SAMS Teach Yourself DVD Authoring

Jeff Sengstack

SAMS 800 E. 96th Street, Indianapolis, Indiana, 46240 USA

Sams Teach Yourself DVD Authoring in 24 Hours

International Standard Book Number: 0-672-32513-6

Library of Congress Catalog Card Number: 2002115933

Printed in the United States of America

First Printing: September 2003

06 05 04 03 4 3 2 1

Trademarks

Warning and Disclaimer

Every effort has been made to make this book as complete and as accurate as possible, but no warranty or fitness is implied. The information provided is on an "as is" basis. The author and the publisher shall have neither liability nor responsibility to any person or entity with respect to any loss or damages arising from the information contained in this book or from the use of the DVD or programs accompanying it.

Bulk Sales

Sams Publishing offers excellent discounts on this book when ordered in quantity for bulk purchases or special sales. For more information, please contact

U.S. Corporate and Government Sales

1-800-382-3419

corpsales@pearsontechgroup.com

For sales outside of the United States, please contact

International Sales

1-317-428-3341

international@pearsontechgroup.com

Acquisitions Editor
Betsy Brown

Development Editor
Jon Steever

Managing Editor
Charlotte Clapp

Project Editor
Elizabeth Finney

Production Editor
Megan Wade

Indexer
Chris Barrick

Proofreader
Gretchen Easterday

Technical Editor
Bruce Nazarian

Team Coordinator
Vanessa Evans

Multimedia Developer
Dan Scherf

Interior Designer
Gary Adair

Cover Designer
Alan Clements

Page Layout
Kelly Maish

Contents at a Glance

Contents

About the Author

Jeff Sengstack has worn many hats: TV news reporter/anchor, video producer, writer focusing on PC technology, radio station disk jockey, music publisher marketing director, high school math and science teacher, and (presently) school board trustee. As a news reporter, he won a regional Emmy and two Society of Professional Journalists first-place awards. He's an Adobe Certified Expert (ACE) on Premiere, is writing a higher-education DV curriculum guide for Adobe, and has written 300 articles and four books, including *Sams Teach Yourself Adobe Premiere 6.5 in 24 Hours*.

Acknowledgments

This book had its beginnings in the fall of 2001. Bruce Bowman, Adobe's "Dynamic Media Evangelist," asked me to evaluate DVD authoring software from Sonic Solutions. That led to a book idea and his offer to introduce me to some technology publishers. I'm grateful for his help.

The folks at Sams welcomed me into their fold and asked that I tackle a book on Premiere first. That project went very well. In particular, it let me reconnect with a lot of my friends in the TV business. I'm gratified to see that *Sams Teach Yourself Adobe Premiere 6.5 in 24 Hours* is selling briskly.

With the Premiere book in stores, my attention returned to DVD authoring. Sonic Solutions had strengthened its place in that rapidly growing market, and a review of competitors' products convinced me to feature Sonic's DVD authoring applications in this book. Sonic's cooperation throughout the creation of this book is greatly appreciated.

And the editorial staff at Sams, as always, has been supportive and helpful.

We Want to Hear from You!

As the reader of this book, *you* are our most important critic and commentator. We value your opinion and want to know what we're doing right, what we could do better, what areas you'd like to see us publish in, and any other words of wisdom you're willing to pass our way.

You can email or write me directly to let me know what you did or didn't like about this book—as well as what we can do to make our books stronger.

Please note that I cannot help you with technical problems related to the topic of this book, and that due to the high volume of mail I receive, I might not be able to reply to every message.

When you write, please be sure to include this book's title and author as well as your name and phone number or email address. I will carefully review your comments and share them with the author and editors who worked on the book.

Email: graphics@samspublishing.com

Mail: Mark Taber
Associate Publisher
Sams Publishing
800 East 96th Street
Indianapolis, IN 46240 USA

Reader Services

For more information about this book or others from Sams Publishing, visit our Web site at `www.samspublishing.com`. Type the ISBN (excluding hyphens) or the title of the book in the Search box to find the book you're looking for.

Introduction

You now can make your own DVDs on your PC.

You can put your vacation videos on a DVD and use DVDs for photo slideshows, your family tree history, or your child's soccer season complete with statistics. You can create them on your PC and play them on your living room TV.

Business people now can use DVDs for marketing, employee training, and catalogs. And event videographers can offer clients wedding DVDs with easy menu access to specific moments. No more wading through a long videocassette to get to "I do."

DVD production can be as simple as recording your old VHS videotapes directly to DVDs or as complex as creating DVDs with multiple menus, subtitles, and extra audio tracks.

Using DVD authoring software, you can create interactive experiences that allow viewers of your DVDs to easily select separate videos, images, and music using menus you design.

DVDs Dominate the Video Market

DVD is the fastest-growing consumer electronics product in history. It has had faster consumer acceptance than TVs and VCRs. There are more than 50 million DVD set-top players and more than 70 million DVD-equipped PCs in the market today.

DVD will soon replace tape as the video publishing format of choice for video professionals and video enthusiasts.

DVDs use high-quality (better than VHS) digital media—images, video, and sound. DVDs are also more compact and more durable than videotapes.

DVDs can store massive amounts of video, images, music, and data; offer amazing versatility; and have near universal compatibility. They are excellent media to archive, publish, and share data and multimedia.

After you make your first DVD, you'll never want to use videocassettes again.

How This Book Will Help You Make Terrific DVDs

Sams Teach Yourself DVD Authoring in 24 Hours opens the door to new possibilities for anyone with a camcorder or still camera and a PC with a Windows operating system.

This book outlines step-by-step instructions on how to create media; edit those videos, images, and audio; and create DVD multimedia projects. It takes you through a straightforward DVD production scenario and then offers details on higher-end techniques.

It covers a wide range of software and hardware products from DVD recorders and DVD authoring software to video and photo editing products.

I've organized the book so each hour builds on what has come before, and I expect most of you will step sequentially through its 24, one-hour lessons. However, if, for example, you already are versed in image editing but need some help with video editing, feel free to take a more self-directed walk through this book.

Or if you want to dive right into DVD authoring itself, skip to Hour 4, "Authoring Your First DVD Project Using MyDVD: Part I," and use the Sonic Solutions MyDVD software and sample media provided with the book to author your first DVD.

How This Book Is Organized

I've organized the book into four sections with several chapters in each. Here's a basic overview.

Part I, "Producing Your First DVDs"

The chapters in this part introduce DVDs and DVD authoring software and let you know what they can do for you. I explain the current state of affairs in the DVD recording hardware business. Finally, I take you through a step-by-step process to burn your first data DVDs and author your first multimedia DVD.

Part II, "Creating Media"

Creating DVDs is more than simply taking some existing media—videos, photos, music, and graphics—and recording them to a DVD. In this section, I offer tips on how to create that media. I also review digital cameras and video camcorders, offer some hands-on tips on their use gleaned from my 15 years in TV news and video production, and show you how to create customized music using the software provided with this book.

Part III, "Editing Media"

In this part, you craft your story. I present a collection of writing and story organizing tips, many of which I picked up from two top-notch TV journalists. Then you use the entry-level and professional video and image editing software included on this book's companion DVD to improve your images and create videos.

Part IV, "Authoring DVDs"

Finally, in this part you tackle the prosumer and professional side of DVD authoring. I review the top DVD authoring software products and then discuss the authoring process in detail using two excellent products from Sonic Solutions: DVDit! and ReelDVD. I cover their features in depth and give you some hands-on experience creating practical DVD applications for personal, business, and professional video and DVD production use.

What's on the DVD?

If you have any doubts about buying this book, the contents of its companion DVD should convince you to make the purchase. We have pulled out all the stops to load the DVD with some top-notch, powerful, and useful products. Most are trial versions—usually full-featured products that time out or expire after a month.

Here's a rundown of what the DVD contains.

DVD Authoring and Movie Playing Software from Sonic Solutions

- **MyDVD**—Consumer- or entry-level authoring
- **DVDit!**—Prosumer-level authoring
- **ReelDVD**—Professional-quality authoring
- **CinePlayer**—DVD movie player software

Image and Video Editing Software

- **Pinnacle Systems Studio 8**—The best consumer video editing software product
- **Adobe Premiere**—The best prosumer/professional video editing software
- **Adobe Photoshop Elements**—Powerful image editing software geared toward entry-level and prosumer users
- **DVIO**—A nifty video capture utility that simplifies transferring video from your digital video camcorder to a PC

Music Composing Software and Royalty-Free Music

- **SmartSound Movie Maestro**—Music creation software that creates engaging tunes that fit your video/DVD production's style and length.
- **Three royalty-free instrumental tunes**—You can use these in any project for any purpose at no charge.

Exciting Times

The timing is just right for this book. The technology—DVD recorders, DVD authoring products, digital camcorders, and video editing software—is finally all in place, relatively easy to use, and reasonably priced.

On a personal level, this book comes at the right time for me. I first proposed a DVD authoring book to Sams Publishing more than a year ago. Back then, the technology had not reached the ease-of-use and accessibility it has achieved today. Understandably, Sams put my proposal on the back burner. That delay gave me the opportunity to write *Sams Teach Yourself Adobe Premiere 6.5 in 24 Hours*.

One feature of Premiere 6.5 is its DVD authoring module. That acknowledgment of this newly emerging DVD technology by one of the world's leading software publishers signaled the tidal shift to DVDs. Numerous other PC industry firms also have begun to see the many opportunities DVDs offer.

Now you are riding that DVD wave. You will find that making your own DVDs is challenging, exciting, and fun. Watching the enthusiastic responses from those who view your DVDs will only add to that enjoyment.

Conventions Used in This Book

This book uses the following conventions:

Text that you see onscreen appears in `monospace type`. Text that you type appears in **`bold monospace type`**.

A **Note** presents interesting information related to the discussion.

A **Tip** offers advice or shows you an easier way to do something.

A **Caution** alerts you to a possible problem and gives you advice on how to avoid it.

PART I

Producing Your First DVDs

Hour

HOUR 1

Discovering What DVDs and DVD Authoring Software Can Do for You

DVDs offer uncountable possibilities. Yet they represent such new technology, few PC owners have begun to exploit DVDs' potential. In this hour I'll offer a taste of what DVDs can do for you.

There are myriad reasons we have arrived at this technological moment. I'll use this hour to bring you up-to-date on how we got here.

Finally, the whole concept of DVD authoring remains shrouded in mystery. I'll demystify that process to prepare you for what's to come in this book's later hours.

The highlights of this hour include the following:

- How converging technologies got us to this point
- Enhancing your media with DVDs
- Delving into DVD project ideas
- What DVD authoring software can do for you

How Converging Technologies Got Us to This Point

Harken back to the multimedia days of yore, back when it was a real struggle to create anything worth presenting in public. Compatibility issues with sound cards and video displays as well as software conflicts fostered uncertainty. Getting your project to run on any PC other than your own was near impossible. And if you used video, it ran in postage stamp-sized windows.

Fast forward to today. Now it's relatively easy to take videos, images, music, narration, and data and build a multimedia project on a DVD. It's not only relatively easy, but also almost guaranteed to play back without a hitch. Drop your DVD into any newer model set-top player, grab the remote, and amaze and astound your viewers.

This dramatic development comes courtesy of several technological advances:

- DVD format
- MPEG-2 video compression
- Digital video (DV) camcorders
- IEEE 1394—"FireWire"—connectivity
- High-speed PC processors
- DVD recorders
- DVD authoring software

DVD Format Adoption

DVDs have their roots in audio CDs. That format emerged in 1982 and single-handedly boosted music industry sales.

By the mid-1990s the movie industry needed a similar jolt. Cable TV and movies via satellite had eroded video sales and rentals. Noting that PC game developers had learned how to put video on CDs, the movie industry sought ways to expand the capacity of those optical discs.

The DVD Logo is a trademark of DVD Format/Logo Licensing Corporation, which is registered in the United States, Japan and other countries.

The result was DVD-ROM. With a capacity seven times that of a CD-ROM, it launched a consumer technological shift never before seen. DVD set-top boxes, launched in the United States in 1997, now number about 50 million. PC DVD players have an even wider acceptance. More than half of all U.S. households own a DVD player.

Blue Laser DVDs: Will They Make Your DVD Recordable Drives Obsolete?

Even though most DVDs can hold up to two hours of high-quality video, that's not enough for some movies. And the ongoing DVD-recordable standards disagreements might continue to fragment the DVD market. I'll cover the competing—some might say combating—DVD recordable standards in Hour 2, "Getting Your Gear in Order—DVD Recorders and Media."

In early 2002, in a move to resolve those issues, a consortium of consumer electronics firms agreed on new DVD recording and playback standards. Those new drives will use blue laser beams instead of red.

Blue beams have a narrower focus and will expand single-sided DVD capacity from their current 4.7GB to 27GB. Double-sided discs will hold an extraordinary 50GB.

Blue laser drives will not be automatically backward-compatible with DVD drives. It'll be up to individual manufacturers to decide whether they want to ship higher-priced drives incorporating both laser technologies.

Does this mean you should wait on the sidelines until these drives ship? The answer is NO.

Those drives are probably years from mass production. Also, the new standards come from the same group that brought us the current set of incompatible DVD recordable standards. There's no telling how well the blue-laser standards will withstand competitive interests as time passes.

By its nature, the technology industry is fraught with change and uncertainty. DVDs are not necessarily here to stay. But by the time blue-laser DVDs hit the marketplace, red laser DVDs will be ubiquitous. It'll take more than a blue light special to change that.

MPEG-2 Video Compression

If movie companies relied on standard analog TV signal data rates, a DVD could hold only three minutes of video. To reach the two-hour figure (most movies are two hours or less), they needed to digitally compress the video signal.

As the movie industry formulated the DVD specifications, it chose to use two data compression technologies for DVD video. Created by the Motion Picture Experts Group, an international image standards setting organization, MPEG-1 (finalized in 1988) and MPEG-2 (finalized in 1994) dramatically reduce video signal data rates while retaining image and audio quality. MPEG-1 approximates VHS video quality, and MPEG-2 matches broadcast quality. Both can play back at varying data rates, but at standard DVD-quality rates, an MPEG-2 video is 1/40 the size of its original TV signal.

Digital Video Camcorders

Digital video (DV) simplifies and streamlines the entire video process. Older video cameras use analog video, and those wave form signals consume massive bandwidth and lead to generation loss during editing and tape copying.

You know that if you dub (copy) a VHS or Hi-8 tape, the copy looks a bit washed out when compared to the original. That's because electronically duplicating a smooth waveform is very difficult. Make a copy of a copy, and the generation loss is even worse.

DV fixes that. DV is a collection of bits—0s and 1s. Any copies merely duplicate those bits, meaning a copy looks exactly the same as the original. Electronic equipment looks only for 0s and 1s. As Figure 1.1 shows, a little noise is not enough to create quality loss, which also means transferring DV from a camcorder to a PC is remarkably easy.

FIGURE 1.1
Signal noise in an analog signal creates visible quality loss but has little or no effect on a digital signal.

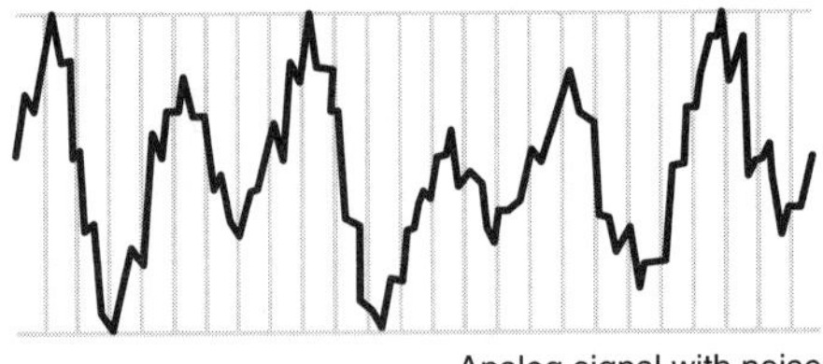

Similar to MPEG, DV is compressed video. A half-dozen flavors of DV are available with varying degrees of compression and quality. The consumer standard is DV25 (25 mega*bits* per second). All flavors of DV have higher data rates than MPEG-2.

Until recently, DV was too expensive for consumers, but not anymore. If you don't have a DV camcorder, go out and get one. I'll offer some buying tips in Hour 9, "Making Videos."

The reason you use DV instead of MPEG-2 video for most video editing applications is that DV offers more precise—frame-specific—editing than MPEG-2 and does not lose quality during editing. MPEG-2's primary function is for playback.

IEEE 1394—"FireWire"—Connectivity

Mac users have long known of FireWire connectivity. Apple Computer invented this data transfer technology in 1986. The IEEE (another standards setting organization similar to MPEG) adopted it as a standard in 1995. IEEE 1394 allows data transfer between devices such as camcorders, printers, scanners, PCs, and hard drives. In addition, it allows direct control—including VCR-style buttons—of connected devices.

Virtually all DV camcorders have IEEE-1394 ports permitting easy connectivity to your PC. I'll explain how you transfer video between your PC and camcorder in Hour 10, "Capturing Video—Transferring Videos to Your PC."

High-Speed PC Processors

Video, even compressed DV or MPEG-2, requires a lot of processor power to achieve smooth playback on a PC. Until recently, the only way to view full-screen, full frame rate MPEG-2 video on a PC was to run it through an expensive video card. With 2GHz processors now commonplace, software playback of MPEG-2 videos is the norm and viewing DVD movies on your PC is routine.

DVD Recorders

In the early days of DVD—from 1997 on—only Hollywood film companies and their DVD development studios could afford DVD recorders. Priced at about $10,000, they fit into a narrow market niche.

Now, thanks to market leader Pioneer Electronics, DVD recorders such as Pioneer's popular DVR-A05 have dropped in price to about $300 while increasing in performance (see Figure 1.2). You probably already have a DVD recorder in your PC. If not, or in case you plan to upgrade, I'll discuss the latest batch of recorders in Hour 2.

FIGURE 1.2
Pioneer's DVR-A05 DVD-recordable drive retails for about $300.

DVD Authoring Software

Coinciding with those high-priced DVD recorders, Hollywood studios purchased very expensive DVD authoring software. Prices for those specialized products frequently exceeded $20,000.

Then along came Sonic Solutions. In the mid-1990s, the former digital audio workstation manufacturer chose to shift to DVD production hardware and software. In 1999 Sonic released DVDit! at the market busting price of $500. By the fall of 2002, Sonic had shipped its two millionth DVD creation application and expects sales to continue sky-rocketing. A mind-boggling array of DVD recorder and video editing products from the likes of Pioneer, Sony, Adaptec, and TDK come bundled with Sonic Solution DVD authoring titles.

I'll feature several Sonic Solution products throughout this book. Sonic owns this market segment, but some competitors are nipping at its heels. I will introduce them to you in Hour 16, "Evaluating Competing DVD Authoring Products."

Taken together, these seven converging technologies have created a DVD revolution. You now get to reap the rewards by making your own DVDs.

Enhancing Your Media with DVDs

DVDs are much more than simply a new type of videocassette or a repository for more than 4GB of data. They have the following benefits:

- They are interactive.
- They readily can handle all types of media.
- They are easily customized.
- The image and audio quality are markedly better than videotape.
- They allow instant access to video segments or other media.

That's what makes this technology so compelling—its versatility.

That characteristic —versatility—has more or less officially crept into the DVD acronym. The movie industry originally said DVD stood for "digital video disc." But when PC owners began using DVDs a few years later, they considered them primarily to be data-only discs. A mini-political battle ensued between the movie and PC industries. The result? DVD ended up officially standing for nothing. Nevertheless, a news reporter suggested giving it the moniker "digital versatile disc," and that has come to be the de-facto descriptor. The DVD Forum, the group behind one of the three DVD recordable formats, now includes "versatile" in its literature.

Put a Hollywood DVD into your living room DVD player, and what happens? First, that oft-ignored FBI warning appears. Then a brief video sequence might play, followed by a menu that might have video playing behind it along with music from the movie. As you navigate your remote through the menu's various choices, each button or text item becomes highlighted in turn.

One button might take you to a set-up screen that allows you to select subtitles, dubbed foreign languages, or Dolby digital audio. Another might open a scene selection menu with buttons displaying thumbnail images of various scenes in the movie. Still another might take you to a collection of "extras": movie theater trailers, behind-the-scenes clips, actor bios, outtakes, deleted scenes, or director's comments. Finally, yet another button starts the movie. Not only will the video and sound be distinctly better than video-cassettes, no wrinkles or drop-out will be caused by worn-out tapes.

DVDs Improve the Viewing Experience

DVDs have changed the way we view movies at home. It used to be that we rented a video, watched it, and returned it. Now, all these extra features make it worth our while to take a second look. Some DVDs' extra features create compelling reasons to *own* a movie. Directors' comments alone can be the equivalent of college filmmaking course-work.

DVDs have made Hollywood filmmakers rethink their entire creative process. The same holds true for home video enthusiasts, professional videographers, and corporate media producers. DVDs have fostered a whole new creative process.

Delving into DVD Project Ideas

As you begin to create DVD projects, you'll quickly discover that this medium's potential is endless. I will scratch the surface of that potential in Part IV, "Authoring DVDs." There I present about a dozen step-by-step guides to creating specific types of DVD projects. The following is a sampling of what I'll cover.

Home DVD Projects

A homemade DVD can be as straightforward as a simple recording of a video or as complex as a family tree project with multiple menus and media. Here are a few examples:

- **Video recording**—Some software and hardware products let you transfer a single video or TV show directly to a DVD. I cover that process in Hour 21.
- **Vacation videos**—As shown in Figure 1.3, you can give your vacation videos some interactivity by using menu buttons to let viewers jump to specific activities. Clicking a button immediately starts a video clip. Viewers can watch to the end of that clip, and the DVD automatically returns them to the opening menu screen. Or, they can exit a video by pressing the Menu button on the remote.

FIGURE 1.3
Add simple interactivity and ease of access to your vacation videos using a DVD menu.

- **Sports**—This approach offers viewers more interactivity. As demonstrated in Figures 1.4 and 1.5, clicking a menu button takes viewers to a *nested* menu (a commonly used DVD authoring technique that puts a menu within a menu) that offers stats for that season plus menu buttons that play season highlights.
- **Family tree history**—The family tree DVD illustrated in Figures 1.6–1.8 show how you can take the nested menu approach a bit further. The opening menu in Figure 1.6 gives viewers options to view an overall family history video or take a tour of specific family lines and so on. The nested menu in Figure 1.7 lets viewers access the portion of the material that applies to that line, and the third menu in Figure 1.8 uses thumbnail images to give viewers access to those family photos.

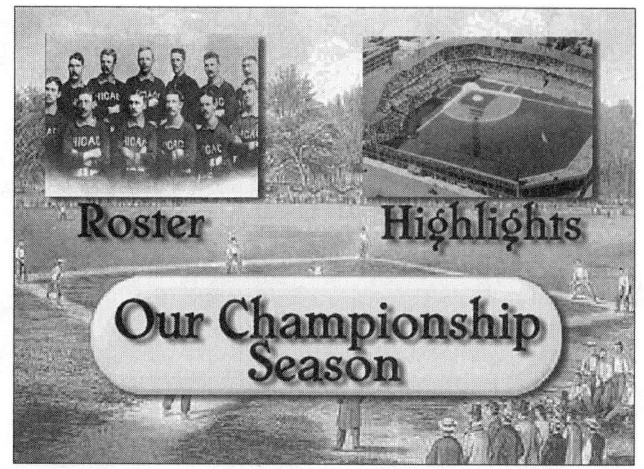

FIGURE 1.4
Sports DVD menus have extra flare when you give viewers options.

FIGURE 1.5
Use a nested menu to offer access to statistics, highlights, and more.

FIGURE 1.6
A family tree history DVD opening menu might look like this.

Figure 1.7
A nested menu, or sub-menu, provides access to a specific branch of the family tree.

Figure 1.8
Accessing the next menu level lets viewer see photos of individuals within that branch of the family tree.

- **Holiday greeting cards**—Using a DVD for those annual family updates sent to friends and relatives adds several new dimensions to what you might now consider to be passé word processor print-outs. Figure 1.9 shows how you can mix videos, photos, and your favorite seasonal music into one appealing, memorable, and long-lasting greeting.

FIGURE 1.9
Use video clips, photos, and music to turn those annual family update letters into holiday multimedia extravaganzas.

Videographer DVD Projects

Professional videographers, many of whom work as one-man bands, need something to give them a competitive edge. DVDs do that. They turn formerly linear event videos—weddings, concerts, and panel discussions—into interactive, enjoyable projects. Here are some DVD ideas that might enhance your product:

- **Event videos**—Weddings are the bread and butter of event videographers, so competition in this market niche is fierce. Adding DVDs to your product offerings makes you stand out above the crowd. Figure 1.10 demonstrates how you might create a wedding DVD. This one gives viewers instant access to specific wedding day highlights and uses a nested menu to jump to individual greetings from the guests. You could create some DVDs using the OpenDVD format (see Hour 24, "ReelDVD Part II: DVD Producer and DVD Trends") to let you or your clients later add honeymoon vacation highlights.
- **Client rough cuts**—Frequently, clients like to play "what if." DVDs let you present customers with all the optional takes or special video effects they request. Place those clips on a DVD with a simple menu, and let clients view them at their leisure. There's no need to cue up a bunch of tapes, play, or rewind. And those clients can view the rough cut DVDs on their home or office TV, in an environment that can duplicate how others will see them.

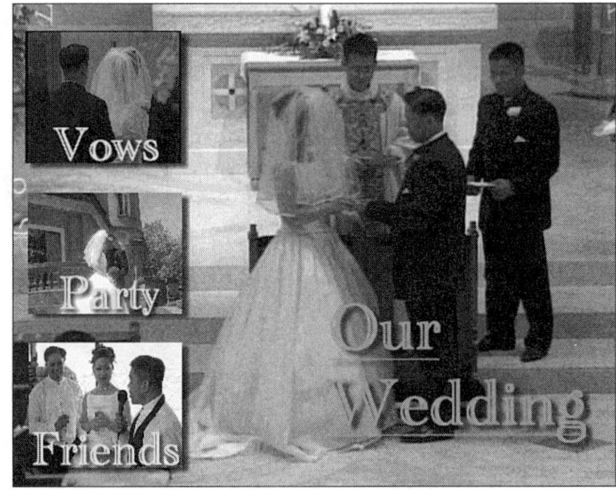

FIGURE 1.10
Wedding videos, the videographers' mainstay, have more sizzle when delivered on DVD.

- **Demo reels**—If you want to give clients a true taste of what you can do, use a DVD. You no longer need to rely on videotape. Besides, few clients will take the time to sit through that kind of linear experience. In addition, if you use tape, you usually must rely on VHS as the lowest common denominator, thus using something that by design does not equal the visual and aural quality of your original material.

Instead, use a DVD. Let your clients choose to view what they want, and give them a variety of material. DVDs send a message that you're on top of things.

Business DVD Projects

DVDs can change the way businesses operate. Instead of the usual PowerPoint demonstrations, arm your roadshow warriors with DVDs loaded with videos, graphs, and data. Promote your products secure in the knowledge that your clients will be able to play your production on their home/office TVs. Here are some business-related DVD ideas:

- **Marketing/Promotion**—No longer do you need to worry whether your laptop will fail. Now marketing presenters need only attach a DVD player to a projector and use the remote to easily and quickly navigate to the marketing material that suits the occasion. Using rewritable DVDs, those same presenters easily can add to or change their presentations. If you can't get to your clients, send them a DVD. Real estate developers who want to show out-of-town customers their properties—complete with model home virtual tours—will find that DVDs make a positive impact.

- **Training and product demonstrations**—DVDs resolve most problems associated with interactive training on a PC. As in the example in Figure 1.11, the videos can run full screen, the menus allow for true interactivity, and you don't need to tie up a PC or worry about compatibility issues.

FIGURE 1.11
Use DVDs to enhance your corporate training videos.

- **Catalogs**—Instead of boring, static images of products in a printed book, send clients a DVD that lets them search for products by category and then view their functionality. This applies to any type of company: Music publishers can excerpt individual works, manufacturers can show their wares at work, and service companies can demonstrate their offerings.

What DVD Authoring Software Can Do for You

To create all these exciting DVDs, you'll need DVD authoring software. Depending on how many features you want to include on your DVDs, that software can range in price from free (bundled with a DVD recorder), to a few hundred dollars for prosumer-quality products, to $20,000+ for Hollywood-style software.

All DVD authoring titles perform the same basic functions. Most importantly, they work within the DVD specification, which is surprisingly configurable, to ensure your finished DVD meets those specs and runs on *most* DVD set-top boxes.

Some older DVD set-top players do not handle all aspects of the DVD spec—thus, the caveat that not all your DVDs will work on all DVD players. Also, three basic types of DVD recordable media exist: DVD-R (dash-R,), DVD+R (plus-R,) and DVD-RAM. The dash and plus formats also offer rewritable (RW) formats: DVD-RW and DVD+RW. Some players work with some formats and not others. I explain more about what some call the "format wars" in Hour 2.

Higher-priced DVD authoring products have correspondingly higher-level functionality that pushes the DVD spec to its limits—allowing things such as games on DVDs. The first *Harry Potter* movie DVD is a good example—it "remembers" your previous answers despite working in a DVD set-top that has no memory.

I will only briefly refer to that very high level of DVD authoring in Hours 23 and 24. For the rest of us, DVD authoring products have some fairly straightforward characteristics. They facilitate

- Menu creation with buttons and text
- Adding special features such as video chapters, extra audio tracks, or subtitles
- Video conversion to MPEG
- DVD burning of single DVDs and masters for mass production

Menu Creation with Buttons and Text

What sets DVDs apart from videotapes is their interactivity and immediate access to media deep within the DVD. DVD menus foster that functionality. Authoring products offer varying degrees of menu creation capabilities.

Lower-priced products limit you to automated templates with few options for menu, button, and text styles or placement. Higher-priced products frequently offer graphics toolsets, letting you create custom menus with animated or video backgrounds and music.

Buttons and text usually are the links to media on your DVD. Depending on the authoring product, they might offer options such as adding text to a button, changing a button's visual characteristics, or using Photoshop-created graphics with multiple layers that alter their appearance depending on user actions (for example, rolling the cursor over a button changes its appearance, as does clicking that button).

Higher-level authoring products let you specify certain behaviors for buttons, menus, and other media. That might seem a bit obtuse, so I'll offer plenty of practical applications later in the book. For now, here's one example: As you create nested menus, you'll want to help users navigate through those menus by noting whether pressing the menu button on the remote takes them to the beginning menu or the previous nested menu. When you create the various menus you can choose to design in those remote-control button behaviors.

Adding Special Features

Some DVD authoring products offer extra features, some of which require detailed explanation. Hour 16, "Evaluating Competing DVD Authoring Products"; Hour 21, "Burning DVDs and Dealing with Mass Replicators"; and Hour 24, "ReelDVD: Part II DVD Producer and DVD Trends," cover these in more detail. Here are three of those features that are less technically detailed:

- **Chapters**—Those scene selection screens in Hollywood DVDs actually do not take you to separate videos of those individual scenes. Rather, clicking a scene's button takes you to that point in the original movie. Some authoring products let you set points like that—called *chapters* in DVD parlance.
- **Extra audio tracks or subtitles**—Higher-quality Hollywood DVDs offer extra soundtracks of varying quality or content: from simple stereo to Dolby digital surround sound, as well as director's comments and foreign language dubs. The *Monsters, Inc.* DVD uses this feature uniquely: Its developers added a sound effects-only track. Higher-priced DVD authoring products let you lay in those multiple tracks. Those same authoring tools let you add subtitles—created in a word processing program or within the authoring software—and place them exactly to match the movie dialogue.

Video Conversion to MPEG

Only two types of video will play back off DVDs on your set-top box: MPEG-1 and MPEG-2. As mentioned earlier, most higher-end PC video editing products do not work with MPEG video. Rather they stick to digital video or computer-based video formats such as Windows Media, Real Media, QuickTime, or the slightly older MOV and AVI files.

Before you begin authoring your DVD, you might choose to *encode*, or convert, all those video formats into MPEG files using third-party software. Or, you can rely on your DVD authoring software to handle this chore sometime during the authoring process. This is no simple task, and some products do that better, faster, and more accurately than others. Some authoring products do not ship with MPEG encoders (they add to a product's cost) and expect you to use third-party MPEG conversion software. I briefly explain MPEG encoding in Hour 21, "Burning DVDs and Dealing with Mass Replicators."

Somewhat surprisingly, the most expensive DVD authoring products sometimes do not offer MPEG conversion tools. The reason is that at that production level, DVD creation studios prefer using very expensive hardware or software MPEG encoders. These perform in-depth analyses of source videos and create sharper-looking and better-sounding MPEG-2 videos than you can get from software-only encoders.

Burning DVDs and Making Masters for Mass Production

All DVD authoring software products *burn* DVDs, meaning they take your media and record it to a recordable DVD.

Some give you extra options such as letting you add data files or create CDs that mimic DVDs when played in a PC CD-ROM drive.

Some even let you take your authored DVD on a test drive before burning it to make sure all buttons lead to their expected destinations. For the most part, this is a fairly straightforward process, but ease of use and quality do vary from product to product.

Until recently, if you wanted to use the services of a mass replicator to make multiple copies of your DVD, you needed higher-priced authoring tools and a digital linear tape recorder (DLT). Replicators now typically use the DVD you create as the recording "master." I offer tips on dealing with replicators, take you through a simple online mass duplication process, and explain basic DVD burning procedures in Hour 22.

Summary

You are about to embark on what I think is an exciting and creative adventure. Finally, all the technological elements are in place to let you create a multimedia experience that meets your imagination and talents.

Low-priced, highly functional DVD recorders and DVD authoring software now give home video enthusiasts, prosumer videographers, and business people unlimited opportunities.

That authoring software lets you make truly interactive DVDs with menu access to video clips, images, music, and data. And burning DVDs is now a remarkably straightforward process.

HOUR 2

Getting Your Gear in Order—DVD Recorders and Media

You can author DVDs to your heart's content, but without a PC DVD recorder, only you can view your masterpiece and only on your PC. If you don't own a recorder, read this chapter and then go out and buy one. If you already have a DVD recorder, check out my overview of the latest industry developments and decide whether you need to upgrade.

The continuing format wars within the DVD recorder industry create confusion over which recordable media is most compatible, fastest, or most useful. I'll do what I can to clear the air and fill you in on one development that might put an end to the bickering.

One way to give your DVD player a test drive is to use it to play a DVD movie. To do that, you need the correct software. In this chapter, I present an overview of the three top DVD software players. Finally, I show you how to use the one included on this book's companion DVD.

The highlights of this hour include the following:

- Clearing up the DVD recording media confusion
- Selecting a PC DVD recorder
- Sony's bid to end the format wars
- Evaluating three DVD software movie players
- Using Sonic Solutions CinePlayer to test drive your DVD recorder

Clearing Up the DVD Recording Media Confusion

Recordable DVD media come in a confusing variety of three flavors: dash R/RW, plus R/RW, and DVD-RAM. The news media call this the *format wars* and liken it to the bloody and brutal VHS versus Betamax quagmire in the early days of home VCRs.

R stands for record-once media, and *RW* means rewritable. It's the same nomenclature used for CD recordable drives.

In that case, Sony's Betamax format lost out to VHS—but only after years of acrimony, bad blood, lawsuits, and marketing miscues.

Sony's Betamax was first to market by a year, momentarily held several technological edges, and was priced competitively. But the company waited more than a year before licensing its technology to another manufacturer—Zenith (eventually seven companies manufactured Betamax VCRs). VHS started with more manufacturers, originally allowed for longer recording times, and by sheer timing was not included in a lawsuit filed against Sony by the movie industry.

The bottom line is that no one can say with certainty why VHS won. Call it serendipity, timing, or luck of the draw. Essentially consumers settled those format wars.

Many misconceptions exist about those Betamax/VHS days. The biggest is that Sony failed to license Betamax. If you want to delve into the history, visit `http://www.urbanlegends.com/products/beta_vs_vhs.html` or `http://lamar.colostate.edu/~dvest/346/project/silos/BEGINNING.HTML`.

As for the DVD recordable format wars—*surprise*—Sony might end up being the peacemaker (see the sidebar titled “Sony’s Bid to End the Format Wars,” later in this chapter).

Dash R/RW Versus Plus R/RW Versus DVD-RAM

Those three DVD recordable flavors currently create consumer confusion. And rightly so. Much of that confusion is a result of the plus R/RW camp that came late to the game promising better performance and a simple upgrade path from rewritable-only drives to rewritable/record-once combo drives.

A 230-company group called the DVD Forum formed in 1997 and supported two DVD-recordable formats: DVD-R/RW (dash R/RW) and DVD-RAM.

DVD-RAM’s original focus was largely massive data backup. DVD-RAM is rewritable up to 100,000 times (versus 1,000 times for DVD-RW), is more expensive than DVD-R/RW, and initially required a special disc holder tray.

The DVD-R/RW format serves multimedia and home/business PC users. Pioneer broke that market open with its $1,000 DVR-A03 DVD-R/RW CD-R/RW drive. The PC and Mac community (the original Mac SuperDrive was the Pioneer A03) embraced the DVD-R/RW format, and the DVD player market followed suit, ensuring its set-top boxes could play back movies recorded to DVD-R discs.

If you plan to have a replication firm mass-produce your DVD project, DVD-R is your best bet.

Most replicators recognize the ubiquity of DVD-R and recently have developed means to use that medium to create duplication masters.

I cover dealing with mass replicators in Hour 22, "Burning DVDs and Dealing with Mass Replicators."

DVD+R/RW—Better But with a Bitter Aftertaste

Philips, Sony, Hewlett-Packard, Ricoh, and others formed a splinter group called the DVD+RW Alliance and developed a slightly different format—DVD+R/RW (plus R/RW).

It promised better performance and improved set-top player backward-compatibility, but this technology not only arrived late to the rapidly growing DVD recordable scene, but also arrived half-baked.

This group’s first drives (virtually all using Ricoh mechanisms) were DVD+RW (rewritable) only and had no write-once DVD+R capability. Early on, customers were led to believe that they could later do a simple firmware update (see the following note) to convert their DVD+RW drives to DVD+R/RW combo drives.

The DVD+RW alliance later discovered it could not use a simple firmware fix to upgrade the drives to +R/RW status. It informed customers that the only way they could make write-once +R discs was to buy a second-generation DVD+R/RW drive that began shipping in mid-2002. That decision created a lot of rancor among early DVD+RW adopters.

Firmware is software electronically inserted into a confusingly named erasable programmable read-only memory (EPROM) or electrically erasable programmable read-only memory (EEPROM) chip within the DVD drive.

In older computing days, a ROM was just that: *read-only* memory. Then along came PROMs, which allow one-time burning of instructions and save manufacturers the expense of building an integrated circuit from scratch.

Nowadays, with easy access to downloadable code on the Internet, most manufacturers use EPROMs or EEPROMs to allow consumers to perform bug fixing or performance-improving firmware upgrades.

Selecting a PC DVD Recorder

Until October 2002, this was a true consumer conundrum. Then clarity arrived in the form of Sony's new dual format, DVD±R/±RW and CD-R/RW DRU-500 drives (see the following sidebar, titled "Sony's Bid to End the Format Wars," for a report on my hands-on evaluation).

By early 2003, no other DVD recordable drive manufacturer had announced a competing dual format drive, but my guess is that other companies will duplicate Sony's bold move. Eventually, DVD-recordable drives might become commodities, not unlike CD-R/RW drives are now. At that point, price, software bundles, technical support, and brand-name awareness might be the only ways to differentiate drives.

Until that situation unfolds, here's a basic rundown of your choices.

DVD-R/RW

This segment's market leader by far is Pioneer. This company's DVD-103 DVD-R/RW drive (or A03), which was also Apple's original SuperDrive, ignited the DVD recordable market after its release in 2001. Its latest, fifth-generation product, the DVR-105 is a solid improvement of the 103/A03 (see Figure 2.1). Retailing for less than $300, it writes to DVD-RW, DVD-R, CD-RW, and CD-R. Despite the improved performance—4x DVD-R and 16x CD-R record speeds—I have an A03 and see no reason to upgrade to the A05 or switch to the new Sony dual format DVD recorder.

FIGURE 2.1
Pioneer's DVR-AO5 is the de-facto industry standard DVD recordable drive.

Toshiba also ships a DVD-R/RW drive, the SD-R2002, with performance and features similar to Pioneer's DVR-AO4. My suggestion is to choose the market leader, Pioneer.

If you own a Pioneer DVD recorder, you probably need a firmware update. (Unlike the firmware update promised but not delivered to owners of the now obsolete Philips/Sony/HP/Ricoh DVD+RW drives, the Pioneer firmware update actually works.)

Pioneer DVD recordable drives have a bug that didn't rear its ugly head until the fall of 2002 when the DVD Forum released specs for high-speed (4X for DVD-R and 2X for DVD-RW) media. The bug is related to the reflectivity of the newer discs.

If you use one of those discs, your Pioneer drive might spin out of control, physically damaging both the disc and the drive.

The firmware fix is fairly easy. Visit `http://www.pioneerelectronics.com/hs/`, and download and run the firmware updater and the version checking software. The version checker tells you whether you need the firmware update. If so, run the update and after a couple minutes, your drive will be capable of handling high-speed discs.

DVD-R DVD-RAM Combo

Panasonic's DVDBurner II Multi drive, released in November 2002 and shown in Figure 2.2, is an attempt to attract both DVD Forum camps—DVD-R/RW multimedia enthusiasts and DVD-RAM data storage folks. After the Sony, it's the second drive to market that offers 4X DVD-R record speeds as well as 3X DVD-RAM and 2X DVD-RW. It also supports CD-R/RW. At $350, it's priced to compete with the Pioneer DVR-104 while offering DVD-RAM and faster record speeds as bonuses.

FIGURE 2.2
Panasonic's DVDBurner II covers all DVD Forum bases.

The only other DVD-RAM drive of note comes from Toshiba, but it's DVD-RAM only. Such a drive is of no use to anyone who plans to create DVD multimedia projects. It offers limited compatibility with DVD players and other PC DVD drives.

DVD+R/RW

Hewlett-Packard, Philips, and Ricoh—all DVD+RW Alliance members—offer so-called second-generation DVD+R/RW drives—write-once and rewritable. Their drives, shown in Figure 2.3, are near mirror images because they use Ricoh's internal mechanism. All that sets them apart are their software bundles. DVD+R/RW drives generally perform better than DVD-R/RW, but consumer acceptance has been slow. Sony used to side with the DVD+RW Alliance but has now ventured off on its own.

Philips DVDRW228

Ricoh MP5125A

FIGURE 2.3
The three principal DVD+RW Alliance drives.

Sony's Bid to End the Format Wars

If you can't beat 'em, join 'em. That seems to be Sony's take on the DVD recordable media format war.

Its solution has been to manufacture a dual RW format DVD recordable drive that supports both the dash R/RW and plus R/RW as well as CD-R/RW. DVD-RAM is not part of this equation, but that format serves only a narrow market niche—data backup.

This is a big shift for Sony. It was one of the original members of the DVD+RW Alliance and, like Philips and Hewlett-Packard, relied on a Ricoh drive mechanism for its first- and second-generation DVD+RW drives.

Now, Sony builds its dual-format drive from the ground up and makes the drive, optics, motor, and key-integrated circuits. The company worked for more than a year to bring this drive to market.

Sony began shipping its DVD±R/±RW and CD-R/RW DRU500 drives, shown in Figure 2.4, in the fall of 2002. The internal ATAPI drive sells for about $350, and the external USB 2.0/i.Link (FireWire) drive sells for about $430.

FIGURE 2.4
Sony's DRU500 might be the olive branch that ends the DVD format wars.

2

Hands-on Testing

I evaluated a pre-release internal drive and came away impressed on several fronts.

The DRU500 looks and installs just like any other internal DVD or CD drive. Just place the jumper on the proper pins (slave or master depending on your PC setup), slide it into a free bay, attach the ATAPI cable and power plug, and turn on the PC. Windows notes the presence of this new drive, identifies it as a DVD-R drive, and gives it a new drive letter.

Sony includes a comprehensive array of bundled software, including Sonic Solutions MyDVD with Arcsoft's ShowBiz video editor, CyberLink's PowerDVD movie player, two data recording utilities, and a music recorder/organizer.

Sony's DVD-recordable drive outperforms all other drives I'm aware of. It's the first to offer a 4X DVD-R record speeds (equivalent to about a 36X CD-R record speed) along with 2X DVD-RW and 2.4X DVD+R/RW record speeds. Its 24X CD-R and 10X CD-RW record speeds put it at the top of that category as well.

You might be wondering whether you should junk your present DVD recordable drive. That depends. I've had a Pioneer A03 drive for about two years, and I see no compelling reason to dump it in favor of the Sony DRU500. Pioneer's DVD-R/RW format is the most compatible of all DVD recordable technologies. I don't swap DVD discs with other users who might have drives that handle only +RW discs, so that is also a non-issue. And I don't burn so many discs that the speed improvement is enough of a selling point.

But, if you are in the market for a new DVD recordable drive or want to replace one of the first-generation DVD+RWs (with no DVD+R write-once capability) then I heartily recommend the DRU500.

By the time this book reaches store shelves, other drive manufacturers might have followed Sony's lead. If so, the format wars will have ended in a tie.

Choosing Between an Internal or External Drive

Few PC owners like to open their PCs to install new hardware, which is understandable. Although newer versions of Windows and motherboards are more forgiving than those from a couple of years ago, trying to get a DVD recorder to fit into the drive bay, making sure the jumper is set properly, and connecting the cables without bending a pin can be nerve wracking.

You probably purchased a PC with a DVD burner already installed. If not, or if you're upgrading, there are two basic approaches you can take: open the case or buy an external drive.

External drives typically cost more but are easy to install. Most simply plug in to a USB outlet on the PC, and then Windows notes the presence of a new DVD drive and gives it a drive letter.

All new internal drives use ATAPI/EIDE connections, like those you probably use for your hard drive. As I mentioned earlier, they take a few more steps to install. Even though they sometimes perform slightly better than external drives, it might be worth it to spend a few more dollars to buy an external unit.

Which Drive to Buy

If you are in the market for a new drive, I suggest you look long and hard at the new Sony recorder—or any others that have taken the multiformat road by the time this book publishes. That said, no real differences exist between the DVD Forum –R/RW drives and the DVD+RW Alliance +R/RW drives. The plus format drives tend to have better benchmarks, but that type of *spec-manship* has little meaning in the real world. On the other hand, plus recordable discs (see the next section, "Selecting Recordable Media") usually cost more. In any event, with prices dropping and performance improving, the main selling point for you might be the drive's software bundle and brand-name reputation.

Selecting Recordable Media

DVD recordable media recently took a giant leap forward. Now 4X DVD-R and 2X DVD-RW media are available. If you have a drive that can handle those improved products, feel free to use them for the time savings they offer.

Meanwhile, choosing among most other DVD media is similar to selecting an audio cassette. You end up buying based on brand identification and price.

In the case of DVD blank media, your choice comes down to drive manufacturer-branded discs from Pioneer, Sony, HP, and so on; name-brand generic media from Verbatim, Memorex, Maxell, Mitsui, and TDK; and *house* brands.

Just about any DVD media retailer on the Internet offers its own house brand of the more popular DVD media formats. User reliability reports vary on these non-name-brand products from noting inconsequential differences between them and name brands to fairly high failure rates for house brand DVDs.

I suggest the middle ground: name-brand generic. Be sure you buy media that match your drive; that is, don't use "plus" R discs in a "dash" R drive. DVD-R media are the lowest cost option, whereas DVD-RAM are the highest.

2

To check the latest disc prices, visit `http://www.mysimon.com/` and search on "DVD blank media".

Evaluating Three DVD Software Movie Players

Most DVD recordable drives wear several hats: DVD-ROM, DVD-recorder, CD-ROM, and CD-R/RW. On the ROM or playback side of things, running DVD movies smoothly is their most fascinating use.

To view movies on your PC, you need DVD movie player software. That software typically includes an MPEG decoder—software that converts an MPEG video data stream into a video signal playable on your PC monitor—as well as DVD remote control-like buttons and extra control features not found in standalone, DVD set-top players.

The three major competing products are Cyberlink PowerDVD XP 4.0 Deluxe, Intervideo WinDVD 4.0 Plus, and Sonic Solutions CinePlayer 1.5. After discussing these, I'll take you through the features of CinePlayer, the movie playback software included with this book.

WinDVD is the industry leader and my top pick of these three DVD software players. But CinePlayer is nearly as feature rich and Sonic Solutions, which is provided a demo copy for inclusion with this book.

Cyberlink PowerDVD XP 4.0 Deluxe

PowerDVD, shown in Figure 2.5, is a worthy competitor to CinePlayer but is my third choice in this group. It offers most of the same features, but its graphical user interface is too small and some of the icons' meanings are unclear.

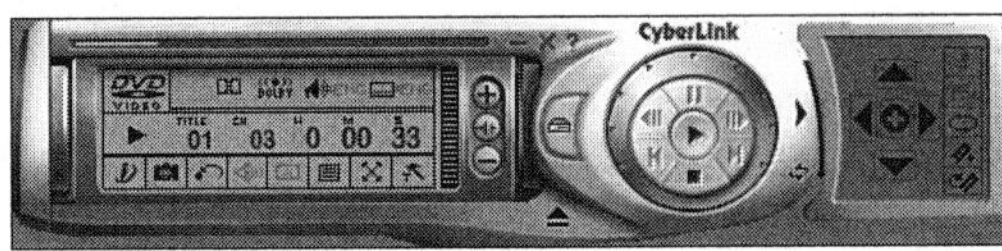

FIGURE 2.5
PowerDVD's interface is too small, and some icons are indecipherable.

The right-click menu offers a time-saving feature. It lets you directly step through items such as subtitles and foreign language dubs instead of making you access those DVD menus to make your choice. But the right-click menu fails to offer the most logical right-click menu item: opening the graphical interface.

One nice feature is bookmarks, which let you easily return to specific scenes. Similar to CinePlayer, PowerDVD has a fly-out number pad, but when I accessed a DVD menu with numbered selections, the number pad inexplicably switched to an arrow navigation pad. And at $75 (versus $50 for CinePlayer), PowerDVD XP Deluxe is overpriced.

Intervideo WinDVD 4.0 Plus

WinDVD, shown in Figure 2.6, is the de-facto industry standard and my top pick. It has every feature imaginable, along with a few surprises.

For instance, not only can you bookmark movie scenes, but you can also open a panel with thumbnail images of your bookmarked scenes. WinDVD's pop-out window offers seven sets of tools, including controls for color/hue; multiple audio options; and easy navigation through chapters, subtitles, and foreign language dubs.

I would, however, prefer a larger interface with more clearly defined icons because it's not immediately apparent how to get from a full-screen movie view to a windowed view. Also, minimizing the interface minimizes the video as well. Niggling inconveniences aside, this $80 product has every bell and whistle imaginable.

Sonic Solutions CinePlayer 1.5

You will get a chance to see CinePlayer's features in detail in a moment. In the meantime, my basic take is that it's a fine product with deep customizability and control over any DVD movie you play. The interface is clean, the icons are large, and the settings and other menus are readily accessible and intuitive.

However, it doesn't have the extra features that set WinDVD apart from the field in this three-horse race. For instance, CinePlayer has no bookmark thumbnail display capability, but it does automatically return to where you left off when last viewing a particular DVD.

At $50, it's the bargain-priced product in this group. Considering that you might get it free in a hardware/software bundle, you can't go wrong with CinePlayer.

FIGURE 2.6
The feature-rich WinDVD 4.0 Plus, with its multipurpose pop-out menus, is my selection as the top DVD player.

2

Using Sonic Solutions CinePlayer to Try Your DVD Drive

PC power users love benchmarks—overclocking motherboards or graphic cards is their forte. But there's not much you can do to change the performance of a DVD recorder other than providing it with the best blank media it can handle.

Nevertheless, just to ensure your DVD drive is up to snuff, you might want to take it for a couple of test runs. This hour covers only the playback side of things. Hour 3, "Burning Data DVDs," covers the recording side of the equation.

> Over the course of several years, I tested dozens of CD and DVD drives for several magazines. In each case I relied on benchmarking tools from TCD Labs (formerly TestaCD Labs). If you want the absolute best DVD benchmark software, which sells for $80, you can buy it directly from TCD Labs' Web site: `http://www.tcdlabs.com/`.

Task: Play a DVD Movie

▼ TASK

This is the first of many tasks you'll encounter in this book. Generally, they are step-by-step introductions to new concepts or practical applications. This will be one of the easier tasks. Go through the following steps to play a DVD movie using the included Sonic CinePlayer software:

1. Install CinePlayer from the disc included with this book. That process will make CinePlayer the default DVD movie player.

> The trial version of CinePlayer included on this book's companion DVD will time out 30 days after you install it. To purchase the retail version, visit `www.cineplayer.com`.

▼

Most people prefer viewing DVD movies on a TV versus a computer monitor. But using a PC has at least one advantage: Monitors have sharper images than TV sets. Movies therefore look crisper and clearer on a PC than on a standard TV. And because TV-out jacks are now available for many video cards and portable PCs, the PC is frequently used as a DVD movie player.

2. Place a DVD movie in your DVD recorder/player. Depending on your version of Windows, either it will automatically open CinePlayer (or some other DVD movie player software you have installed on your PC) or you'll see the screen shown in Figure 2.7.

In Windows XP, Microsoft's Media Player (visible in Figure 2.7) can play DVDs only after installing a third-party plug-in such as Sonic Solutions' CineMaster Decoder Pack (which sells for $15). But no plug-in matches the performance of a standalone DVD movie player.

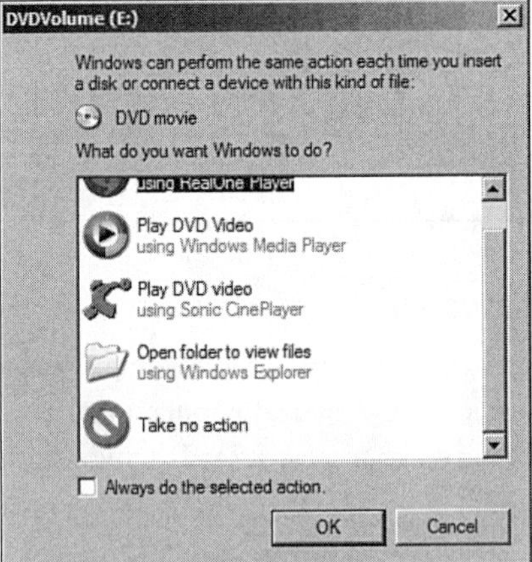

FIGURE 2.7
The Windows Autoplay feature pops up this screen when you insert a DVD movie into your DVD drive.

3. If you see the screen in Figure 2.7, select Play DVD Video Using Sonic CinePlayer. If some other DVD player software starts, close it and open Sonic CinePlayer by double-clicking its desktop icon or locating it in the Start menu.

You should now see the interface in Figure 2.8. It has the instantly familiar look and feel of a set-top DVD player. Most of the buttons are self-explanatory, but I'll explain some special features in a moment. For now, use Figure 2.8 as a guide to try the standard DVD playback buttons.

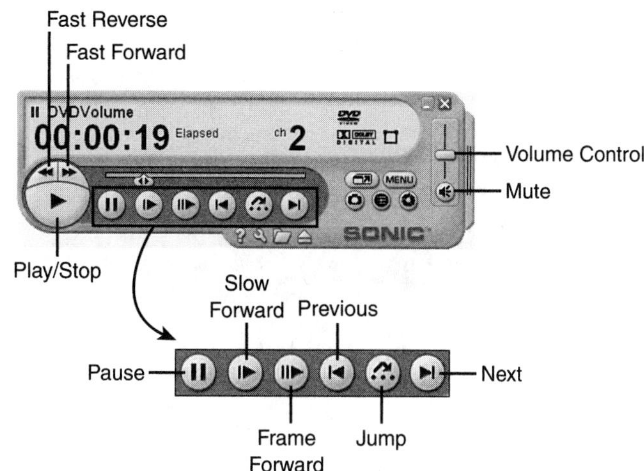

FIGURE 2.8
Sonic CinePlayer's user interface has the look and feel of a set-top DVD player.

2

In particular, check out the Fast Forward and Fast Reverse buttons. Note that with each click, the speed increases—to 1.5X, 2X, 4X, 10X, and 20X.

The Slow Forward button also operates in increments: 1/4x, 1/2x, 3/4x, and full speed.

Previous and Next move from one part of the DVD to another—that is, from a menu, movie, graphic, or still image to another such element on the DVD, depending on how the DVD was authored.

Jump opens a submenu that lets you move directly to a specific chapter or to a time within the current video segment.

Task: Examine the CinePlayer's Extra Features

TASK

Most PC DVD movie players offer extra features beyond those available in a standard set-top, standalone DVD player. CinePlayer's offering tops this category. Follow these steps to get a taste of the options at your fingertips:

1. Using Figure 2.9 as a guide, click the Question Mark icon to open the Help menu. The opening help screen explains all the buttons in Figures 2.8 and 2.9. Sonic Solutions uniformly has the best help screens I've encountered for any software. Feel free to explore them. They cover CinePlayer in depth, so I will touch on only a few items.

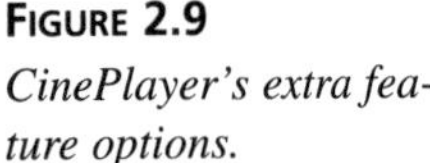

FIGURE 2.9
CinePlayer's extra feature options.

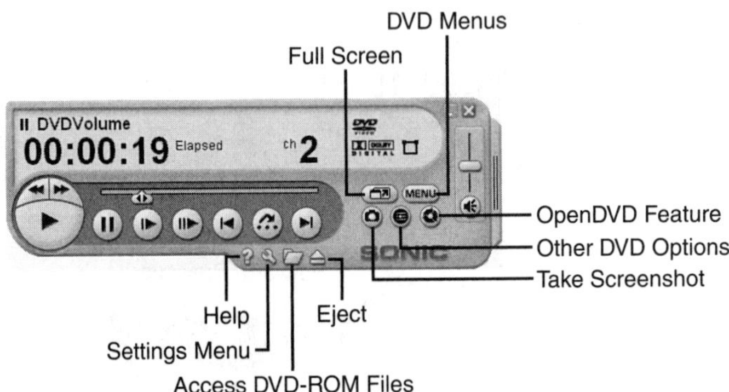

2. Close out of the help screen, click the Menu button, and select your DVD's Main or Chapter menu. Click the Expand/Collapse button circled in Figure 2.10 to open the Navigation/Keypad Button menu. These buttons duplicate the functions of a remote control and let you navigate around the menu buttons or type in a chapter number. However, simply using your mouse and clicking an onscreen button or chapter is much easier.

FIGURE 2.10
CinePlayer's pop-out Navigation/Keypad Button menu is only for those who are mouse-click challenged.

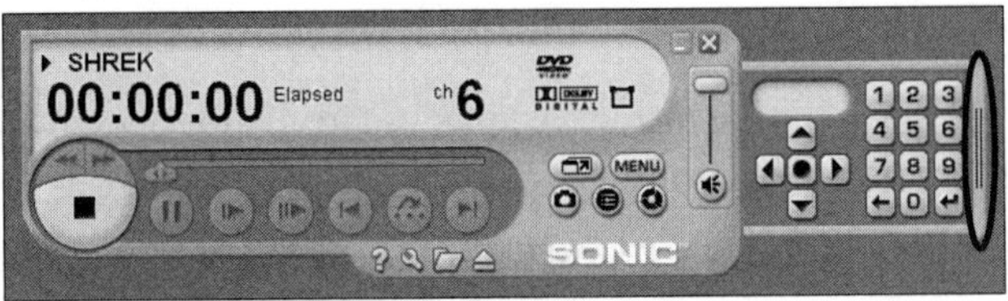

3. Click the Wrench icon to open the Settings menu. As illustrated in Figure 2.11, you have more options than you probably imagined would be available in a simple DVD movie player.
4. Select the Video Settings tab and click the Picture Controls button at the bottom of the screen. Depending on your video card's capabilities, some or all will be adjustable.

Because DVD movies are made for display on TV screens, they can appear dark on a PC monitor. Use Gamma to adjust the brightness for all colors except black and white, and use Brightness to make all the colors, including black and white, brighter or darker. Contrast changes the difference between dark and light, Hue shifts all the colors, and Saturation changes the color intensity.

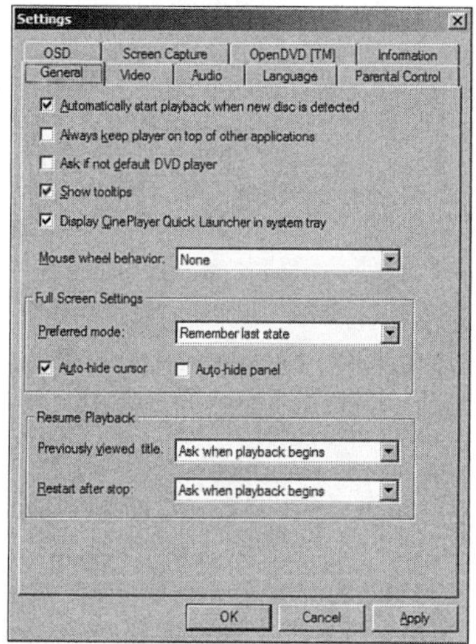

FIGURE 2.11
CinePlayer's Settings menu accesses myriad options.

5. The Decoding section gives you two options: Enable Hardware Acceleration and Use VMR (Video Mixing Renderer). Normally, you want to take advantage of your graphic accelerator, so keep that enabled. VMR is available only for Windows XP users. If you have XP, click VMR and click OK.
6. CinePlayer pops up a box asking whether it's okay to close your movie to switch on VMR or Hardware Acceleration. Click OK. Moments later you'll return to where you left off in the movie.
7. VMR might not make an obvious difference in image quality. However, one clear change is that it lets you adjust the Closed Captions Opacity option. Reopen the Settings interface (using the Wrench icon), select the Video tab, and note that the Opacity slider in the Closed Captions section is no longer grayed out. With your DVD movie playing, select Always Show "Closed Captions" if Available, move the Opacity slider, and note how the captions change in real-time onscreen.

A handy way to get a fix on your PC's hardware setup and current DVD software is to select the Information tab in the Settings window.

8. Now select the Screen Capture tab, which lets you grab any frame from any DVD movie. Some image editing products don't make this very easy because DVD movies run in a DirectX Overlay. CinePlayer, however, makes it simple. Select Bitmap (BMP) or JPEG image files, check the Correct Aspect Ratio box, and select a hard drive file storage folder.

JPEG is a picture format from the Joint Photographic Experts Group. It uses a sliding scale of 0–100 to define picture quality. Anything less than 50 is unacceptable to most people, so the CinePlayer slider, although it doesn't have a numeric scale, actually ranges from 50 to 100.

Most DVD movies are enhanced for wide-screen TVs, meaning they're stored on the DVD in a 4:3 standard TV aspect ratio and then are stretched to fit a 16:9 format. Selecting Correct Aspect Ratio guarantees your screen grab will look similar to what you see on the monitor.

9. Close the Settings interface and use the Take Screenshot camera icon to capture an image from your movie. If you selected the Audio Feedback on Success check box, you'll hear a camera shutter sound effect. Your screenshot will be in the designated file folder.
10. Finally, reopen the Settings interface and select the On Screen Display Options illustrated in Figure 2.12. This lets you add a time display or change the characteristics of the Events display: the words that confirm user actions, such as "Play," "2X FastF," "Pause," and the like.

Use CinePlayer's handy right-click menu. Right-click anywhere on the video screen or control interface and up pops a menu similar to the one in Figure 2.13. (The options that are accessible and those that are grayed out depend on features available on the DVD movie you've selected.)

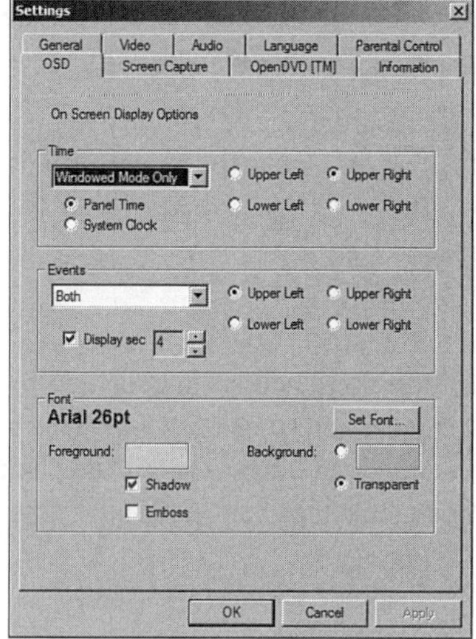

FIGURE 2.12
CinePlayer's On Screen Display Options interface lets you fine-tune CinePlayer's visual feedback.

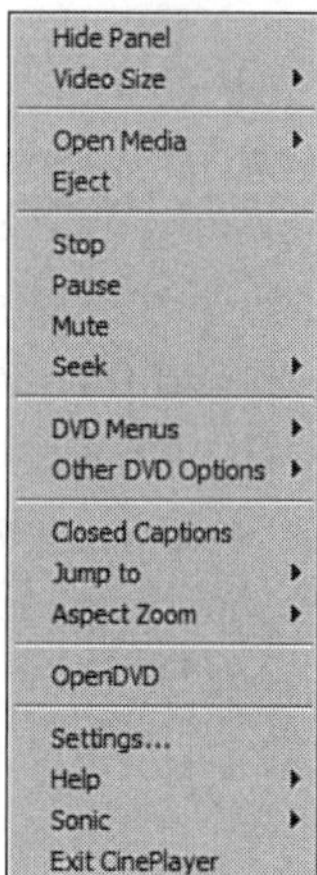

FIGURE 2.13
CinePlayer's right-click option provides rapid access to most items in the player interface.

Summary

The DVD format wars are still fresh in many DVD users' minds. But clarity has arrived in the form of a cross-format drive from Sony. If other manufacturers follow suit, that might signal the end to this issue. If so, as a multimedia producer, your choice is clear: DVD±R/±RW.

Choosing media is easier. Stick with generic name brands or better, and don't buy 4X DVD-R-rated media when your drive can handle only 2X.

One of the benefits of a PC DVD recorder is its capability to play DVD movies on your PC. Not only does most DVD movie playing software perform MPEG decoding, but it also offers features no set-top DVD player can provide.

Workshop

Starting with this hour, and for most other hours, I'll conclude with a set of how-to or what-if questions and answers, a quiz, and some exercises. This hour didn't cover many technical topics, so this will be something like a review-"lite." In any case, do take a few moments to review the Q&A and tackle the short quiz and exercises.

Q&A

Q I have a first-generation DVD+RW drive, and it doesn't record write-once DVD+R. What should I do?

A Replace the drive. DVD+RW is too narrow a format with a too limited compatibility to use as your only DVD recorder. Because you're essentially starting from scratch, you should use a DVD±R/±RW drive. At the moment, your only option is the Sony DRU500. By the time this book ships, other choices might be available.

Q When I capture an image from a DVD movie and then view it using my image editing software, it looks vertically stretched. What's going on?

A You have captured an image in its raw form as it exists on the DVD. Your DVD movie player software converts that tall image into the wider, anamorphic image you see on your PC monitor. You need to set your DVD movie player to capture in the proper aspect ratio.

Quiz

1. For Windows XP users, DVD movie subtitles can be obtrusive. Depending on how they're displayed, they might obscure the video too much. How do you reduce their opacity and let the movie shine through?
2. How do you use CinePlayer to capture bitmapped screen images and store them in a specific file folder using the name of the currently selected movie?
3. What's the fastest way to change the audio playback format using CinePlayer?

Quiz Answers

1. In the CinePlayer interface, click the Wrench icon to open Settings, select the Video tab, click Use VMR, click OK and let the movie restart, reopen the settings Video tab, select Always Show "Closed Captions" if Available, let the movie play so you can view the closed captions, and move the opacity slider to the desired position.
2. Select Settings, click the Screen Capture tab, select Bitmap (make sure Correct Aspect Ratio is checked), type in the movie name in the Custom Name window, browse to and select your file folder, click OK, play the movie, and click the Camera icon when a scene you want to capture appears.
3. This is something you might have discovered on your own. Use CinePlayer's right-click menu, select Other DVD Options, select Choose Audio, and select the audio format that fits your needs. The choices can run from several flavors of Dolby audio to various language subtitles.

Exercises

1. Take a critical look at the layout and organization of several DVD movies. Note what happens at the start of each DVD. Does it go directly to a menu, to the FBI warning, or through some introductory material? Can you skip that opening material, or does the DVD refuse to budge until you've seen it all? Does it have any extra features: subtitles, language dubs, or extra angles? How all these elements work within a DVD is the result of DVD authoring decisions and tools. You'll delve into all these topics in later hours.
2. Using a full-featured DVD movie, step through all of CinePlayer's right-click menu options.
3. Change the OSD—On-Screen Display—options in CinePlayer's Settings menu.

Hour 3

Burning Data DVDs

We'll start with normal DVD (and CD) data recording. This is not DVD *authoring* per se, but rather simple file transfers.

This hour starts with a hands-on run-through of Windows XP's built-in optical media disc burning module. I'll take you step-by-step through a well-known DVD creation product that's equal to, or at least very similar to, any you're likely to find bundled with a PC DVD recorder. Finally, this hour details the new features of the latest update to the de-facto DVD creation software industry-standard product.

The highlights of this hour include the following:

- Using Windows XP's My Computer to burn a CD
- Using bundled software to burn a DVD
- Taking a quick tour of Roxio's Easy CD & DVD Creator 6

Using Windows XP to Burn Data Files to a CD

My goal in writing this book is to help you create full-featured, interactive, multimedia DVDs. In this hour you'll take the first steps in that direction, beginning with burning data files, first to a CD and then to a DVD.

It might seem a bit mundane simply to use a DVD for normal data storage, but DVDs, with their 4.7 gigabyte (GB) capacity, are excellent data backup media.

If your PC's DVD recorder is of recent vintage, it also can record to CD-R and CD-RW discs.

We'll discuss those discs first because Windows XP's built-in optical disc recording software *cannot* handle DVDs. Those of you with older versions of Windows might want to skip ahead to the next section, "Using Bundled Software to Burn a DVD, "because Windows Me, 2000, 98, and earlier do not have built-in CD-recording capability.

Task: Burn Data Files to a CD-R or CD-RW Disc

TASK

With Windows XP, your PC DVD recorder can behave much like a massive floppy disc drive: You simply copy and paste files to it. If you use a CD-RW (rewritable) disc, you can erase files or update them. Here's how:

1. You'll need a CD-R or CD-RW disc. Insert a blank, writable CD into your DVD recorder, and the screen in Figure 3.1 pops up. You can select the Open Writable CD Folder option or click Cancel. Click Cancel because you don't need to go to that folder just yet.

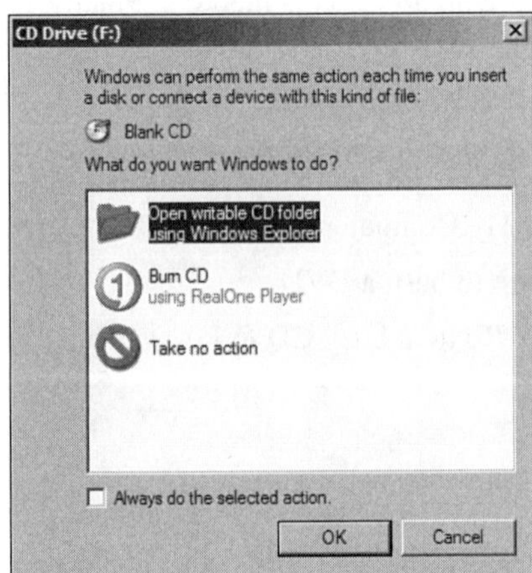

FIGURE 3.1
The Windows Autoplay feature notes when you've inserted a recordable CD into your DVD drive and asks what you want to do with it.

2. Open My Computer by double-clicking its icon. Depending on whether you've adjusted any My Computer Views settings, it'll look similar to Figure 3.2. Note that after you've inserted a CD-R or RW disc, the Total Size and Free Space values will be equal. In the case of my CD-RW disc, both those values are 656 megabytes (MB). A typical CD-R disc holds about 702MB.

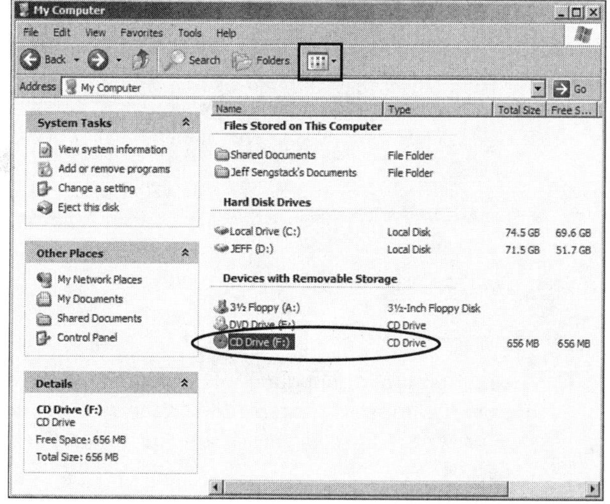

FIGURE 3.2
How My Computer should look after you've inserted a recordable CD into your DVD drive.

I do not like the default Windows My Computer icon-oriented settings; I much prefer the Details view. To switch to this format, click the Views icon (highlighted in Figure 3.2 at the top of the screen) and select Details from the drop-down list.

Windows lags behind the DVD technology curve. My Computer demonstrates that.

Before inserting a CD-R or CD-RW disc into the DVD-R/RW drive, My Computer refers to that drive as `DVD-R Drive (F:)`—*`CD Drive`*. As highlighted in Figure 3.2, after inserting a CD-RW disc, Windows updates the My Computer display by renaming the DVD-R/RW drive `CD Drive (F:)`.

If you insert a DVD disc into your DVD drive, Windows XP refers to it as a CD. If you try to write to a DVD recordable disc, you'll get an error message similar to the one in Figure 3.3.

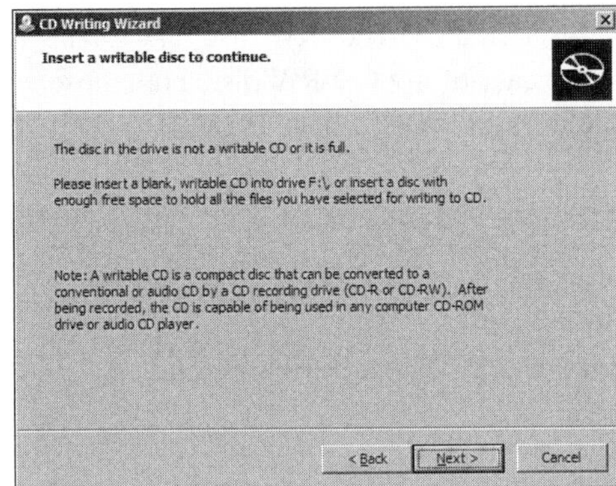

FIGURE 3.3
Using My Computer to try to write to a DVD recordable disc leads to this error message. The message should read, `My Computer can not write to recordable DVDs.`

3. Select the files you want to add to your recordable CD.

> To select groups of contiguous files, click the top file and then Shift-click the bottom file. To select scattered files in the same window, click one file and then use Ctrl-click to select each subsequent file. The same process works for folders.

4. Now you must copy each file, group of files, folder, or group of folders. I like to use right-mouse click shortcuts, so here's how I do it. As illustrated in Figure 3.4, right-click one of the selected files/folders (if you have more than one selected, right-clicking one applies to all those you've selected) and select Copy from the drop-down menu.

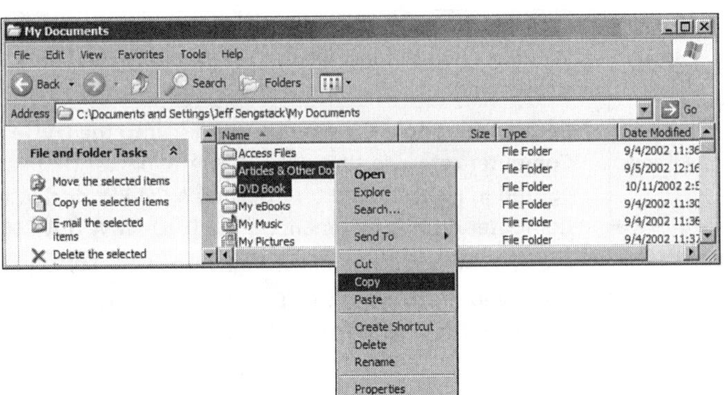

FIGURE 3.4
Use the right-click window to simplify the file copying process.

5. Navigate back to the opening My Computer screen (use the Backspace key as a shortcut). Right-click the DVD-recorder drive and select Paste. Depending on the size of the selected files, a pause might occur as Windows creates temporary files.

If you do not care for the right-click approach to Windows file management, feel free to use the menu-driven approach. Here's how: After you've selected some files or folders to copy, the Files and Folders Tasks window in the upper-left corner of My Computer will offer an option labeled Copy the Selected Items. Click that to open the Copy Items dialog box. Select the DVD drive, and then click Copy. When you're ready to burn your files or folders, go to step 6.

6. When you've copied/pasted all the files and folders you want to transfer to the recordable CD in your DVD drive, double-click its icon. Your screen should look similar to Figure 3.5. Note the highlighted window in the upper-left corner. Click Write These Files to CD.

Make sure your selected files don't exceed the capacity of your CD disc, which is usually about 650MB–700MB. You're on your own in this regard because Windows does not display your total file size as you add files and folders to the queue for later writing to the CD.

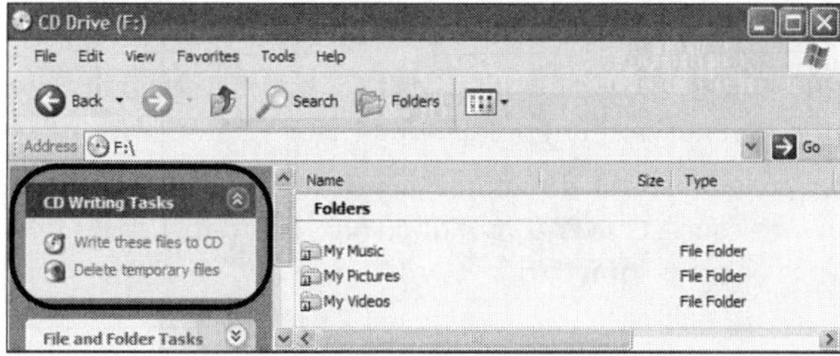

FIGURE 3.5
Use the CD Writing Tasks window to copy selected files and folders to your recordable CD.

7. The CD Writing Wizard shown in Figure 3.6 opens. Name your CD, click Next, and watch as Windows burns the files to your CD. When it's done, your CD will eject.

   Pop the CD back in to make sure all went well. It should have the name you gave it in the CD Writing Wizard.

3

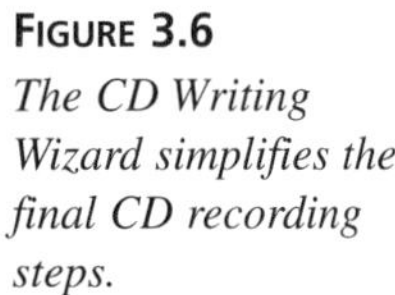

FIGURE 3.6
The CD Writing Wizard simplifies the final CD recording steps.

Getting Burned

In 1994, I was the principal writer for the now defunct *CD-ROM World Magazine*. (*Multimedia World Magazine* bought it out in 1995, and *PC World* absorbed *Multimedia World* shortly thereafter.)

CD-ROM World's parent company organized a trade show that fall, and I was one of the presenters. One of my sessions covered the then breakthrough technology of recordable CDs. Knowing how finicky they could be, I gave the trade show computer supplier detailed specs for my demonstration PC.

They almost got it right; the only minor deviation from my specs was a slightly different model video card that created an irresolvable conflict that killed my demo. I still could go through all the steps to prepare the data for recording to the CD-R drive, but when it came to actually burning a disc, the CD-R drive refused to function.

I let the audience know well in advance that it was not going to work. Many in attendance had had similar experiences, though, and I noted a sea of sympathetic expressions.

These days, bugaboos still abound; they just take on different forms. For instance, you might run into conflicts over competing DVD recordable standards—dash-R, plus-R, and DVD-RAM. It's always something.

Using Bundled Software to Burn a DVD

PC DVD recorder drives typically ship with a collection of reasonably good software: movie player, DVD authoring, and file transfer. Although some variations exist, data management products from Veritas are very common.

I'll use one of its offerings to demonstrate how to write data files to a DVD. Retail products such as Roxio's Easy CD & DVD Creator (see the next section) have a similar look and feel but throw in more features and functionality.

Prassi PrimoDVD—The Bundling Favorite

Prassi started in Italy more than 25 years ago and moved to California in the mid-1990s. It created Easy-CD Pro (later acquired by Adaptec) and DVD-Rep (bundled with the early Pioneer DVD-R drives), and these days its PrimoDVD (purchased by Veritas which in turn was bought out by Sonic Solutions in December 2002) is bundled with several DVD recorders.

Some drive makers have begun switching to Veritas's newer DVD data recording products—Record Now and DLA—but their functionality is similar to PrimoDVD. Even if you don't have PrimoDVD, whatever DVD recording product you do have will have similar processes.

Task: Use Prassi PrimoDVD to Record Data to a DVD

PrimoDVD uses wizards or lets you dive right in. I'll show you both methods. Here's how to use PrimoDVD to record files to your DVD:

1. PrimoDVD's wizards, shown in Figure 3.7, give you three choices: Disc Copy, Audio Disc, and Data Disc. You'll use Data Disc in this example. Click that button to open the Data Disc window.

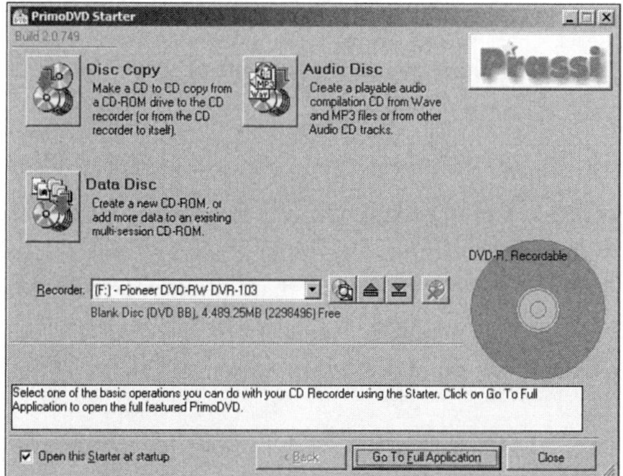

FIGURE 3.7
Prassi PrimoDVD offers three wizards to step you through the DVD creation process.

As with Windows, PrimoDVD has failed to completely make the leap to DVD. You'll notice that the wizards all refer to CDs, not DVDs. Nevertheless, PrimoDVD handles DVDs just fine.

2. As shown in Figure 3.8, you can add individual files or folders. Click the Add File or Add Folder button, and select from the displayed lists. The process is a little klunky. You can select only one *folder* at a time, but you can select more than one *file* at a time by using Ctrl-click or Shift-click. Click OK to place the files or folders in the Data Disc list.

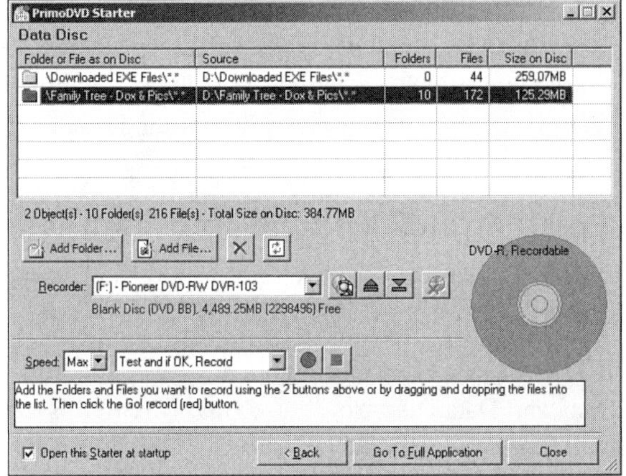

FIGURE 3.8
PrimoDVD's Data Disc creation interface lets you select multiple files but only one folder at a time.

You don't have to use the Add File or Add Folder button. You can drag and drop files and folders from My Computer directly to the Data Disc list.

3. When you're ready to record, check a few things first. Make sure you've selected your DVD recorder (PrimoDVD usually easily finds it), and be sure the default speed is set to Max (don't change that unless you have slower-rated media). In the drop-down list next to the Speed window, select either Record or Test, and If OK Record. Then click the red Record button.

These days it's less critical to test before doing the actual DVD (or CD) recording. Most DVD recorders are fast enough and have large enough data buffers (Sony's new DRU500 has an 8MB buffer) to avoid those nasty buffer underrun errors that have created many useless coasters. So, I recommend skipping the test phase and thus saving a big chunk of time.

4. Depending on whether you chose to test, your DVD drive will start recording. When completed, the DVD drive will eject. Insert the newly recorded disc again and check in My Computer to make sure it recorded properly.

Testing Your DVD Drive's Speed

As you make data CDs and DVDs, you might want to test the performance of your DVD recorder.

In PrimoDVD, before clicking the red Record button, note the Total Size on Disc figure in MB or GB. Start recording and note two times: how long it takes to get to 100% and how long it takes to finalize the disc.

Pioneer's specs for my AO3 DVD-R/RW say its top recording speeds are 8X for CD-R and 2X for DVD-R.

These *Xs* can be confusing. 1X for a CD means 150KB per second (KBps). That's the original speed for audio CDs.

On the other hand, 1X for a DVD is 1.25MBps, or more than eight times the data transfer rate of a 1X CD. Double that rate is the speed necessary for smooth playback of MPEG-2 videos.

Data is packed much more tightly on a DVD disc, so when the DVD drive reads a DVD, it does not need to spin as fast as it does when reading a CD to get the same data throughput. My Pioneer has a 24X CD-ROM read rating and a 6X DVD-ROM read rating. But a 6X DVD rate is the equivalent of a 50X CD read speed in data throughput.

Testing my DVD write performance with a CD-R, I found it records 1.47MBps. Dividing that by .15 (150KB), I discovered my Pioneer CD-R speed is about 10X, or slightly better than its specs. If I factor in the 40 seconds it took to finalize my CD-R, it ends up hitting that predicted 8X mark.

You can do the same testing with a CD-RW, DVD-R, or RW. RW speeds will be slower than their R counterparts.

Keep your decimal places in mind. Your total file size might be some number of MB or GB even though you might be dividing by KB. Note that GB = billion bytes, MB = million bytes, and KB = thousand bytes.

Task: Use Prassi PrimoDVD's Full Application

TASK

PrimoDVD gives you the option to bypass the wizards and go directly to the full application. If you use PrimoDVD more than once, you'll probably opt for this route. Here's how to do it:

1. Open PrimoDVD; then click Go to Full Application to open the interface shown in Figure 3.9.
2. Using the icons circled in Figure 3.9, select the type of disc you want to make. In this case, select Data Disc, the icon in the middle. That opens a new window within the main interface, shown in Figure 3.10.

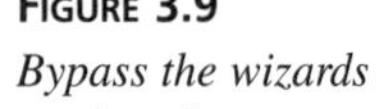

FIGURE 3.9
Bypass the wizards to go directly to PrimoDVD's full application.

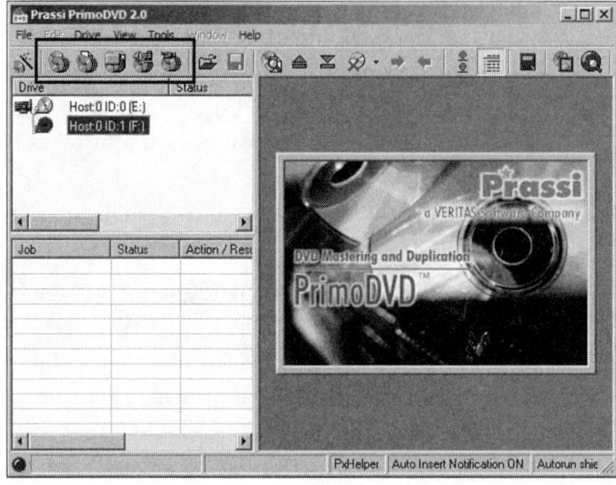

FIGURE 3.10
Take the direct route to PrimoDVD's Data Disc interface to expedite DVD creation.

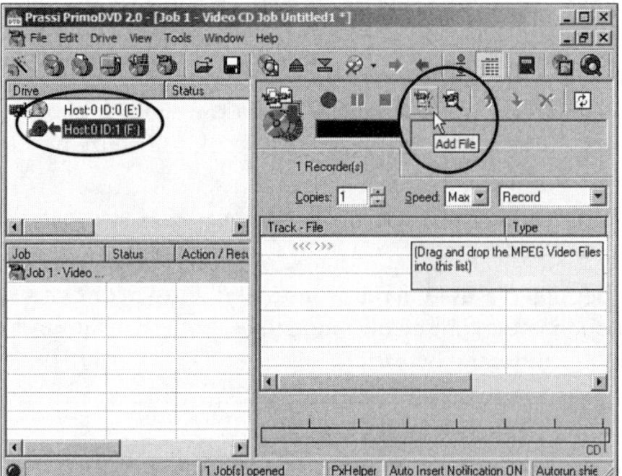

3. Select a recorder by double-clicking the drive icon in the upper-left of the screen in Figure 3.10. You'll know you've completed that task when a red arrow appears next to the drive.
4. Drag and drop files and folders to the data list, watching your total data figure in the bar at the bottom of the page.

Alternatively, you can use the Add File button circled in Figure 3.10 to access files much the same way you did when you used the Data Disc wizard.

5. When you're ready to record, click the red Record button. As before, when PrimoDVD finishes that task, it ejects the disc. Remember to check the disc to ensure that all went well.

PrimoDVD and most other optical disc recording software let you make VideoCDs. These use video files recorded in MPEG-1 format (they have a lower quality and lower data rate than MPEG-2). They play on most PC CD and DVD drives and many newer DVD set-top players. I explain how to capture, or *transfer*, video from your camcorder and convert it on-the-fly to MPEG files in Hour 10, "Capturing Video—Transferring Videos to Your PC."

PrimoDVD has some other features of note: disc-to-disc copy and audio CD creation. Disc-to-disc works well with CDs. Even if you have only one drive, PrimoDVD copies the contents of a CD or data-only DVD (see the following caution) to a temporary location on your hard drive, prompts you to place a recordable CD or DVD in the DVD recorder, and then records to it.

If you try to record a DVD movie, you'll probably run into some roadblocks. Hollywood studios have come up with copy protection schemes for most of their DVD movies, and PrimoDVD does not let you make duplicates of your favorite movies.

Making audio CDs (DVDs won't play in your music CD player) is easy. The only issue is knowing how to *rip*, or copy, songs from your personal music CD collection to your hard drive. I explain this in detail in Hour 8, "Acquiring Audio."

Taking a Quick Tour of Roxio's Easy CD & DVD Creator

Roxio, a spinoff from Adaptec, is the industry leader in optical disc creation software. Its Easy CD Creator is the de-facto industry standard. Microsoft uses Roxio software in the Windows CD recording module you might have tried at the beginning of this hour.

Roxio's Easy CD Creator 5 treats DVDs as an afterthought. Recognizing it needed a stronger DVD emphasis, Roxio has updated and renamed its flagship product, releasing

Roxio's Easy CD & DVD Creator 6: The Digital Media Suite. I tested a beta of that product. Here's my quick take.

Version 6 is a collection of loosely connected modules. As shown in Figure 3.11, its opening screen lists five tasks. Two are only marginally connected to DVD/CD creation: Audio Central and Photosuite.

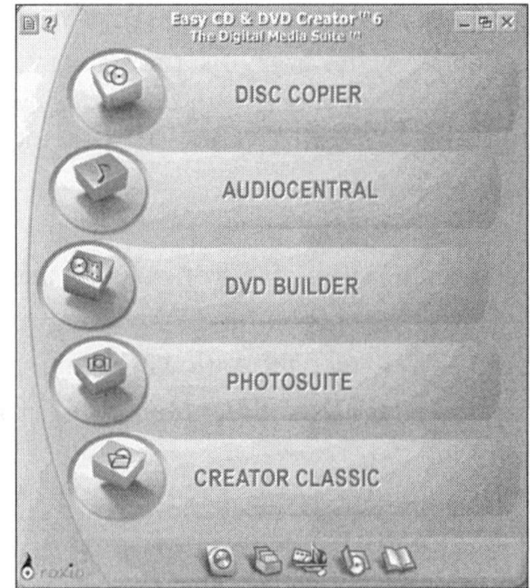

FIGURE 3.11
Roxio's Easy CD & DVD Creator 6 is a collection of five loosely connected program modules.

AudioCentral, shown in Figure 3.12, mimics the Windows Media Player while offering less functionality. Its primary role is music playback and selecting tunes to rip and record to a CD. PhotoSuite is a rudimentary image file management and touch-up and tool with only marginal usability.

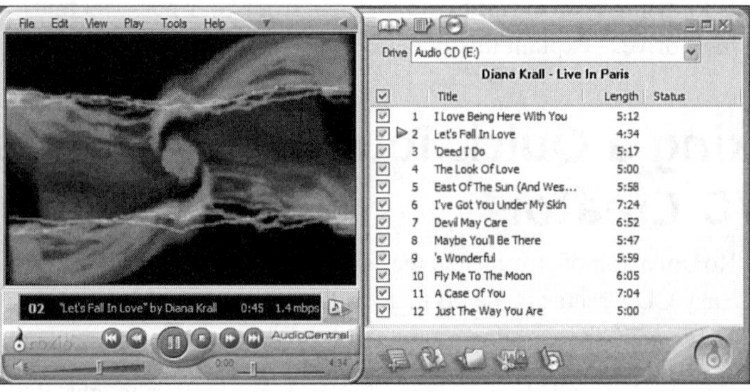

FIGURE 3.12
AudioCentral has a similar look and purpose as Windows Media Player with less functionality.

The core of Easy CD & DVD Creator remains the optical disc tools: Creator Classic and Disc Copier. The version 6 update stands out from its predecessor because of the acknowledgement that you can use those tools to make DVDS as well as the inclusion of a barebones DVD authoring module.

Disc Copier couldn't be easier to use. As shown in Figure 3.13, you simply open the Disc Copier module, tell the program the source and destination drives, and then click the Roxio logo at the bottom-right to start copying. You can copy using one DVD/CD drive or two. It's easy, just as the name says.

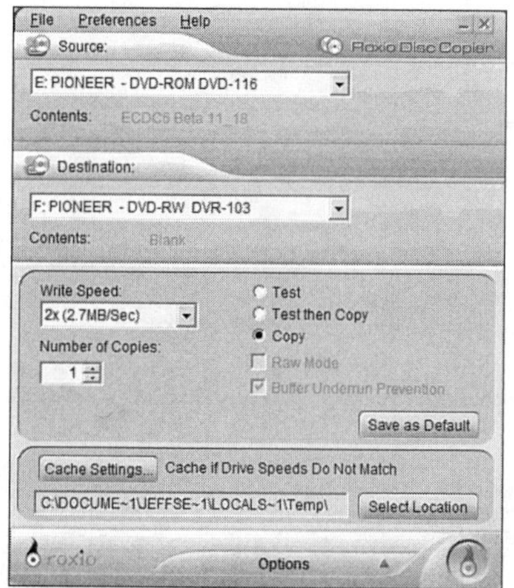

FIGURE 3.13
Roxio's Disc Copier is simple and effective.

Creator Classic, shown in Figure 3.14, is the real workhorse of this five-module collection. Its clear and remarkably easy-to-use, drag-and-drop interface and simplified setup for a wide variety of disc types makes working with this module a snap. I really like that the project size display at the bottom notes when you've exceeded the capacity of a disc but then simply lets you know that Creator Classic will put that excess capacity on another disc. That is a very clever twist on the old `out of disc space` message other products use.

Finally, DVD Builder, illustrated in Figure 3.15, is Roxio's attempt to get its foot in the DVD authoring door. My take is that it's not much more than a toehold. Its interface is confusing, its menu button and design options are too limited, and its controls are awkward.

You would be better off using a DVD authoring module included in a video editing product or, better still, a standalone DVD authoring product. I cover DVD authoring modules—those that ship with video editing software—in Hour 11, "Crafting Your Story and Selecting Video Editing Software." review standalone, entry-level, and prosumer DVD authoring products in Hour 16, "Evaluating Competing DVD Authoring Products."

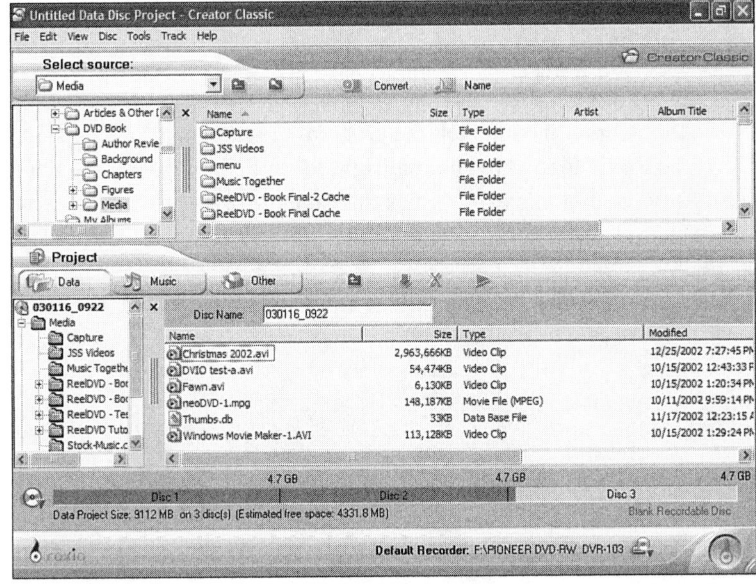

FIGURE 3.14
Roxio's Creator Classic makes the other Easy CD & DVD Creator modules look like afterthoughts.

FIGURE 3.15
DVD Builder is a less-than-adequate first stab at DVD authoring from Roxio.

Summary

My goal for this book is to help you produce full-featured, multimedia DVDs. But to get there, I think it's a good idea to take care of a few fundamentals. Topping that list is having you use your DVD recorder to create data CDs and DVDs. Windows XP has a built-in CD recording module created by optical recording industry leader Roxio. But that module does not work with DVDs.

Recording data to DVDs requires DVD creation software. Most PC DVD recorders come bundled with rudimentary DVD burning software. Veritas's (now Sonic Solutions) PrimoDVD is a commonly used product and readily dispatches most DVD recording chores. Moving up to a full-featured product such as Roxio's Easy CD & DVD Creator means added functionality and features.

Workshop

Review the questions and answers in this section to reinforce your data DVD creation techniques. Also, take a few moments to tackle the short quiz and the exercises.

Q&A

Q I put a DVD-R disc in my drive and nothing happens. Windows XP fails to display that pop-up interface asking me whether I want to open a writable folder. I can copy and paste files to that drive, but when I click Write These Files to a CD, I get an error message.

A Windows XP's new optical disc creation software does not work with DVDs. If you want to use that basic file management software, simply rely on CD-R/RW discs instead of DVDs.

Q How can my PC DVD recorder have a 6X DVD read speed and a 24X CD read speed?

A DVDs have seven times the capacity as CDs. DVD data are packed much more tightly than CD data. So, if a drive spins at the same rotational speed, data flows off a DVD about seven times faster than off a CD. But all that DVD data flow does create a bottleneck at higher speeds. Therefore, it's not a straight 7:1 ratio at all rotational speeds. Nevertheless, a 6X DVD rate equals 7.5MBps throughput and a 24X CD rate equals 3.6MBps. DVDs can have lower rating numbers but decidedly faster true data transfer rates.

Quiz

1. Windows XP's My Computer CD recording module offers two ways to tell it which files to copy. What are they?
2. How do you test your DVD recorder's DVD and CD record speeds?
3. How do you ensure you don't try to burn too much data to a disc?

Quiz Answers

1. I like the copy/paste and right-click approach. Select the files you want to copy, right-click them, select Copy, right-click your DVD recorder drive letter, and select Paste. Method #2 uses the Files and Folders Tasks window in My Computer. Select the files you want to copy; then, in the File and Folder Tasks window, select Copy the Selected Items. Finally, in the pop-up Copy Items window, select the DVD drive and click Copy.
2. Before you click the Record button, note the total size of the files you're copying and check the time. Then, click Record. When you're finished, calculate the elapsed time and divide that into the total file size. For a CD, divide that answer by 150KB, the standard 1X rate. For a DVD, divide by 1.25MB, which is equivalent to 1X in DVD parlance.
3. If you rely on Windows XP's CD creation module, you need to do the addition yourself. My Computer will note how much free space the recordable CD has, and you need to be sure you don't exceed it. PrimoDVD (and other bundled optical recordable media tools) tracks that for you as you add files to the queue.

Exercises

1. Make a backup disc of all your critical files. If you think those files will exceed 700MB, use a DVD. You might consider using rewritable discs to do backups on a regular basis or, as you update files, simply replace the old ones on the disc.
2. Make a CD-to-CD or DVD-to-DVD copy using PrimoDVD or whatever bundled CD/DVD creation software came with your DVD drive. If you have more than one drive, that simplifies things, but try it using only one drive. Try to copy a DVD movie and see how that works. Depending on the copy protection scheme, you will probably get a disc read error.

HOUR 4

Authoring Your First DVD Project Using MyDVD: Part I

You're undoubtedly itching to create a multimedia DVD—one with menus, buttons, videos, music, and still images. In this hour you start that process, and in the next two hours you finish it.

To do this DVD authoring, you'll use a trial version of MyDVD 5 and some tutorial assets—video, stills, and audio—included on this book's companion DVD. Your project will sport some Hollywood flare, such as animated menus with music accompaniment by video buttons.

In this first hour, we'll limit the lesson to some getting-acquainted fundamentals. We'll tackle the real DVD authoring—video editing, menu creation, and DVD burning—in the next two hours.

The highlights of this hour include the following:

- Introducing the authoring process
- Checking out the MyDVD 5 interface and feature set
- Using MyDVD's scene detection module

Introducing the Authoring Process

Authoring a DVD can be as simple as transferring a video directly from your camcorder to a recordable DVD or as complex as creating multiple menus with dozens of buttons, several audio tracks, and subtitles.

The process varies from authoring project to authoring project, but the fundamentals remain the same.

Creating, Acquiring, and Editing the Media

You first create, acquire, and edit the media you want to include on your DVD. You typically use a camcorder to make videos, a digital or film still camera for photos, a scanner to load documents and other images onto your hard drive, and graphics software to create menu backgrounds and buttons.

You can choose to use the basic video editing software included on most consumer DVD authoring products to simply select some scenes from your video and add transitions—dissolves, wipes, fades, and so on—between those scenes. Or you might step up that video production quality by using higher-end video editing software. I'll cover video editing techniques and scanning tips in Part III, "Editing Media."

Authoring Your DVD

Authoring is when you bring together all the elements of your DVD. You typically begin by assembling all the media assets into one readily accessible location; then you build menus. Menu backgrounds simply can be static (a graphic, video still image, or photo) or, depending on the sophistication or feature set of your DVD authoring software, you might opt to use a video or animated look.

Consumer DVD authoring products offer a plethora of menu templates, and some even let you build your menus from scratch using graphics supplied with the software. In this and the next hour's first-time DVD authoring process, you'll rely on the wide range of graphics and templates provided with MyDVD. A product covered later in the book, DVDit!, comes with a full range of more customizable graphics and numerous menu creation options.

You also might want to add audio to your menus, such as music or a brief narration, and have it repeat (or *loop*) itself until the viewer takes some action.

Adding Buttons and Links

The viewer action usually entails using the remote control to click an onscreen menu button. Entry-level products such as MyDVD let you choose a button style from a wide variety included with the program. Other higher-level products expect you to create the buttons and other graphic elements on your own, outside the authoring environment.

You add buttons plus text to those menus. Frequently, the buttons are just thumbnails of video scenes placed in a frame style of your choosing—that's MyDVD's approach. Other products like DVDit!, however, let you use custom-made graphics and text as buttons.

Finally, you'll connect (or *link*) those menu elements to other menus or media. In MyDVD, this takes place automatically. Higher-level authoring products give you detailed control over this so-called menu navigation.

Checking Out the MyDVD 5 Interface and Features

MyDVD 5 is the best consumer-level DVD authoring product. I reached that conclusion after testing five entry-level products for Hour 16, "Evaluating Competing DVD Authoring Products."

MyDVD lets you create DVD projects with ease. It has built-in templates and wizards that step you smoothly through the DVD authoring process. The available menu and button styles give you all the options you need to cover most of your DVD production needs.

As you gain expertise, you can substitute your own graphics, images, and audio during menu creation. Using a *plug-in* (a mini-program inserted into another program to perform a specific function), you can use Adobe Photoshop to create menus with built-in buttons and import them directly into MyDVD. (I cover that in Hour 22, "Creating Custom MyDVD Templates with Style Creator.") As you increase your skills, you might want to build customized DVD templates and post them on the Sonic Solutions Web site (more on that in the next hour).

Downloading, Installing, and Starting MyDVD 5

You probably know how to install software, but in case you need a little assistance, here's a brief explanation.

As we went to press, Sonic Solutions had not created the trial version of MyDVD 5. They will make it available for download at `http://support.sonic.com\downloads.asp`. There will be a link to download MyDVD for readers of this book at that site. After downloading that file, double click it to install it.

The default installation folder is `C:\Program Files\Sonic\MyDVD`. Move the tutorial assets from the DVD to that folder by opening `My Computer`, right-clicking the `MyDVD Tutorial Assets` file folder on the book's DVD, selecting Copy, navigating in Windows Explorer back to `C:\Program Files\Sonic\MyDVD`, right-clicking that file folder, and selecting Paste. It takes a few minutes to transfer the files from the DVD to your PC's hard drive.

Task: Tour MyDVD's Interface

MyDVD's interface is logically laid out and very user friendly. Here's a basic run-through to give you an idea of its features and what's to come:

1. Start MyDVD by double-clicking its icon on your desktop. Alternatively, if the installation process failed to place an icon there, select Start, Sonic, MyDVD, Start MyDVD. Doing so opens the interface shown in Figure 4.1.

Figure 4.1 *MyDVD's opening interface is the first indication of how streamlined and simplified this entry-level DVD authoring process can be.*

2. Your choices are fairly straightforward. You can author a DVD or put a video on a CD. Within each of those two options, you can create or modify a project, transfer video directly to a DVD or CD, or edit an existing DVD or Video-CD project. For this task, select DVD-Video and then click Create or Modify a DVD-Video Project.

If you want to explore a bit, select Transfer Video Direct-to-DVD, which opens the screen shown in Figure 4.2. This wizard steps you through a simple process to transfer (or *capture*) your video straight from your camcorder to a DVD, giving that DVD a simplified opening menu if you select that option. I cover this direct-to-DVD process along with several other DVD recording topics in Hour 22, "Burning DVDs and Dealing with Mass Replicators."

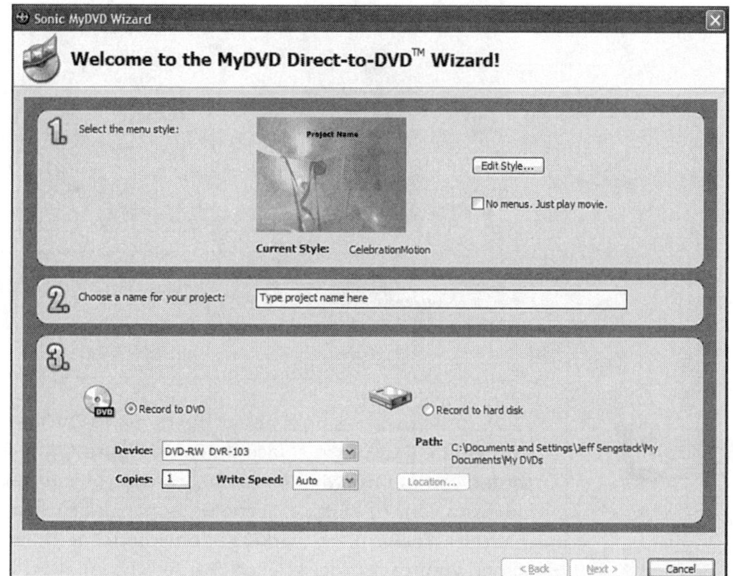

FIGURE 4.2
MyDVD's Direct-to-DVD option lets you make a straight copy of your video to a DVD.

3. The main MyDVD user interface opens. As shown in Figure 4.3, the central part of the screen represents the DVD menu viewers of your DVD will see when they first insert your DVD into their set-top players or PCs. The icons down the left side let you capture or transfer video from your camcorder (I cover this topic using other software products in Hour 10, "Capturing Video—Transferring Videos to Your PC"), get movies (video files) from your hard drive, add still images to a slideshow, and add a submenu (or *nested* menu). The buttons across the top—Edit Style, Edit Video, and Edit Chapters—come into play in the next hours' tasks. For now, click Get Movies.

FIGURE 4.3
MyDVD's main user interface menus and icons are in a logical workflow layout.

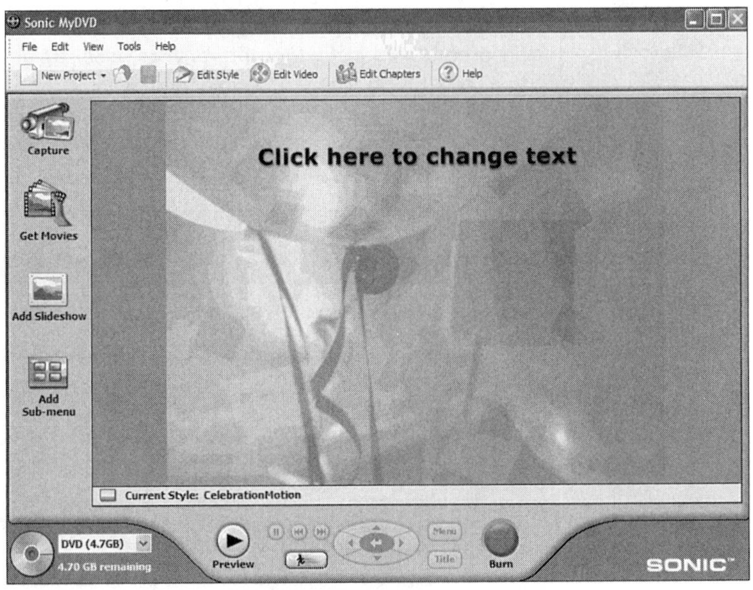

As with the previous note about the direct-to-DVD process, feel free to explore the Capture interface shown in Figure 4.4. If you have a DV camcorder, you can use MyDVD to directly control your camcorder and copy videos to your hard drive. Scene Detection, a new feature in MyDVD 5, automatically creates place markers at obvious scene changes (or where you turned your camcorder's Record/Pause button), making it easier to edit the video later. You also can perform scene detection with videos already stored on your hard drive. I explain that in this hour's next section, "Using MyDVD's Scene Detection Module."

4. The Add Movies to Menu window opens. Navigate to the `MyDVD Tutorial Assets` file folder. Later in this and the next Hour you'll select one or more files but for now, click Cancel to return to the main user interface.

In Figure 4.5, I opted for the Thumbnail view (displaying the opening frame of each video) versus the default List view. To switch to the Thumbnail view, click the View Menu icon (in the upper-right corner) and select Thumbnail.

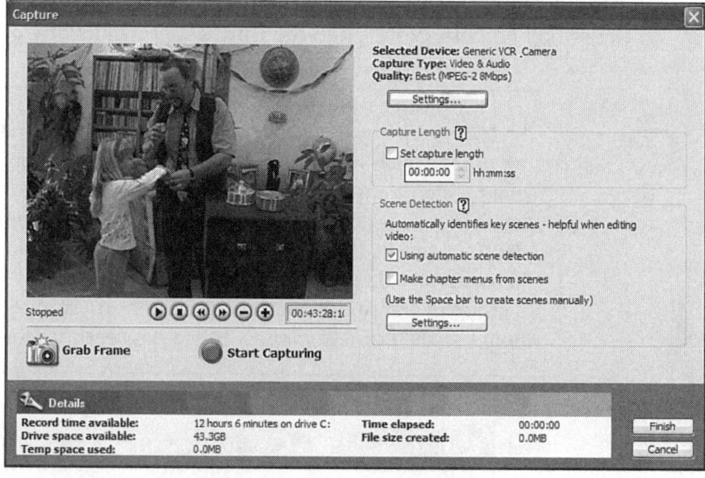

FIGURE 4.4
The Capture interface lets you transfer video from you camcorder to your PC's hard drive, marking scene changes in the process.

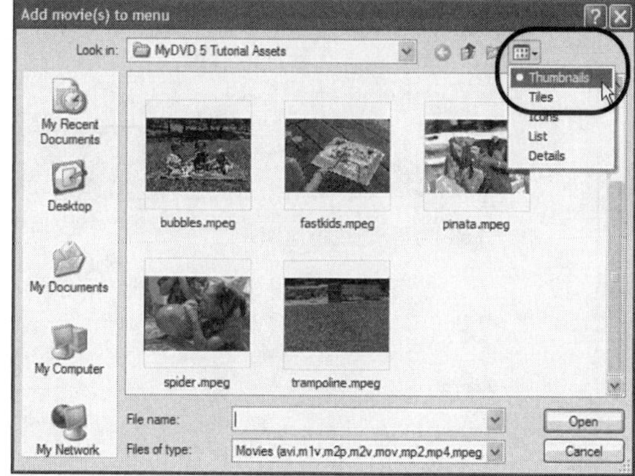

FIGURE 4.5
Use the Add Movies to Menu window to select videos for your DVD project.

5. Before you do any video editing or authoring, take a look at one other interface. Select File, Preferences.
6. Most of the Preferences tabs (shown in Figure 4.6) are self-explanatory, but we'll look at two. Select the Video Editing tab.
7. Here you can have MyDVD place the same transition between all video segments and select the duration for all transitions (whether added automatically by MyDVD or manually by you). In addition, you can set the display time for each slide in a slideshow and set the color for the letter box (the gap between the edges of the slide and the TV screen for slides that don't exactly match a TV's standard 4:3 aspect ratio). Make only one change: Click the Duration drop-down list and change

the default transition duration to 1 Seconds. Leave all the other items at their default settings—no automatic transition and a 5-second slide display with black letter box borders. Now, go to the Burn DVD window by clicking its tab.

MyDVD has many excellent transitions, and I recommend experimenting a bit to see which works best for your video scene changes. That said, my general take on transitions in a video is that you should minimize their use because they can be distracting and garish. Video editing is discussed in depth in Part III. If you still want to automatically add the same transition between all video scenes, use the top drop-down list in Figure 4.6 to display a group of transitions and then use the next list to select the specific transition. Use the Preview window to see how each transition looks.

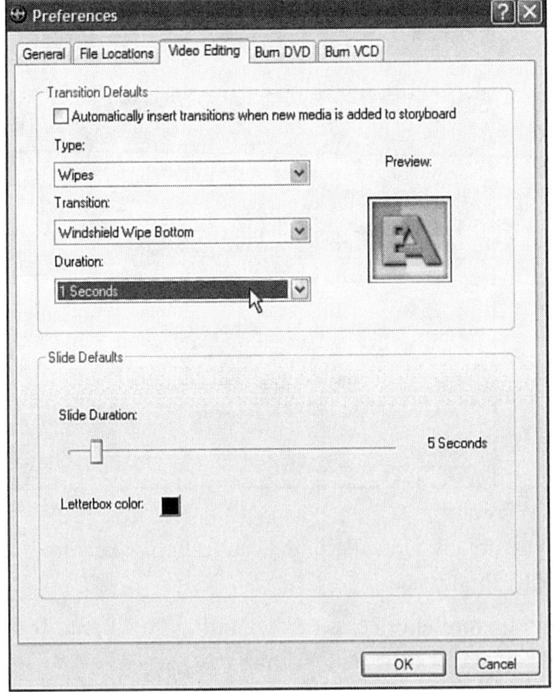

FIGURE 4.6
MyDVD's Preferences interface gives you control over some automated processes, such as video transitions and converting video files into MPEG files.

8. The Burn DVD window, shown in Figure 4.7, lets you set the quality of how MyDVD converts your video files into MPEG—encoding, or *transcoding*, in MyDVD parlance. Faster transcoding times lead to lower video quality and smaller files. The Audio Options area lets you use PCM versus Dolby Digital. Again, accept the default settings by clicking Cancel.

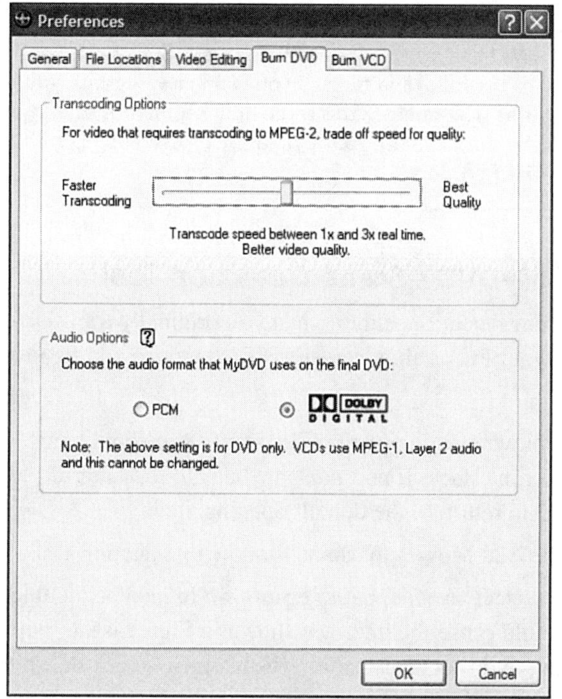

FIGURE 4.7
The Burn DVD window lets you select the overall quality of the MPEG video as well as the audio format.

Choosing Dolby Digital over PCM is generally a no-brainer because Dolby Digital is the de facto DVD standard, so all DVD players should support it. Plus, its file size is generally smaller than equivalent quality PCM files. PCM is at the core of most digital audio signals, including music CDs. Some authoring products let users create lower-quality PCM files, thereby reducing the file size. However, MyDVD does not offer that option, preferring to give the user better sound quality.

Some DVD authoring products don't include Dolby Digital. However, MyDVD has it, so take advantage of it.

Using MyDVD's Scene Detection Module

New to version MyDVD 5 is its video editor. Built from the ground up by Sonic Solutions, it replaces an editor in version 4 licensed from ArcSoft. Although it's still geared to the consumer market, this new editor has plenty of bells and whistles, starting with a feature not found in any previous version of MyDVD: scene detection.

We have provided video files on this book's companion DVD for you to use in the following tasks. If you want to work with your own video files, that's fine. The goal for this three-hour segment is to author a DVD from start to finish. If you use your own assets, you'll end up with a much more personalized product.

Task: Use MyDVD's Scene Detection Tool

▼ TASK

Relying on scene detection, either when you originally transfer (capture) video from your camcorder to your PC or after, can simplify the video editing process. Here's how it works:

1. When you wrapped up the previous task, you should have ended up at MyDVD's main user interface. If not, navigate back to that interface or close and restart MyDVD to return to the default opening view.
2. Open the Add Movies to Menu window by selecting Get Movie.
3. You can select any file, but to ensure we're in sync, double-click `Go-Karts.avi`. That should cause the interface shown in Figure 4.8 to pop up asking you whether you want to do scene detection. (If that alert screen doesn't appear, read the following caution and follow its instructions.)

Sometimes, for a number of reasons, MyDVD does not automatically display that scene detection alert shown in Figure 4.8. Typically, it's because you previously added the selected file to a different project. In any event, MyDVD simply returns you to the main interface. In that case, click the newly created menu button (in this case `Go-Karts`) to select and highlight it and select Edit Video from the taskbar at the top of the main interface. Then, when you're asked if you want to go to scene detection, click Yes.

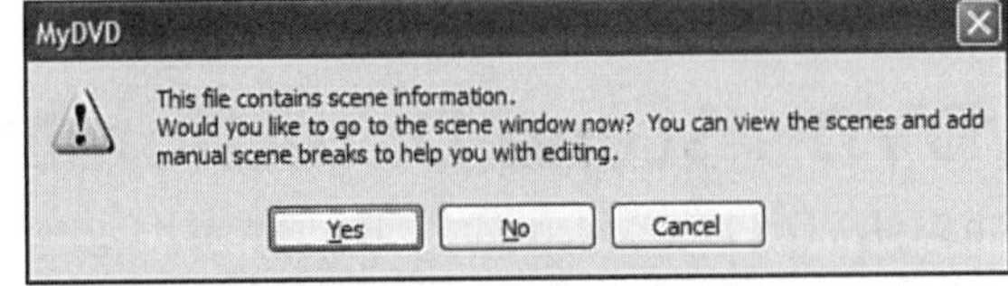

FIGURE 4.8
By default, each time you first add a video file to a project, MyDVD asks whether you want to use its scene detection feature.

4. Click Yes on the '!' Scene Detection message interface to open the Mark Scenes for Editing window shown in Figure 4.9.

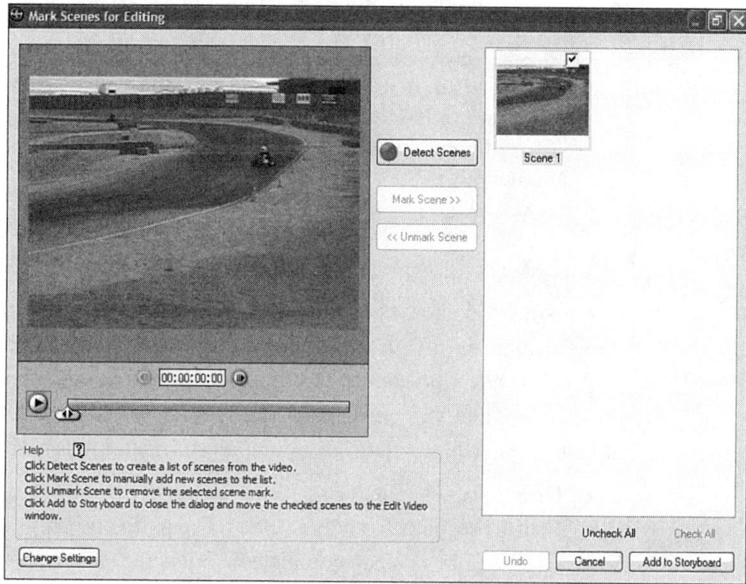

FIGURE 4.9
Use this feature to automatically detect scene changes, making it easier to trim individual scenes and later add transitions between them.

5. You can accept the defaults and click Detect Scenes, or you can change one option. Click the Change Settings button at the bottom of the screen to open the Scene Detection options dialog box shown in Figure 4.10.

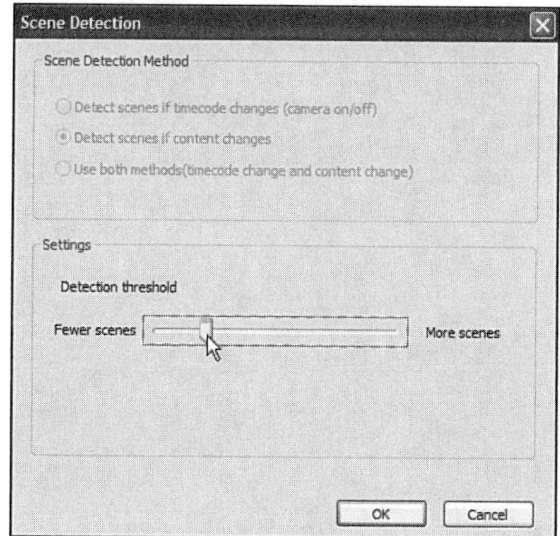

FIGURE 4.10
When using scene detection on a video file, your only option is Settings: Detection Threshold.

The reason the Scene Detection Method section at the top of the window is *grayed-out*, meaning you cannot access those options, is because it applies only when you use MyDVD to *transfer* video directly from your camcorder to your PC. In that case, the section's options are available to you and you can choose to have MyDVD mark scenes each time you turned the camcorder on or off when making the original videotape and when the scene content changes.

6. You have only one option—Settings: Detection Threshold. Moving that slider right or left simply decreases or increases the scene detection sensitivity, but it is an inexact science. In general, trial-and-error is your best bet here because different video clips have different types of scene breaks. It uses a numeric setting that displays only when you click the slider. Setting it to 2 is your best bet in this case, so click the slider button and make that adjustment. Click OK.
7. Click Detect Scenes. The `Go-Karts.avi` clip is 2 minutes long, and it takes MyDVD only about 15 seconds to analyze it and display thumbnails of what it thinks are the scene change points. When completed, your screen should look similar to Figure 4.11.

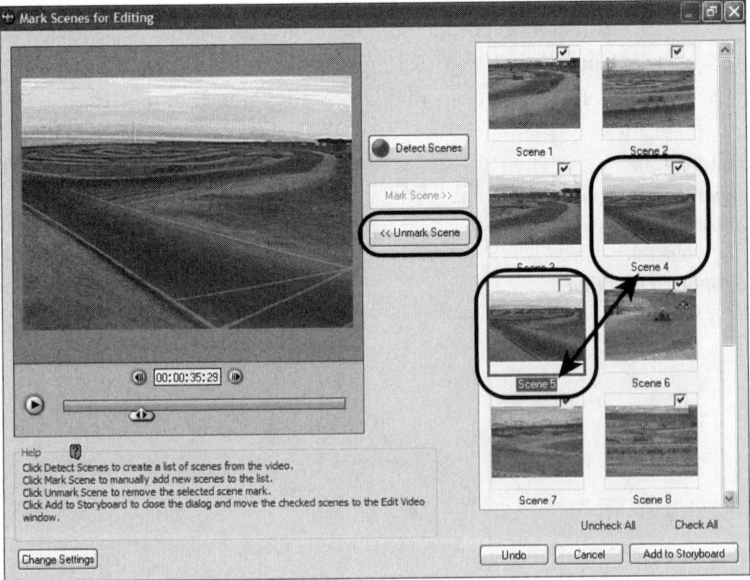

FIGURE 4.11
Using scene detection creates a collection of thumbnails representing scene breaks.

To prepare this chapter, I worked with a beta version of MyDVD 5 that was still being tweaked. The scene detection algorithm was a work in progress. The trial version of MyDVD 5 included with this book should have nearly flawless scene detection. Therefore, you can view step 8 as simply a means to see how to add missed scenes.

8. If the scene detection module fails to detect a scene change, you can fix that minor miscue by manually adding that one scene. To do that, scrub through the video—click and drag the double-arrow slider highlighted in Figure 4.12—looking for the unselected scenes. When you find it (I've noted one in Figure 4.12), click Mark Scene. MyDVD adds it to the collection of thumbnails images in its sequential location.

The double-arrow scrubber generally takes you to the vicinity of a scene change, but you might want to use the nudge tools (the triangles next to the time code display) to move on a frame-by-frame basis. Or, if you know the time code (in this case, it's 36 seconds and 12 frames, or 36:12), you can click that time code display to highlight it and then type in the exact time code.

You can hold down the Ctrl key while clicking the Next and Previous buttons to jump in 1-second increments. This is useful for getting to the vicinity of the scene change without having to step through every frame.

MPEG videos (as opposed to AVI or QuickTime MOV files) have inherent technical issues that sometimes lead to less-than-accurate manual scene selection. The fault does lie with MyDVD, but the reason lies in how MPEG files are compressed and encoded. That's why, as you scrub through an MPEG video, you might notice the Mark Scene button switch on and off. Only when you've scrubbed to a selectable or editable frame can you actually mark a scene change.

9. Click the Add to Storyboard button to open the MyDVD video editing interface and place all those scenes on a timeline.

FIGURE 4.12
Scrub through the video to find an unmarked scene change, and then mark it.

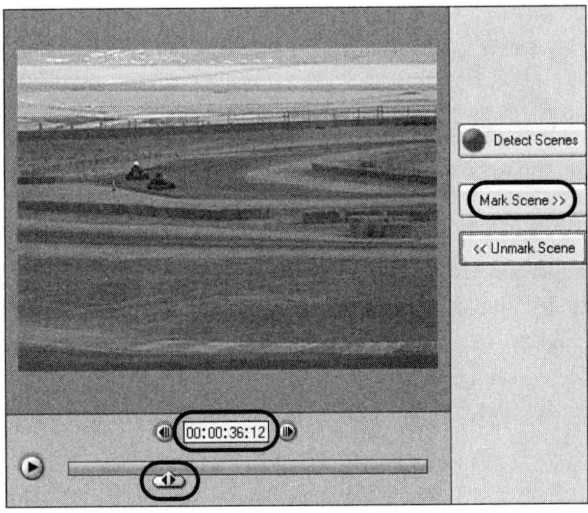

Summary

This hour introduced you to the DVD authoring process. The first steps include gathering and editing your media assets. I cover the remaining steps—video editing, menu creation, linking assets and menus, and DVD burning—in the next two hours.

You've taken a close look at MyDVD's features. It's much more than a straightforward consumer-level DVD authoring product; it gives you a video capture option with built-in and customizable scene detection.

In the next hour, you'll find that MyDVD's easy-to-use video editing module offers a multitude of transitions and video effects, an additional audio track, and a text function. Its authoring and menu creation tools greatly simplify those processes while giving you more options than other products in this category and price range.

Workshop

Review the questions and answers in this section to reinforce your introductory MyDVD skills. Also, take a few moments to take the quiz and do the exercises.

Q&A

Q I have some video files already on my hard drive and want MyDVD to do scene detection only at the points when I turned my camcorder off and on. But I don't see any way to do that. How does that work?

A It's too late for that option. After you've transferred a video to your hard drive, that camcorder on/off scene detection option is no longer available. You might note that that section in the MyDVD Preferences is grayed-out, meaning it's unavailable as an option in this case.

Q I selected a video from the Add Movies window and wanted to do scene selection, but the automatic alert message that normally pops up to give me that option failed to display. How do I start scene selection?

A The alert message probably failed to appear because you've accessed this file before. To apply scene detection to this video, click its button in the main interface menu screen to select it and select Edit. This causes a different alert window to appear asking whether you want to go to the Scene Detection window. Click Yes.

Quiz

1. You used the scene detection process, and it failed to mark several scene changes. How do you increase the scene detection sensitivity?
2. You want to add a transition between all scenes in your video. To keep those scene changes from being too obtrusive, you want to use a cross fade. How do you do this?
3. You've used the scene detection process and ended up with several duplicate scenes. How do you remove them?

Quiz Answers

1. In the Scene Detection window, click the Change Settings button and slide the Detection Threshold slider to the right.
2. Set the default transition *before* you do scene detection. Do that by selecting Edit, Preferences; going to the Video Editing tab; and finding Cross Fade under the Fades & Dissolves group. I'd suggest setting the Duration to 1 second.
3. One at a time, click each check box above the scene's upper-right corner; select the scene by clicking its thumbnail, then click Unmark Scene.

Exercises

1. Several other video files are available in the MyDVD Tutorial Assets file folder, including a set of `Birthday Party Picnic` AVI files. Use scene detection on them. Because some are edited clips (instead of video transferred directly from a camcorder), they have transitions in them (typically cross fades or dissolves). That

makes things difficult for scene detection because no obvious breaks exist. Experiment with the scene detection sensitivity slider settings to see how it handles this situation.

2. In the next hour, you will embark on the menu creation process so now is a good time to see how it all works from a viewer's perspective. Take a look at some feature film DVDs, noting how their menus work. As you use your remote to move from button to button, take a look at how the button colors (highlights) change when you move the cursor over them or click them. Check out how the menus are interconnected and give you options to return to previous menus or go on to other menus.

HOUR 5

Authoring Your First DVD Project Using MyDVD: Part II

It's time to author a DVD. In this hour you complete a series of tasks using MyDVD 5 to edit a video, assemble your assets, and create menus and buttons. You'll place chapter points in a video with associated menu buttons and add a slideshow to your DVD. In the following hour, you'll edit the menu style, learn how to share your customized menu templates online, and finally burn your project to a recordable DVD.

The highlights of this hour include the following:

- Editing video with MyDVD
- Assembling assets and building menus using MyDVD
- Setting chapter points and adding a slideshow

Editing Video with MyDVD

MyDVD offers a full slate of entry-level video editing tools. You can trim any scene (change its in- or out-points), remove (delete) a scene, or move a scene out of its original sequential order. MyDVD offers a solid collection of scene transitions, such as fades, dissolves, and wipes. It lets you choose from some very cool special video effects—such as embossing, adding a sepia tone, or turning a video scene into a mosaic—that can change the look of your video. Also, you can add music or a narration to the entire video.

Task: Edit Video with MyDVD

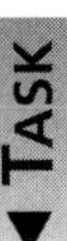

This entirely new editing module created from scratch for MyDVD 5 is remarkably easy to use. Here's how it works:

1. At the completion of the previous hour's task, you should have ended up in the Edit interface shown in Figure 5.1. If not, click the `Go-Karts` menu button in the main interface and click Edit Video. You'll see a message asking whether you want to go to scene detection. Click Yes—your previously selected scenes should show up—and then click the Add to Storyboard button.

Skipping the Scene Detection option (clicking No on the message), even after previously going through the detection process, cancels any scene detection work you've done. Only after you've done some editing will MyDVD retain your scene detection information.

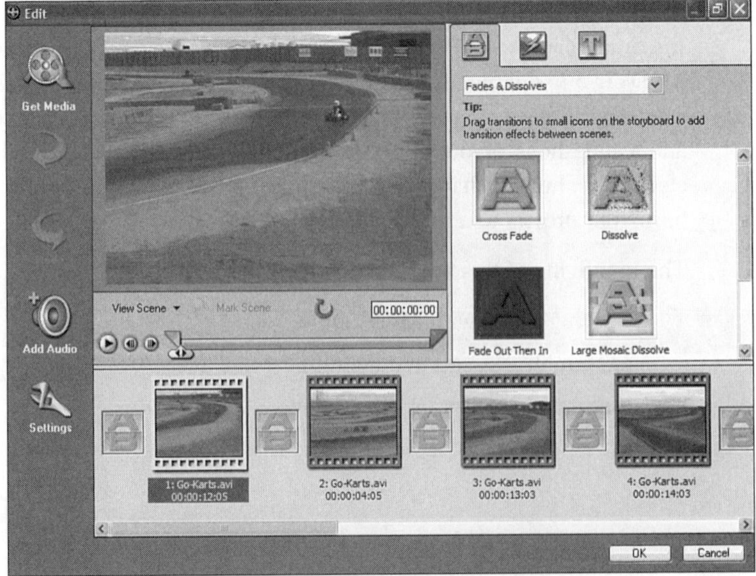

FIGURE 5.1
MyDVD's Edit interface lets you add transitions between clips, apply special effects to the video images, superimpose text on clips, and add audio.

2. First, you need to move a clip. I sometimes like to start a video with a wide, establishing shot. `4: Go-Karts.avi` fits that category, so click and drag its filmstrip icon ahead of the first scene thumbnail, `1: Go-Karts.avi`. As you drag the icon, you'll see a vertical bar (shown in Figure 5.2) showing where it will go if you release the mouse button. The moved scene should now show up in the Video display window.

As you drag scenes to new locations on the timeline, you'll note that they change their scene numbering to match their new locations. That is, if you drag the fourth scene ahead of the first scene, the fourth scene label becomes `1: Go-Karts.avi` and the previous first scene becomes `2: Go-Karts.avi`. To ensure my explanation is clear, from this point on I'll refer to scenes by their ordinals—that is, third, fourth, fifth, and so on.

As you make changes to your video, you might notice that the Undo button becomes active. If you don't like what you've done, simply click Undo to back up one step for each click. This is a very helpful tool. Conversely, if you undo something and realize it was okay the way it was, click the Redo button.

FIGURE 5.2
When moving a scene, a vertical black line notes its new location. After you make any change to your video, the Undo command becomes active.

5

3. I think that newly placed opening shot is too long. Trim it to 3 seconds by clicking the red triangle at the right side of the viewing window and dragging it until the time code display reads `3:00` (more or less—this isn't brain surgery). This process is shown in Figure 5.3.

Here's a scene length rule of thumb: Unless a scene has some compelling action or some other reason to keep it onscreen for a while, 3 seconds is generally a comfortable length for a scene.

You can trim your clip using the red triangle, the green triangle, or both. If you do any trimming from the beginning of the scene (the green triangle) and want a specific clip length, you'll need to do some calculating. Click the green and red triangles to see their respective time codes; then do your subtraction. Higher-end video editing products do this calculation for you as you change a scene length.

FIGURE 5.3
Use the green and red triangles to trim a clip.

If you play the trimmed scene, it displays as if no trim took place. To see the trimmed version, click View Scene (shown in Figure 5.3) and select View Entire from the drop-down list. Now, when you click the Play button, the trimmed scene runs for 3 seconds and the video continues to the second scene.

4. I think the edit from the third scene to the fourth is abrupt. Instead of a cut edit (the most frequently used edit), use a transition to smooth that scene change. Click the A/B box at the top of the screen, and then look at the drop-down list shown in Figure 5.4. Six transition groups are available, each with 4–25 different transitions. In this

case, keep things simple and select Cross Fade from the Fades & Dissolves group. Click and drag it to the transition placeholder between the third and fourth scenes.

Feel free to check out all the cool transitions included with MyDVD. To see them in action, simply click one and watch its animation. If you want to change a transition you've already added to a video, simply drag the replacement transition on top of the existing transition.

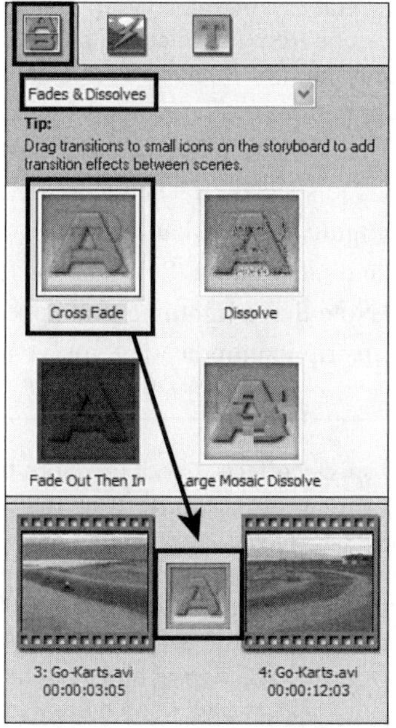

FIGURE 5.4
Add a transition by selecting one from the drop-down list groups and dragging it to the transition icon between two scenes.

5. Check your handiwork by clicking the newly placed transition and clicking the Play button. If you changed the default transition time when completing this hour's first task, the dissolve lasts for 1 second. If you did not change the default time, it lasts for 2 seconds.

Whenever you add a transition in MyDVD's editor, it shortens the overall length of your video by the duration of the transition. Why? MyDVD (and most other entry-level video editing products) overlaps the end of one scene and the beginning of the next to blend the two scenes during the transition (I explain this in greater detail in Hour 12, "Editing Video with Pinnacle Studio 8"). For the most part, this isn't an issue, but if you want to add a narration, wait until you finish editing the video so your commentary can match the scenes.

5

6. A good way to start a video is by fading up from black. Likewise, a good way to finish is to fade out to black. To add a black scene, go to the Edit window, click Get Media, select the black still, and click Open. Doing so adds that black scene to your timeline. Move it to the beginning or end of your video and change it to 5 seconds (or so) in length. Then drag the Cross Fade transition to the transition placeholder between that black scene and the selected scene.

A black scene lets any fade-up or fade-out last longer and is a great background for a title or credits. However, an even easier way to add credits to your DVD is to choose one from the selection provided with MyDVD. They say things such as "Starring...," "Directed by ...," and so on.

7. MyDVD has a full suite of video special effects. Open the collection by clicking the lightning icon shown in Figure 5.5. You can choose from four groups: Filters, Frames, Image Adjustment, and Orientation. Select Emboss from the Filters group and drag it to the last scene. Note that a lightning bolt appears in the storyboard/timeline scene filmstrip icon indicating you've applied an effect.

Feel free to try other special effects. Check them out by simply double-clicking them in turn and noting their effects on the selected clip. A couple are worth special note: Sepia gives a nice, old photo feel; using the Black & White, Blue, or Brightness option with a negative numerical value works well when you want to emphasize text placed over a clip (I cover adding text in step nine); and Sharpen can give the appearance of clarity to slightly out-of-focus video. One other note: As a sign of how far video editing software has come, until fairly recently editors paid good money to buy individual special effects such as Scratchy Old Film. Now you get it at no extra charge.

You can replace an effect simply by dragging a new effect to a scene. If you want to remove an effect, right-click the scene and select Remove Effect.

8. Going from a scene with no special effect applied to it to a scene with a special effect such as Emboss can startle most viewers. Ease viewers into the closing embossed scene by dragging a cross fade (or other transition) between it and the previous scene.

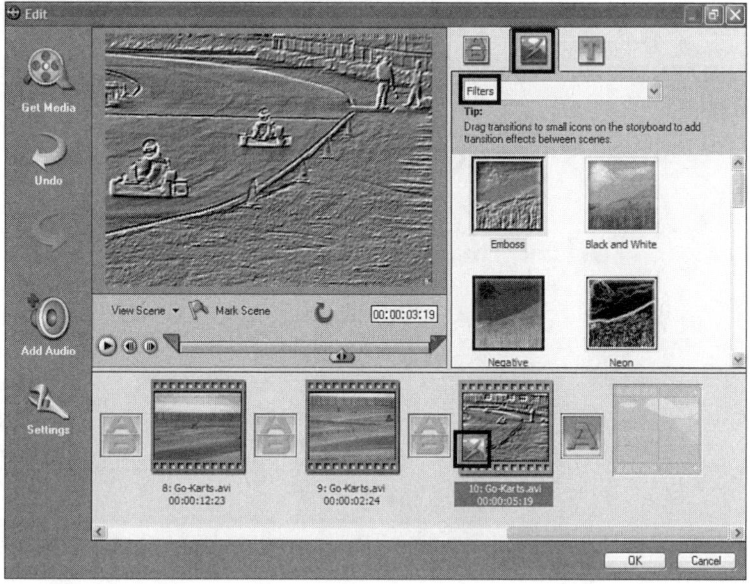

FIGURE 5.5
Use a special effect to give your video some extra visual pizzazz.

9. Superimpose an opening title—place text on top of the first clip—by selecting that scene, clicking the Text icon, and typing a message in the text window (see Figure 5.6). You can select a typeface from the drop-down list, a color from the color box, and other typical font features from the various icons in that window.

Some text limitations do exist. All the text characteristics—font style, size, bold, italic, underline, drop shadow, and color—apply to the entire message instead of individual words. That is, you cannot underline part of the message or use multiple colored letters. You can, however, click the text and drag it in its entirety to the best location onscreen.

Text can be a valuable addition to your DVD production. You can add a title, date, or location *super* (TV news parlance for superimposed text) at the start of your video. If you interview someone, insert that person's name and title at the bottom of the screen. I discuss working with text in videos during Hour 13, "Adding Audio, Tackling Text, and Improving Images."

5

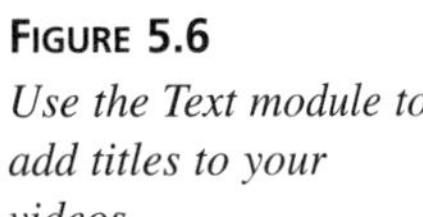

FIGURE 5.6
Use the Text module to add titles to your videos.

10. Add background music to your video by clicking the Add Audio button, navigating to the MyDVD Tutorial Assets folder, and selecting the `GO-Karts Music.wav` file. An alert screen pops up telling you that you are about to add background audio to the entire video. Click OK, which returns you to the Edit interface.
11. Click the Settings icon (the wrench in the lower-left corner) to open the interface shown in Figure 5.7. The Audio Levels slider at the bottom of the screen lets you adjust the *mix* (the blending of the two tracks) of your video's original audio and the added music track. Slide it toward the middle and click OK to return to the Edit interface.

You might need to experiment to find the best mix: Move the slider, click OK, and play the video to check the sound mix. If you need to make any fixes, return to Settings.

I created the `Go-Karts Music.wav` file in about 1 minute using SmartSound Movie Maestro. It's a fun program, included on this book's companion DVD, that fashions music to fit a mood and a specific video length. I cover Movie Maestro in detail in Hour 8, "Acquiring Audio."

FIGURE 5.7
Use the Audio Levels slider to create the best mix of your video's original audio and the added audio track.

12. View your video by clicking View Scene, selecting View Entire, and clicking the Play button. Your edited video should fade up from black, display the title over the first clip, have a hard rock musical score along with the go-kart engine noise and squealing tires, incorporate any transitions you added, and conclude with a cross fade to the embossed view of the finish line and a fade to black as the music ends.
13. Save your edited video by clicking OK to return to the main interface. Select File, Save and then select a file folder and project name (I'd suggest `Go-Karts` to keep things simple).

Assembling Assets and Building Menus Using MyDVD

In this and the next hour's remaining tasks you will create a DVD with a full range of features. You'll work with the video you just edited, as well as a set of still images from that video and a collection of short videos about a birthday party picnic. The purpose is to give you experience in several DVD authoring processes so you can apply what you've learned to any media you put on a DVD.

As shown in Figure 5.8, your DVD will open with a main menu offering two button choices: A Day at the Go-Kart Track and Birthday Party Picnic.

Clicking the Go-Karts button will take users to a submenu with buttons for the video you just edited, chapter points within that video, and a slideshow of stills from that video.

5

Clicking the Birthday button will take users to a screen with button links to six short videos.

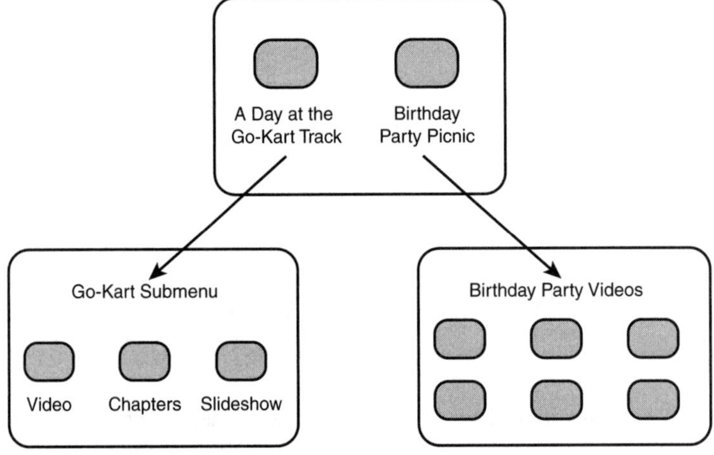

Figure 5.8
The basic design of the DVD you will create in this hour and the next includes three menus, chapters, and a slideshow.

Task: Organize Your Assets into Menus

TASK

You build your DVD project by creating menus and then adding assets to them. We'll start by creating the three-menu structure shown in Figure 5.8. Here's how:

1. Start MyDVD. Open your saved Go-Karts project by selecting File, Open. Then navigate to the Go-Karts folder (it should be in the default Open window) and double-click `Go-Karts.dvd`. Doing so places a button on the default opening menu, as shown in Figure 5.9.

> An easier means to open a previous project is to select File, Recent Files and then select the project from that drop-down list.

2. You want the button that accesses the Go-Karts video to be on a submenu, not on this opening menu. So, create that submenu by clicking the Add Sub-menu icon. (see Figure 5.10). It adds a button named Untitled Menu 1 to the opening menu next to the Go-Karts button.

> At this point, you are working in the default menu, with its associated buttons, text style, and music. In the next hour you'll give this project a different look, add descriptive text, and change the menu music.

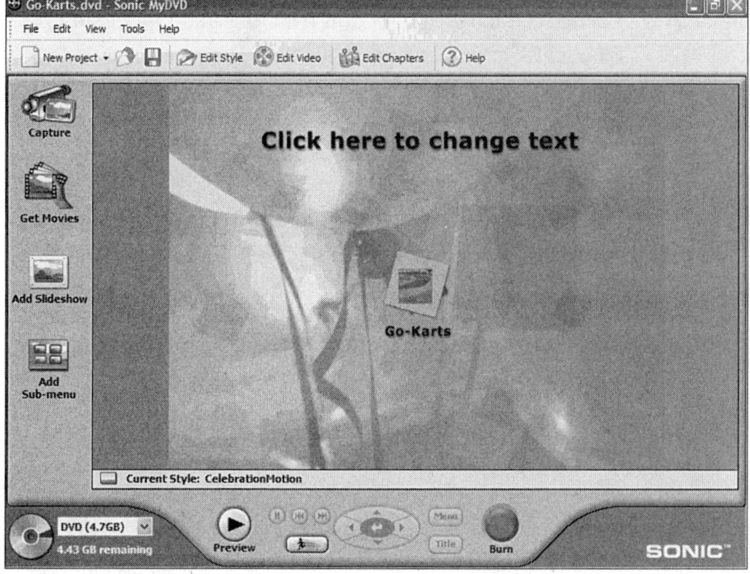

FIGURE 5.9
Reopening the video you edited in the previous hour presents you with this menu interface.

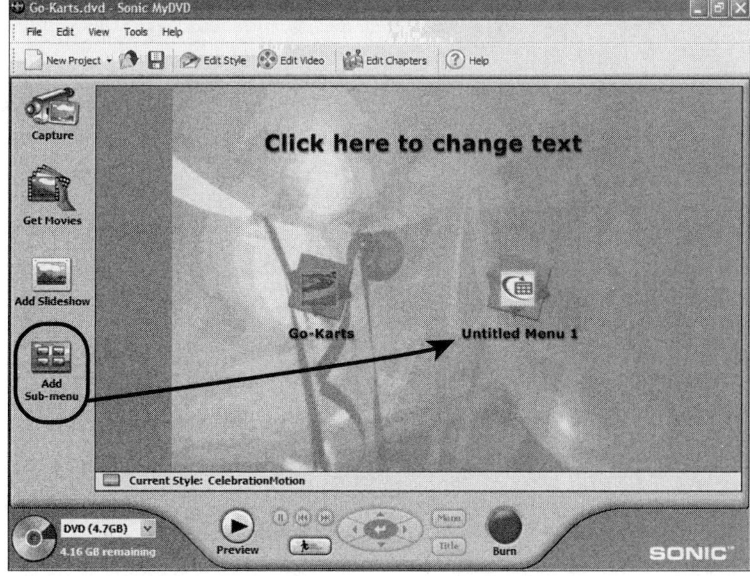

FIGURE 5.10
Clicking Add a Sub-menu places a new menu button on your project's opening menu.

5

3. To move the Go-Karts button to the Untitled Sub-menu is a three-step process. Start by right-clicking the Go-Karts button and select Cut, as shown in Figure 5.11. Doing so removes the button from the main menu and puts it in the Windows Clipboard so you can paste it.

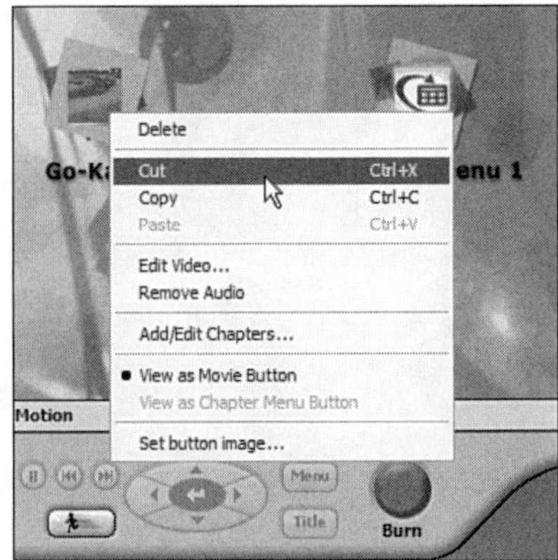

FIGURE 5.11
To move a button to another menu, right-click it and select Cut.

4. Double-click the Untitled Sub-menu button to open it. As shown in Figure 5.12, you'll see a blank menu with two menu navigation buttons in the lower-left corner.

The two icons in the submenu are built-in DVD navigation aids. When viewers click the Home icon, they return to the main menu, and when they click the Return curved arrow icon, they are taken to the previously selected menu.

5. As shown in Figure 5.12, right-click anywhere on the Untitled Sub-menu window and select Paste to place the Go-Karts button in the submenu window.
6. Return to the opening main menu by double-clicking the house icon in this submenu.
7. Add another submenu to the main menu by clicking Add Sub-menu. A button named Untitled Menu 2 appears next to the other submenu button. Go to that newly created menu by double-clicking its icon.
8. You will add the six Birthday Party Picnic videos to this menu. Start the process by clicking the Get Movies icon, which opens the Add Movie(s) to Menu window.

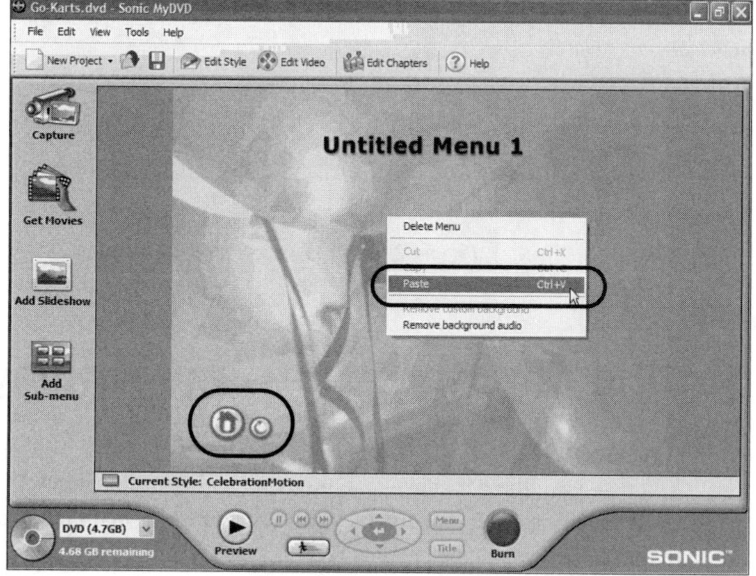

FIGURE 5.12
Use MyDVD's right-click menu to paste the Go-Karts button in this submenu.

9. Navigate to the MyDVD Tutorial Assets folder. Select all six Birthday Party Picnic videos using the Shift+click method (click the first file, `Birthday Party Picnic-1.avi`, and then press Shift before clicking `Birthday Party Picnic-6.avi`). Click Open. As shown in Figure 5.13, after MyDVD does some processing, six buttons appear in Untitled Menu 2.

10. This would be a good time to save your project. Select File, Save.

5

The Shift+click method of selecting a group of files is a standard Windows shortcut. Here are two more: You can select several files that don't fall neatly into a group by using Ctrl+click on each file. And you can select a group of adjacent files by clicking slightly to the left of the first file and dragging the mouse over the other clips. Doing so highlights the group of files. In addition, you can Ctrl+click files in that group, one by one, to deselect them.

FIGURE 5.13
Adding several video clips to your project automatically creates buttons individually linking each clip.

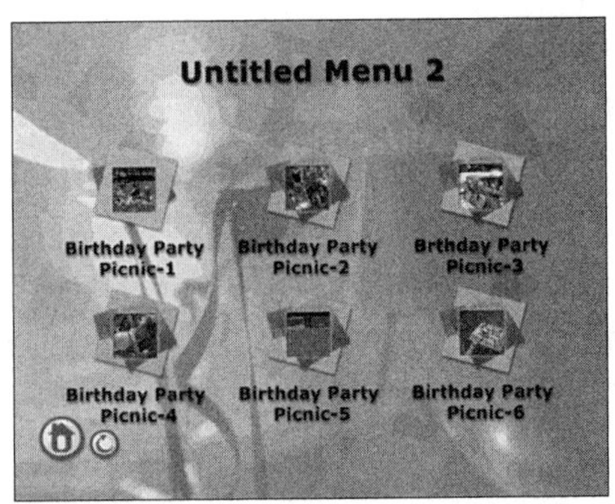

Setting Chapter Points and Adding a Slideshow

You now have three menus—a main menu and two submenus. When viewers put your DVD in their set-top players, they'll see the opening main menu and then use their remotes to access the submenus.

In this task you'll give viewers something in addition to only a couple of videos to look at.

Task: Work with Chapters

TASK

Chapters give viewers immediate access to points in your video, such as a scene selection option in a Hollywood feature film on DVD. In this task, you'll set chapter points in the Go-Karts video. Here's how to do that:

1. Continue where you left off in the previous task, with the submenu, Untitled Menu 2, open. If you closed your program, simply restart it and open the Go-Karts project.
2. Return to the opening menu by double-clicking the house icon in the lower-left corner of the submenu.
3. Go to the Untitled Menu 1 by double-clicking its button, which returns you to the submenu with the Go-Karts button shown in Figure 5.14.
4. Click the Go-Karts button to select it (the frame should change color); then click the Edit Chapters button (see Figure 5.14). That opens the Chapter Points interface shown in Figure 5.15 with the Go-Karts video open.

FIGURE 5.14
With the Go-Karts submenu open, click the Edit Chapters button. Doing so takes you to the Chapter Points menu and creates a sub-submenu.

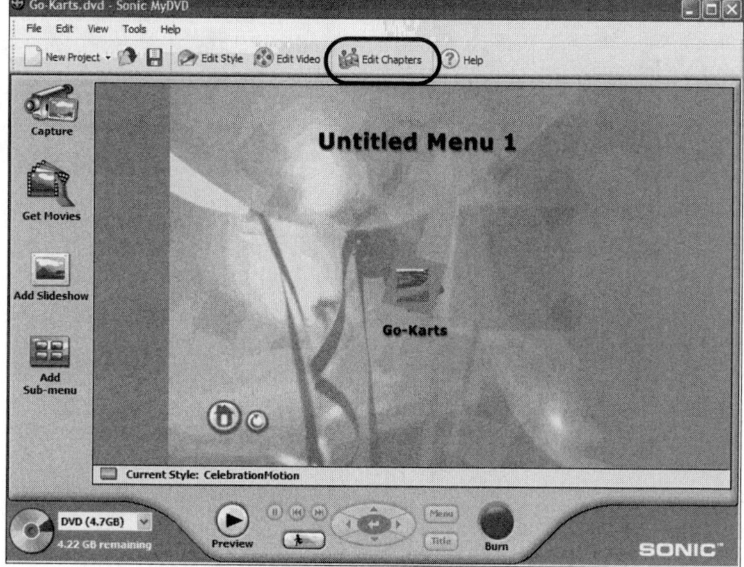

5. MyDVD tracks the original detected scenes and any editing changes you made to them any rearranging you did to their display order. You can use the Chapter Points menu to automatically put a marker at the beginning of each of the scenes by clicking the Add Chapters at Scenes button shown in Figure 5.15. Several little orange markers show up along the timeline.

> When you work with longer videos, chapters are a great way to let viewers jump to a particular point of interest. For example, if you have highlights of a baseball game, you can use chapters to let viewers jump quickly to individual plays or innings.

6. You can delete, move, or add chapter markers. As shown in Figure 5.16, you can delete a chapter by clicking a marker and then clicking the Remove Chapter button.

> When two chapters are very close together, MyDVD displays a double chapter marker icon, as shown in Figure 5.16. To delete one of them, click the double marker icon to open a drop-down menu, select one, and click the Remove Chapter button.

FIGURE 5.15
The Chapter Points interface lets you flag scenes to give viewers easy access to your video.

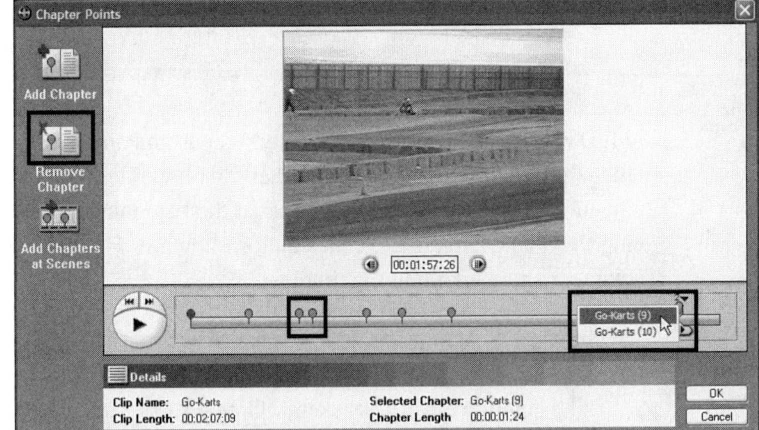

FIGURE 5.16
Delete a chapter marker by selecting it and clicking Remove Chapter. Use the drop-down list for markers that are very close together.

7. Move a chapter by clicking an icon and dragging it while watching the video display to select a new location.
8. Add a chapter by scrubbing to a new location in the video and clicking the Add Chapter button.

When viewers click a chapter point to jump to a particular scene, that scene plays but does not stop at its conclusion—it continues on beyond the final frame. This is the default behavior for most feature film DVDs. The assumption is that most viewers use scene selection to jump to a section in a film to watch from that point onward as opposed to watching only that selected scene.

9. When you have finished setting, moving, deleting, and adding chapter markers, click OK. A message pops up asking whether you want to add a chapter menu button (see Figure 5.17). Click Yes to return you to the main MyDVD interface and the Go-Karts submenu. The menu's button text will have changed to `Go-Karts Chapters`.

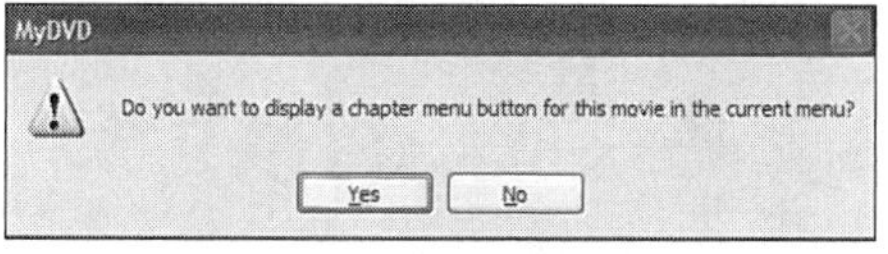

FIGURE 5.17
After adding chapters, MyDVD asks whether you want to add a chapter button to your menu.

MyDVD's designers have purposely simplified the menu creation process by allowing only one button per video—either to access a chapter menu or to play the video from start to finish. I prefer two buttons, one for each function. Steps 10–14 use a work-around to accomplish this.

10. Right-click the Go-Karts Chapters button and, as shown in Figure 5.18, select the View As Movie Button. Doing so changes the button back to its original state.
11. Right-click the now renamed Go-Karts button and select Copy.
12. Right-click anywhere on the main menu and select Paste. As shown in Figure 5.19, this adds another button called Copy of Go-Karts.

If you do not first change the state of the Go-Karts Chapters button to a View As Movie Button, MyDVD will not allow you to copy/paste it to the menu.

13. Right-click the Go-Karts button and change its state back to View As Chapter Menu Button. This updates its name to Go-Karts Chapters.

5

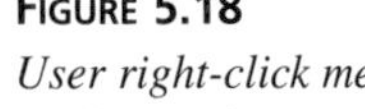

FIGURE 5.18
User right-click menus to change the state of a button and copy/paste a new button onto the menu.

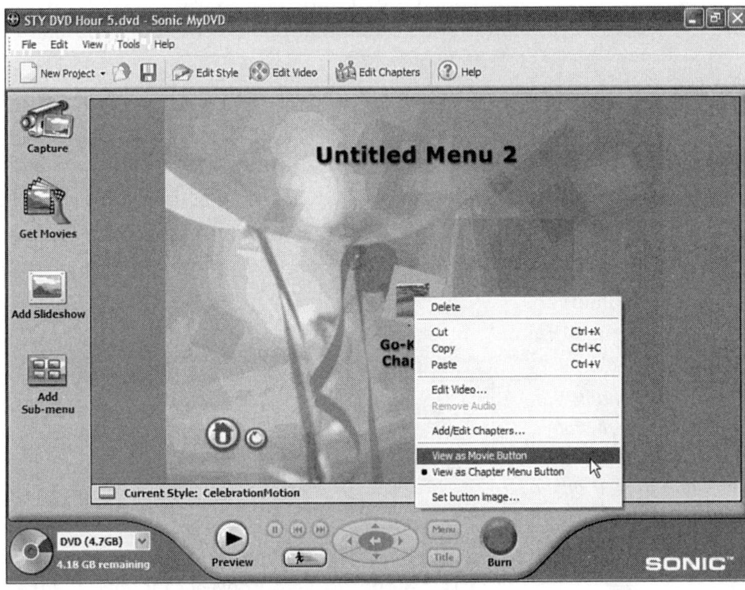

FIGURE 5.19
Add an extra button to the menu to offer viewers direct access to the entire video or its chapters.

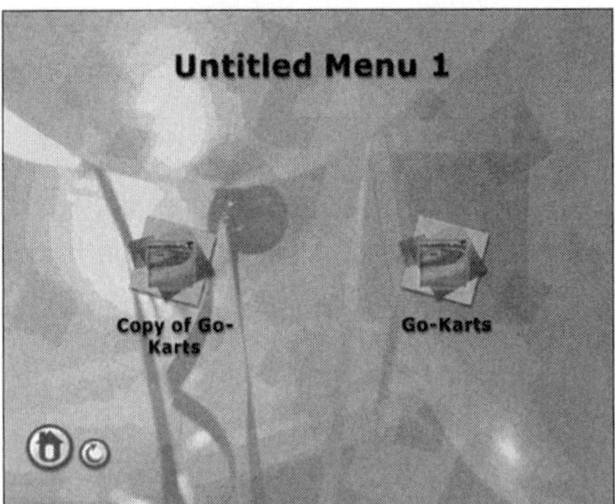

14. Click the words `Copy of Go-Karts` to highlight them, type in `Go-Karts Video`, and press Enter. Your menu should now look similar to Figure 5.20.
15. To check out your handiwork, double-click the Go-Karts Chapters button to access that sub-submenu. As shown in Figure 5.21, by default it displays up to six chapter buttons. If you created more than six chapters, MyDVD adds a small triangle (the Next button) in the lower-right corner of the chapter menu, giving viewers access to the next group of chapter buttons (most MyDVD menu templates permit a maximum of six buttons per menu).

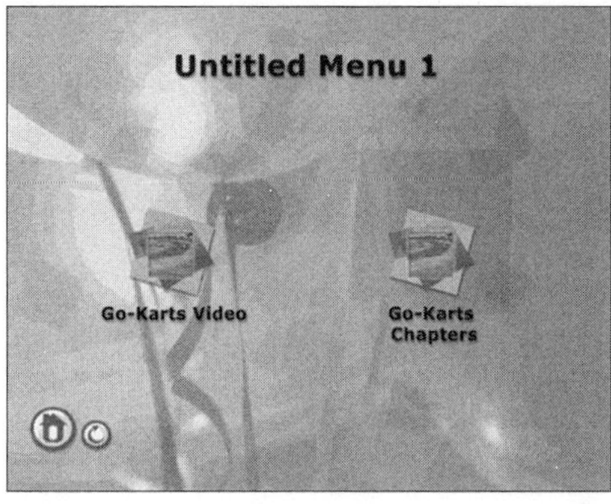

FIGURE 5.20
How your Go-Karts submenu should look after adding a chapters button.

16. Save your project by selecting File, Save.

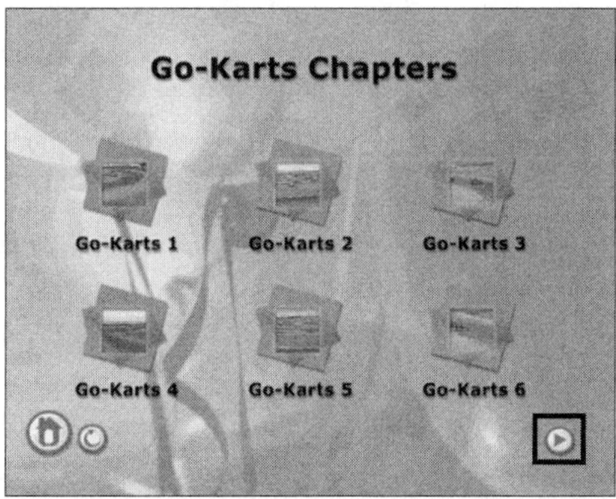

FIGURE 5.21
The Go-Karts Chapters sub-submenu displays six of the chapter points.

5

Task: Add a Slideshow to Your DVD Project

TASK

A slideshow is a collection of still images that can play with background music or a narration. In this task, you add a slideshow of Go-Kart stills to the Go-Karts submenu and a musical accompaniment. Here's how:

1. Return to the Go-Karts submenu by double-clicking the little curved arrow icon in the lower-left corner of the Go-Karts Chapters menu.
2. Click the Add Slideshow button in MyDVD's main interface, which opens the Create Slideshow interface shown in Figure 5.22.

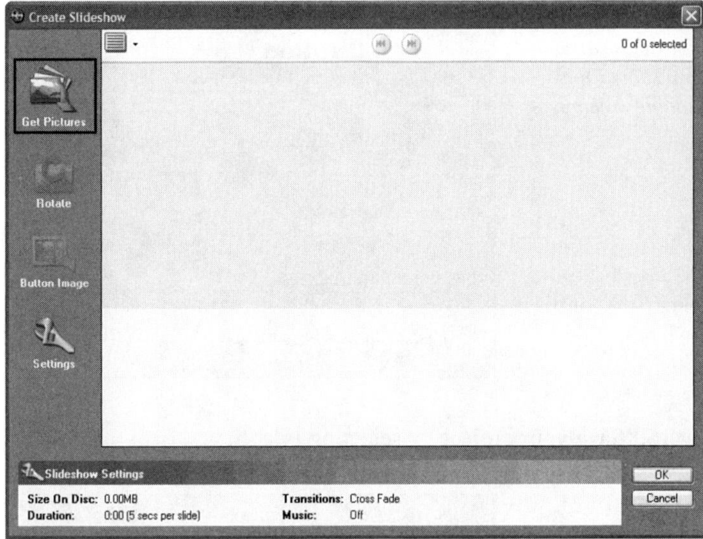

FIGURE 5.22
Use the Create Slideshow interface to add a collection of stills to your DVD.

3. Click the Get Pictures button and navigate to the MyDVD Tutorial Assets file folder.
4. As shown in Figure 5.23, use the right-click and drag method to select all 10 Go-Kart still files. Right-click to the left of `Go-Kart Still 01.bmp` and drag the mouse until you've highlighted all of them. Then click Open, which returns you to the Create Slideshow interface.

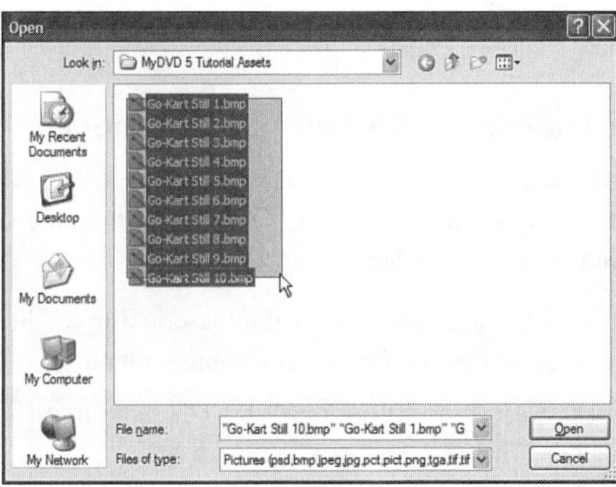

FIGURE 5.23
Use the right-click and drag method to select this group of adjacent image files.

5. Rearrange the display order of the slides by clicking and dragging images to other locations in the slide timeline. Break up the string of similar shots (stills 4–6) by dragging wide shots of the track between them.
6. Change the image that will display as a thumbnail on the menu button by clicking one of the stills—I recommend #3—and then clicking the Button Image icon shown in Figure 5.24. Doing so adds a thumbs-up symbol (or thumbnail, take your pick) next to the newly selected image.

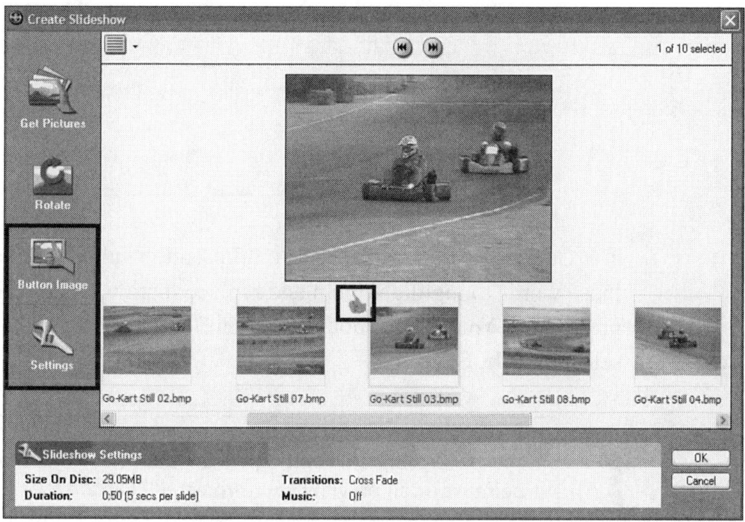

FIGURE 5.24
Use this interface to rearrange images, change the menu button thumbnail image, and add a soundtrack.

7. Open the Settings interface by clicking its button, shown earlier in Figure 5.24. As shown in Figure 5.25, under the Basic tab you can change the slide duration (keep it at 5 seconds for this task) and add audio. In this case, click the Audio Track check box, click Choose, navigate to the MyDVD Tutorial Assets folder, and select the `Go-Karts Slideshow Music.wav` file.

The Settings Advanced tab offers several options: You can change the transition between the slides, change the letter box color (in this case, the slides exactly fit the screen so there will be no letter box), or save the slides as data files on the DVD to make the slides accessible to PC users. That latter feature is one of several that make MyDVD a cut above the entry-level DVD authoring application crowd. You can make changes here if you care to. Click OK to return to the Create Slideshow interface, and then click OK to return to the main MyDVD interface.

5

FIGURE 5.25
Slideshow settings has five options, including altering the slide duration, adding audio, and three others, in the Advanced tab.

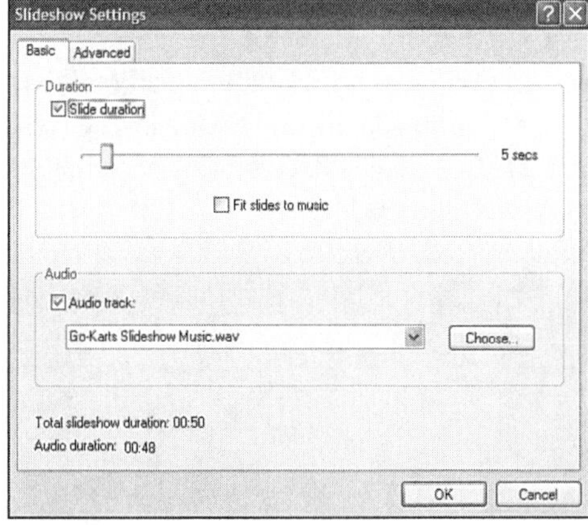

8. The Go-Karts submenu now has a third button called Untitled Slideshow. Click those words to highlight them and replace them with **`Go-Karts Slideshow`**; then press Enter. Your menu should look similar to Figure 5.26. Save your project by selecting File, Save.

You have completed setting up the basic menus and project flow. At this point, you can preview your project. I'll explain that process in detail in the next hour, but if you want to experiment a bit now, feel free. Simply click the Preview button at the bottom of the main MyDVD interface and click menu buttons to see what happens. You should be able to navigate around the various menus and view the videos and slideshow.

FIGURE 5.26
The Go-Karts submenu with all buttons in place.

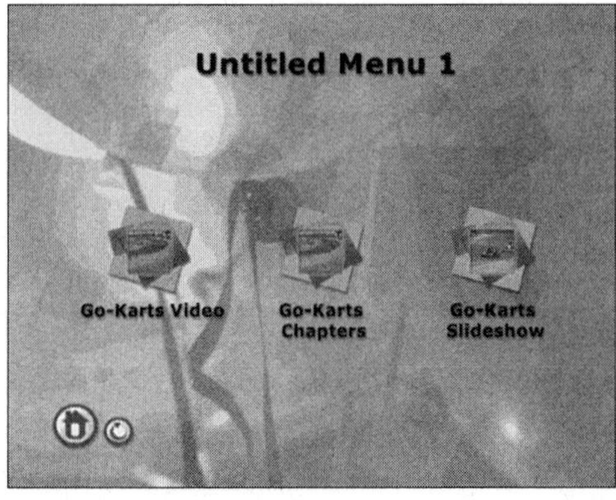

Summary

In this hour you continued the DVD authoring process, finally getting to something that really feels like DVD authoring: building menus.

You began by editing a video using MyDVD's Edit interface, which lets you trim scenes; add transitions, text, and audio; and apply special video effects.

You added chapters to your video along with a separate chapter sub-submenu. And you created a slideshow with its associated menu button.

Workshop

Review the questions and answers in this section to reinforce the MyDVD video editing and authoring techniques. Also, take a few moments to take the quiz and do the exercises.

Q&A

Q When I finished the scene detection process and went to the Edit interface, each scene had the same transition ahead of it. That seems like overkill. What should I do?

A Uncheck a box in the Preferences window. To do that, select File, Preferences; go to the Video Editing tab; and uncheck the Automatically Insert Transitions when New Media Is Added to Storyboard button. However, this won't fix your current video, only future videos. Remove any or all of those transitions by right-clicking them individually and selecting Delete.

Q When I add audio to the video in the Edit window, all I hear is that audio, not the original audio. What's going on?

A You need to adjust the audio mix. In the Edit interface, click Settings and move the Audio Levels slider toward the center. You might have to use some trial-and-error—going back and forth between the Edit window to listen to the video and the Settings screen to fine-tune the mix—to find the sweet spot.

5

Quiz

1. How do you use MyDVD's Edit module to start your video with a fade up from black and end it with a fade out to black?
2. You want to start your video by giving it an old-fashioned sepia-toned look and then gradually change to full color. How do you do that?
3. You've added a slideshow to your project, but the button thumbnail image is of the wrong slide. How do you fix that?

Quiz Answers

1. Drag a transition to the transition placeholders before the first scene and after the final scene. You can use any transition but, a cross fade works best for this use. If you figured out how to add a black clip, place one at the beginning and end of your video.
2. Use a long transition. Increase the transition duration by going to MyDVD's main interface and selecting File, Preferences. Then click the Video Editing tab and change the Transition Default Duration to however long you want the gradual change from old to new to last (3 seconds works well). Return to the Edit module, drag the Sepia effect (in the Filters group) to the first scene, and add a cross fade transition after that scene to the next scene. The one caveat is that you cannot set transition durations on a transition-by-transition basis. All other transitions in this video will run for the time you selected for this gradual time shift.
3. Go back to the Add Slideshow interface by double-clicking the Slideshow menu button, clicking the slide you want to use as the menu button, and clicking the Button Image icon.

Exercises

1. Experiment with the Go-Karts video. Trim scenes, rearrange them, add titles, try the special effects, throw in some transitions, and add your own audio.
2. Use MyDVD to create your own edited video: Use MyDVD's capture module to transfer video from your camcorder to your PC; use scene detection to break it up into workable chunks; and then open the Edit module to add text, transitions, and special effects. You get extra credit if, when you finish editing, you record a narration using your PC's microphone and then add that audio file to your finished piece.
3. Create a slideshow using your own images. Either import them from your digital still camera or use a scanner. If they have a similar color scheme (old sepia-toned photos, for instance), change the letter box color (select Create Slideshow, Settings and then click the Advanced tab) to complement that color. Add music and a transition to match the mood.

HOUR 6

Authoring Your First DVD Project Using MyDVD: Part III

In the previous hour, you created a basic DVD menu structure, built a chapter menu, and added video clips and a slideshow. In this hour, you'll give your DVD project a unique look and feel and then burn it to a recordable DVD.

You'll start by testing the DVD navigation—using your mouse as a remote control to click to various menus and media—making sure things work as you'd expect. Then you'll edit the appearance of the menu backgrounds, buttons, and text. Finally, you'll take your DVD on one last test drive before burning it onto a DVD. With the newly recorded disc in hand, you'll then run down the hall, place it in your set-top DVD player, and view your work.

The highlights of this hour include the following:

- Previewing your work-in-progress
- Editing the menu style

- Creating templates and sharing them on the Web
- Burning your DVD project to a recordable disc

Previewing Your Work-in-Progress

MyDVD lets you test your project, reproducing the experience viewers will have when they play it in a set-top DVD machine. The only difference is that you get to use your mouse to click buttons onscreen as well as buttons on a simulated remote control.

Task: Preview Your DVD Project

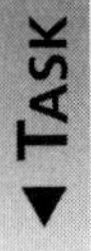

As you build menus and add elements, you might lose track of how everything is connected. Use the preview mode to clarify that. Here's how to do it:

1. Return to the main MyDVD interface. Open your DVD project by selecting File, Open. Then navigate to the Go-Karts (or whatever name you gave your project) file folder, open that folder, and double-click the `Go-Karts.dvd` file.
2. Preview your project by clicking the Preview button at the bottom of the main MyDVD interface (see Figure 6.1).

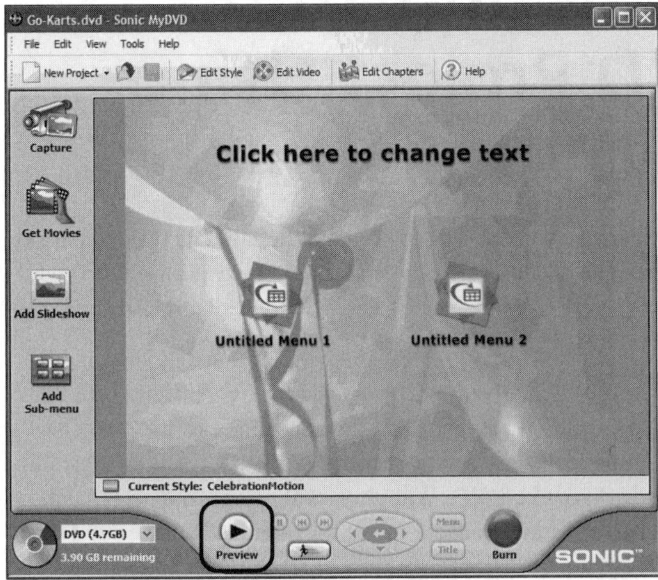

FIGURE 6.1
Use MyDVD's Preview feature to take your DVD on a preliminary test drive.

3. The Information screen, shown in Figure 6.2, pops up and notes that animated menus will not display. Click OK, which takes you back to the main screen

During preview you can see the animated/video menus, but it requires some time to *render* (combine the menus and animated buttons into new video clips) those screens and buttons. To see the motion menus and animated button in action, click the Build Motion Menu button (the little runner) next to what was previously the Preview button but is now the Play button.

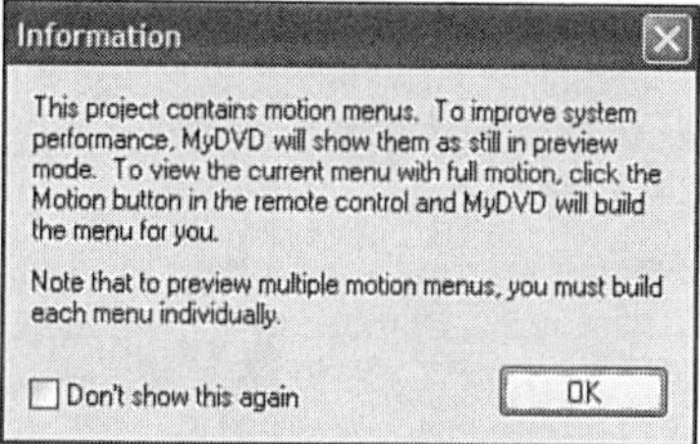

FIGURE 6.2
Before entering Preview mode, MyDVD lets you know that any animated menus and buttons will remain static.

Upon entering Preview mode, your MyDVD main interface changes a bit. As shown in Figure 6.3, the previously grayed-out remote control buttons are now activated, whereas the various menu-building and video-editing buttons at the top of the screen are grayed out. Also the main menu music plays in a continual loop.

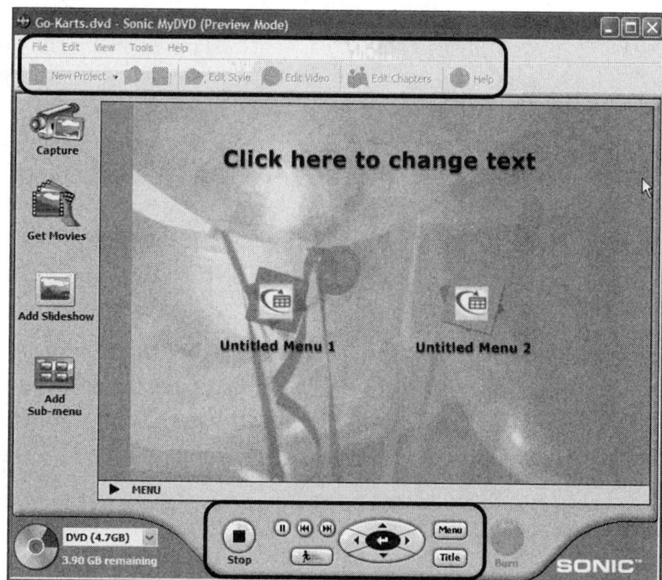

FIGURE 6.3
After clicking the Preview button, the MyDVD main screen changes, activating the remote control buttons and graying out the menu/video editing icons.

4. Before beginning the preview process, take a look at Figure 6.4. It identifies the various buttons you can use to simulate a remote control. You can, however, forgo the remote control buttons and click directly on the DVD menu.

If you click the Stop button, you exit Preview mode and return to the main interface.

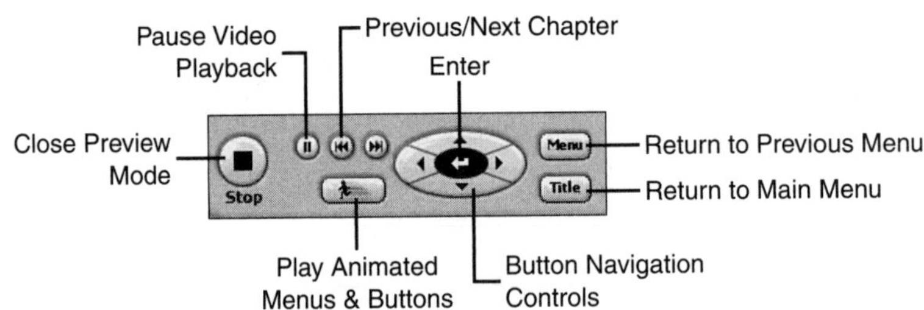

FIGURE 6.4
Use these controls to take your DVD for a test drive.

5. Start previewing your DVD project by clicking directly on Untitled Menu 1. Or use the Button Navigation Control buttons to highlight that menu and then click the Enter (crooked arrow) button. Either method takes you to the Go-Karts submenu with its three buttons.

Unlike when building a menu, in Preview mode you don't need to double-click to go to another menu. A single-click is all you need, just as when using your remote control.

6. Click the Go-Karts Video button and that video plays. You can let it play to the end, in which case it automatically returns you to the Go-Karts submenu. To speed up that process, click the Menu button (clicking the Title button takes you back to the Main DVD menu).
7. Click the Go-Karts chapters button to access that six-button sub-submenu. Click a chapter button, and you are taken to that point in the video. Stop the video playback (the default behavior is to play the entire video from that chapter point) by clicking the Menu button in the remote control section. Doing so takes you back to the Go-Karts submenu.

If you click the small triangle-shaped Next Menu button in the lower-right corner of the chapter menu, you are taken to the second chapter sub-submenu with the remaining chapter buttons. And, in this and all other sub-menus, clicking the small curved arrow Return button in the lower-left corner takes you to the previous menu.

8. Click the Go-Karts Slideshow button. The music should change, and the still images should display in succession for 5 seconds each. Let the slideshow play to the end; it returns you to the Go-Karts submenu.
9. Return to the main menu by clicking the house icon in the lower-left corner of the menu or the Title remote control button.
10. Click Untitled Menu 2 to move to the Birthday Party Picnic videos menu; then click any of the six party video buttons. The video plays and returns you to the Birthday Party submenu at its conclusion.
11. Click Stop to terminate the preview mode and return to the main MyDVD interface.

Editing the Menu Style

Until now, you've been working with a default menu style—background, audio, buttons, and text—as well as *placeholder* wording: `Click here to change text`, `Untitled Menu 1`, and so on. It's time to move beyond default settings and personalize your project.

Task: Make Menu Style Changes

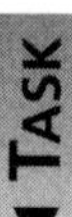

In MyDVD's Edit Style section, you can change text characteristics, menu and button graphics, menu music, and a couple other items. Here's how:

1. At the conclusion of the previous task, you ended up in the main MyDVD interface. If you're not there at this point, return to that location. As shown in Figure 6.5, click the Edit Style button to open the interface shown in Figure 6.6.

Take a quick look around the interface shown in Figure 6.6. Clockwise from the upper left, it offers drop-down lists to change the Menu backgrounds, menu title and button text characteristics, animated menu/button options, and customized menu/button selections.

6

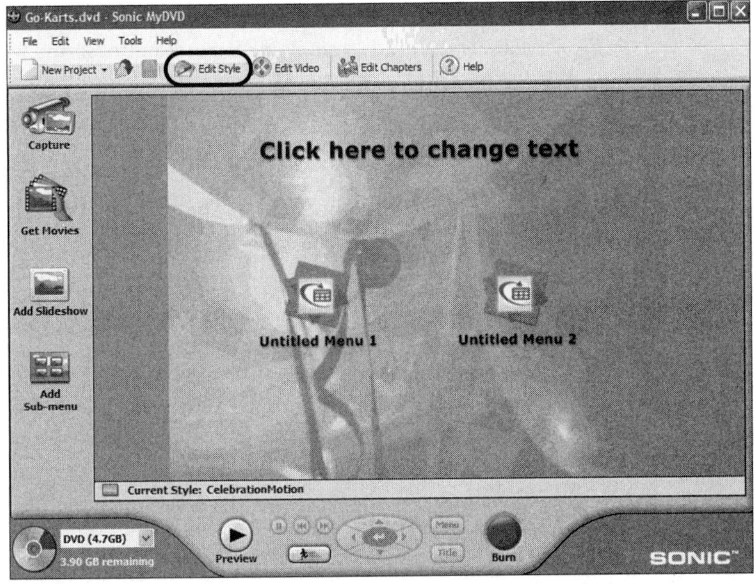

FIGURE 6.5
Click Edit Style to open the interface shown in Figure 6.6.

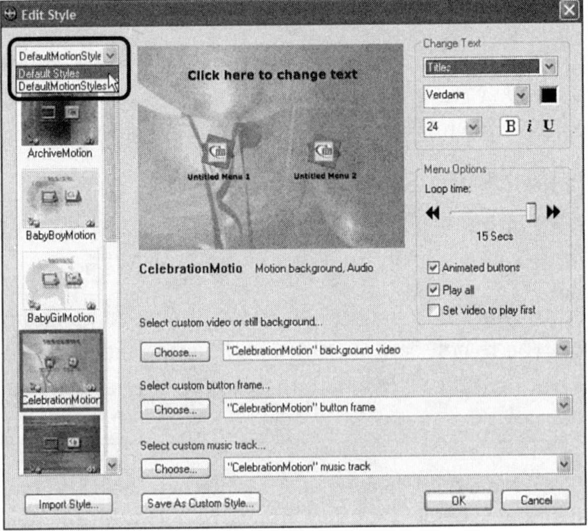

FIGURE 6.6
The Edit Style interface lets you change the look and feel of your DVD.

2. Click the Menu background drop-down list, circled in Figure 6.6, and select Default Styles. As shown in Figure 6.7, this updates the scrollable collection of

menu templates so it now offers more than 25 menu styles that use static backgrounds. Scroll through them and click some to get an idea of how they look. As shown in Figure 6.7, when you select one, it appears in the main menu window.

After selecting a Default Style static background menu, note how the wording in the interface changes. Beneath the main menu display MyDVD notes that this menu is a still image with no audio. By default, it also turns off Animated Buttons, but you can turn them back on. You can also add audio. (I cover both topics later in this task.)

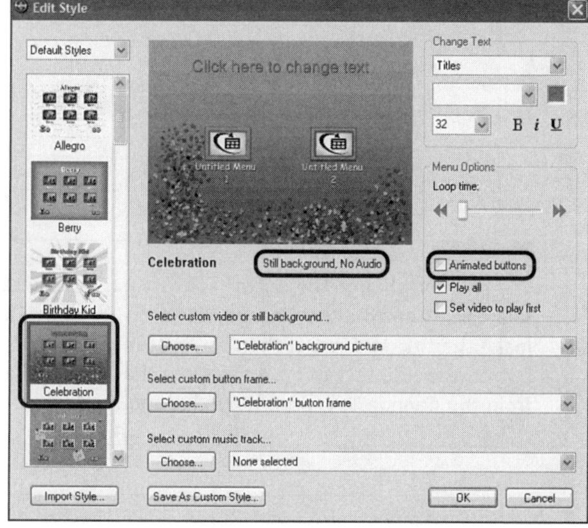

FIGURE 6.7
Selecting the Default Styles menu group displays more than 25 static menu background templates with their own button and text styles.

3. Switch back to the DefaultMotionStyles menu backgrounds, scroll through the 10 motion menu backgrounds, and click SportsMotion (see Figure 6.8).
4. Change the title text style for the entire project (you cannot change text on an individual menu basis) by selecting Titles from the Change Text drop-down list and then selecting whichever font, size, and other features (such as bold, italic and underline). In the example shown in Figure 6.9, I chose Team MT font style (MyDVD displays all the fonts you have installed on your PC), 36-point type, and Bold. As you make changes, they appear in the menu window.

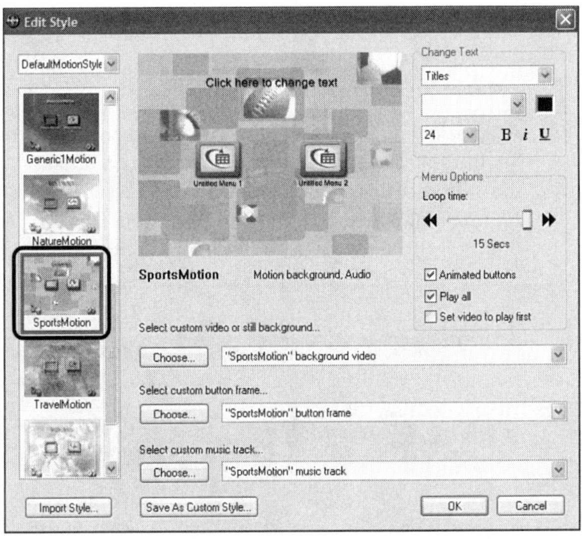

FIGURE 6.8
Choose SportsMotion as a best-fit template for this hour's DVD project.

You also can change the button font style using the same process. To do so, select Buttons from the Change Text drop-down list. Normally you'll want to make the title text larger than the button text, but if you want both the button the title text to have the same characteristics and size, select All Text from the Change Text drop-down list and make your changes.

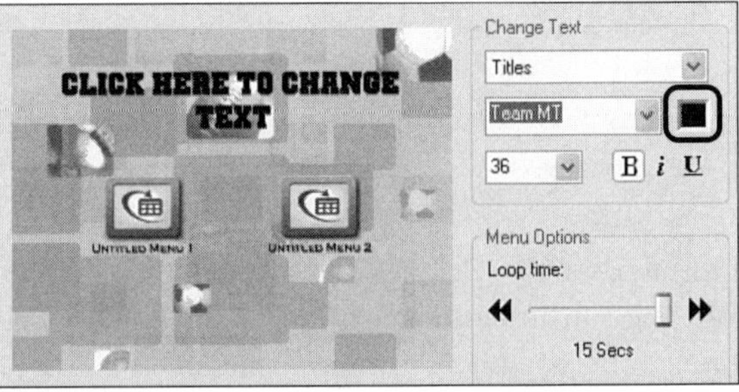

FIGURE 6.9
Use the Change Text features to alter your text's appearance. Use the highlighted Color selection window to change your font color.

5. Change the font color by clicking the black box circled in Figure 6.9 to open the Color selection window shown in Figure 6.10. This is the standard Windows color selection tool. Click Define Custom Colors to gain a wider range of color options. Use the custom palette to select a color that suits you and click OK.

Experiment with the Color selection palette. Click around in the large box on the right to select a hue and then drag the slider on the right to set its brightness (or luminance). Note how the numeric values in the six boxes change as you move the slider and crosshairs around. After you've selected a new color, click an empty Custom Colors box and then click the Add to Custom Colors button. Your new color will appear in the selected Custom Color box. Click OK to set that as your new font color.

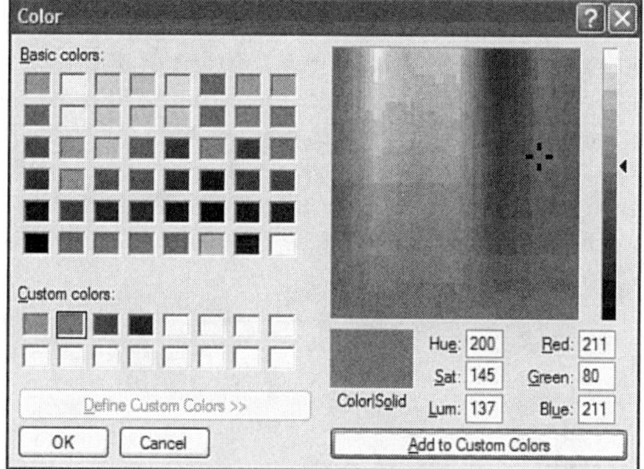

FIGURE 6.10
The Color selection box lets you select from a default palette of colors or create your own color.

6. Check out the Menu Options section shown in Figure 6.11. Here's a quick run-through with my recommended settings:
   - **Loop Time**—How long the menu animation will play before restarting. Keep it at its default maximum setting of 15 seconds.

The menu animation default time is 15 seconds only because that is the length of the video used in MyDVD's default motion menus. If replace a motion menu background with your own video or add a video to a static background menu, the menu and button animations will run for the length of the video background clip—up to one minute.

   - **Animated Buttons**—Checking this box means all the buttons created with video thumbnails will play when this menu is open. Keep this in its default, checked condition.

6

- **Play All**—Checking this box means all video clips on the DVD will play in order without the viewer clicking them separately. While this automated playback may appear to defeat the purpose of the menus, viewers can always opt-out by clicking the Menu button on their remote. In any event, for this our purposes you can uncheck this box.
- **Set Video to Play First**—This plays whichever video MyDVD determines is the first a viewer will encounter in your project. Leave it unchecked.

The purpose of the Set Video to Play First check box is very simple—you are just telling the DVD player that, instead of displaying the menu when you insert the disc, that it should jump straight into the video (so mom doesn't have to worry about how to operate the remote control—she can just pop the DVD in and watch the video). The one caveat is that if you have more than one video on your DVD, MyDVD selects the first one on your project's opening menu as the so-called First Play video. So keep that in mind if you choose to use this option.

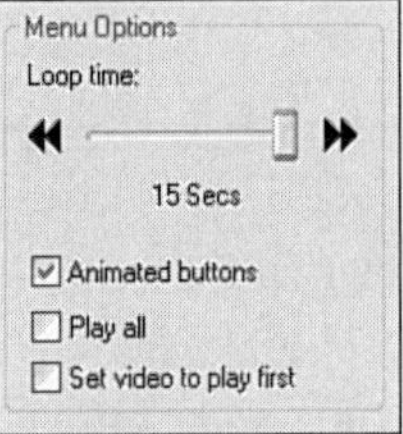

FIGURE 6.11
The Menu Options section gives you an extra level of control over your menu display.

7. At the bottom of the Edit Style interface are the three drop-down lists shown in Figure 6.12. Here's a list of what you can do with them (when you finish examining them, click OK to return to the main MyDVD interface):
   - **Select Custom Video or Still Background**—This lets you choose a menu background still image or video from your own collection.

If you use a still image, it should meet certain size and aspect ratio (proportion) characteristics for best results. Refer to the MyDVD Help section—under Authoring DVDs, Creating Video, Audio, and Graphics—for specific details. Basically, for NTSC projects, menu stills should be 720×480 pixels.

- **Choose Custom Button Frame**—This offers 15 button frames that ship with MyDVD. As shown in Figure 6.13, you can use any of them to replace the frames used throughout your DVD project.
- **Select Custom Music Track**—This lets you select any compatible audio file (MP3, WAV, MPA, WMA, ABS, AIF, or AIFF) and use it with your menus (static or animated).

As you work on projects, the drop-down lists display items you've added to them by selecting the respective Choose button. At the moment, they probably have only the SportsMotion default items.

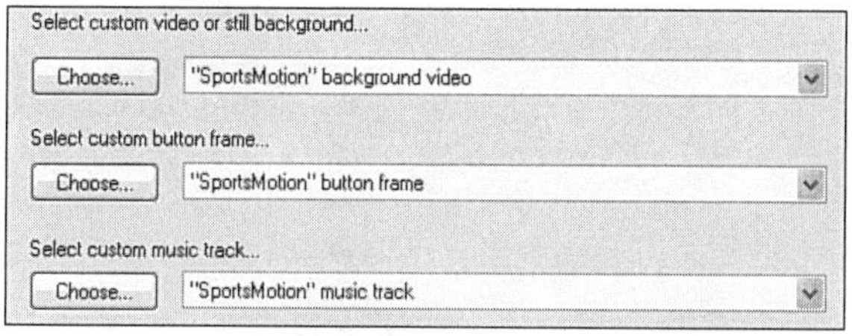

Select custom video or still background...
Choose... "SportsMotion" background video
Select custom button frame...
Choose... "SportsMotion" button frame
Select custom music track...
Choose... "SportsMotion" music track

FIGURE 6.12
Use these three options to further customize your project.

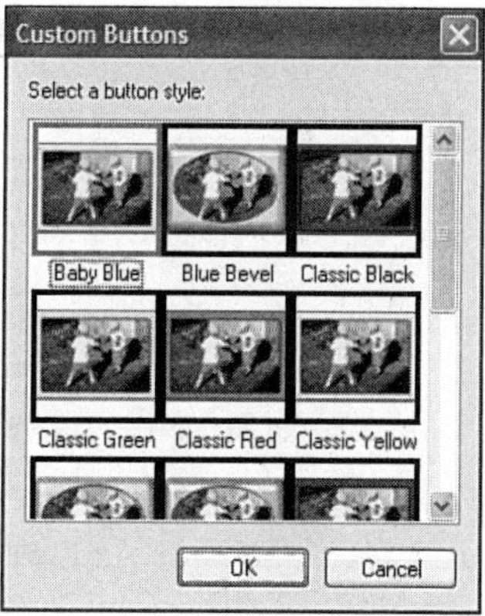

FIGURE 6.13
You can replace the default button style with one of these custom frames.

Fixing Other Project Features

Stepping out of the task mode, I'll wrap up the menu editing process by taking you through three additional items: changing thumbnail images, rearranging menu buttons, and changing text copy.

Selecting Different Thumbnail Images

All the menu buttons that lead to videos use the first frame of each video as the thumbnail image on the button. However, you can select different frames that better represent each video.

Go to the Birthday Party submenu (Untitled Menu 2). As shown in Figure 6.14, right-click a thumbnail (my example is Birthday Party Picnic-1) and select Set Button Image.

FIGURE 6.14
Right-click a video button and select Set Button Image to change the thumbnail.

The screen shown in Figure 6.15 opens. Move the slider to a video frame that suits you (I chose the frame at 17:01) and click Set. This returns you to the submenu with the new thumbnail image in place. Feel free to do that for some or all of the video clips and chapters in your project.

You might have noticed that buttons connected to submenus, such as the ones shown in Figure 6.16, have the same default thumbnail image: a swooping arrow pointing to a collection of buttons. You cannot change that thumbnail within MyDVD. You can, however, change the frame around it in the Edit Style interface. And you can change the thumbnail graphic by using Sonic Solutions Style Creator. I introduce that product later this hour in "Creating Templates and Sharing Them on the Web." And I explain how to use it in Hour 22, "Creating Custom MyDVD Templates with Style Creator."

FIGURE 6.15
Use the slider to find a more representative button thumbnail image; then click Set.

FIGURE 6.16
Buttons linked to submenus have unchangeable, default appearances.

Rearranging Menu Buttons

Rearranging menu buttons is very easy. If you want buttons to appear in a different order, simply click the button you want to move and drag it ahead of its new location. As Figure 6.17 shows, a transparent image of that button and a vertical line highlight the new location. Release the mouse button to drop it in its new location.

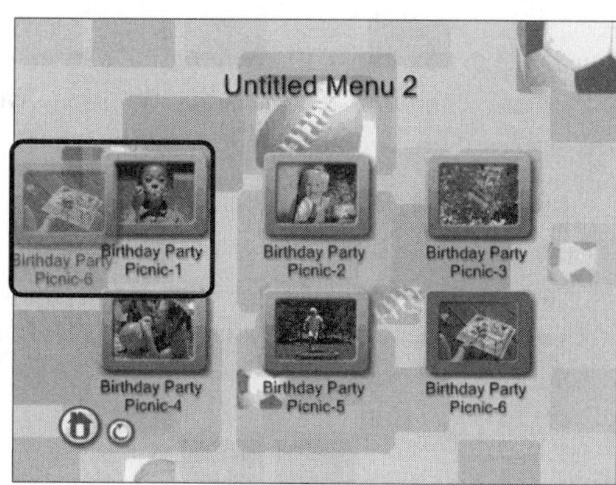

FIGURE 6.17
Drag and drop buttons to rearrange them.

Changing Text

Changing text is also easy. As shown in Figure 6.18, you can change any text line by clicking it to highlight it (the best spot to click is at the bottom of the text line) and then typing in new text. If you exceed the length of the space allotted for the text, MyDVD automatically adds another line.

If you edit the title of a submenu, MyDVD automatically renames the button that links to that submenu, and vice versa, to match the new name. You cannot have different wording for a submenu button text and its submenu title.

FIGURE 6.18
Changing text is easy; simply select the text and make the changes.

Creating Templates and Sharing Them on the Web

As you increase your DVD production skills, you might want to create your own DVD menu styles and share them with others on the Sonic Solutions Web site. I'll give you a brief overview of the process here and go into greater detail in Hour 22.

In general, to create a custom style, you add a plug-in (a small program within a program) called Sonic Solutions Style Creator to Photoshop ($600)—or its much less expensive younger sibling Photoshop Elements ($99)—that gives you access to a style creation template. You use that to build menus and buttons and, if you choose, post them as templates to the Sonic Solutions Web site.

I use the trial copy of Photoshop Elements included on this book's companion DVD in my explanation of Style Creator in Hour 22.

Visit the Sonic Solutions MyDVD Styles Web site at `http://styles.mydvd.com/default.asp`, shown in Figure 6.19, to get a feel for how this works.

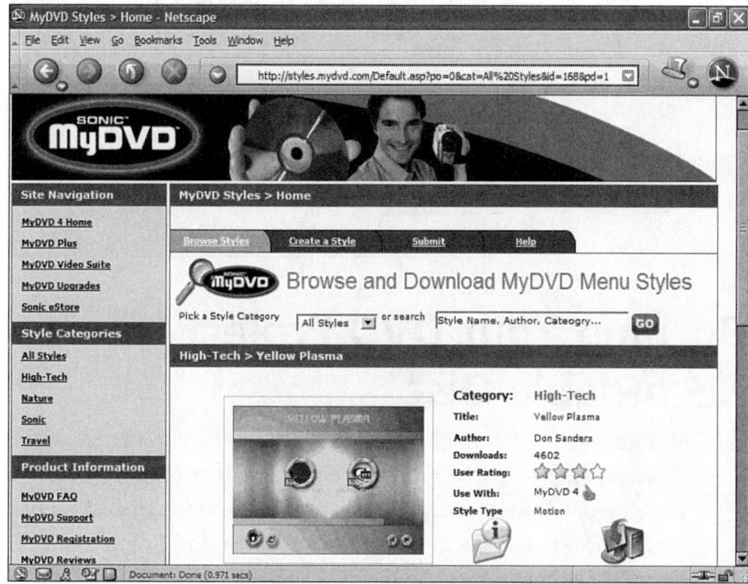

FIGURE 6.19
You can download MyDVD user-created menu templates at the Sonic Solutions MyDVD Styles Web site.

Even if you don't post any templates, you can download all the styles here at the MyDVD Styles Web site at charge. To add a downloaded style to your MyDVD collection, simply go to the Edit Style interface and click the Import Style button (in the lower-left corner of Figure 6.20). You need to create a new style file folder (MyDVD won't let you add styles to the two default style groups) and then locate the downloaded style on your hard drive and click Open.

As shown in Figure 6.20, the new template shows up on the left side ready for immediate use.

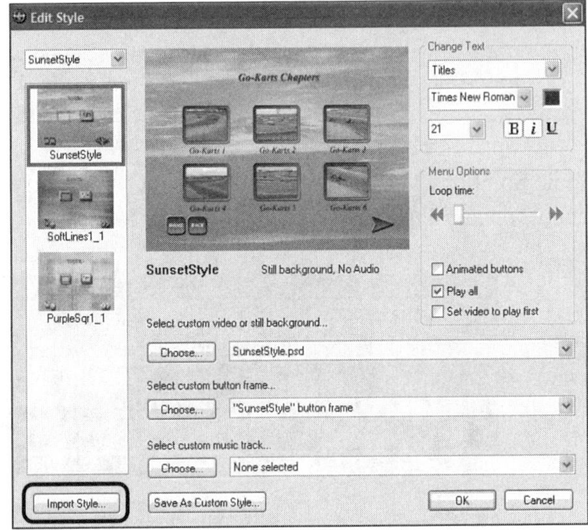

FIGURE 6.20
Use the Edit Style, Import feature to add new menu templates.

Burning Your DVD Project to a Recordable Disc

Before tackling the final step in the DVD production process, take your DVD on one more preview test drive. This time you might choose to render some or all of the animated menus. To do that, you need to return to the main MyDVD interface, navigate to a menu you want to see in action, and click Build Motion Menu. Depending on the number of video buttons on the menu, it can take from about 30 seconds to 2 minutes to complete the process.

During the preview you'll note that in a rendered motion menu all the video buttons play for whatever duration you selected for the loop time in the Edit Style interface. This is one of the extra features that sets MyDVD apart from its competition.

The time to render depends on the length of the loop time. If you use a long video clip along with several video buttons, it will take much longer than 2 minutes to render your motion menu.

Task: Burn Your DVD

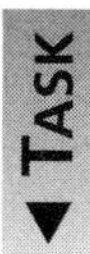

When you're satisfied that all your menus look and act as you expected them to, it's time to burn your DVD project to a recordable disc. Here's how:

1. Insert a recordable DVD into your DVD recordable drive. That likely will open the window shown in Figure 6.21. Click Cancel.

FIGURE 6.21
Inserting a recordable DVD into your DVD recordable drive usually pops up this window.

2. Return to the main MyDVD interface and click the Burn button (it's the big red button at the bottom of Figure 6.22).

FIGURE 6.22
Click the Burn button to start the DVD recording process.

▼

3. The Make Disc Setup window, shown in Figure 6.23,opens. You have very few options (higher-level DVD authoring products typically offer several more). In this case, you probably will want to accept the defaults and click OK.

If you have more than one DVD recordable drive, you'll need to tell MyDVD which one to use. And if you want to make more than one DVD, you'll need to change that number. If you do make more than one, MyDVD prompts you when it's time to insert a new, blank DVD recordable disc.

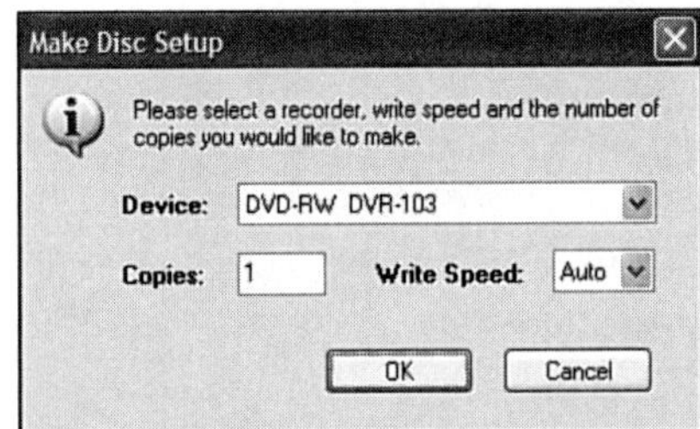

FIGURE 6.23
The MyDVD Make Disc Setup window has very few options.

4. You can watch your progress at the bottom of the main MyDVD interface, and you can stop the recording by clicking Cancel.

In a few minutes you'll have a finished DVD. When completed, insert it into your nearest DVD set-top player and see how it has turned out. Unless you encounter a rare compatibility problem with an older player, your DVD should play back without a hitch. The first time I created a DVD and saw how smoothly it worked on my home TV, I realized that this was "it." Suddenly the possibilities were (and are) endless.

Summary

This hour took you through the final stages in the MyDVD DVD authoring process. You started by previewing your project to see how the menu buttons link and viewing the video and slideshows in action.

Then you used the Edit Style interface to fine-tune your project's look and feel. Even in this consumer-level product, you have detailed control over the menu backgrounds, button frames, text styles, and menu music.

Outside the Edit Style interface, you edited your menu title and button text, changed the button thumbnail images, and rearranged the button placement.

Finally, you burned your project using the streamlined and simplified two-step process in MyDVD. You successfully (I'm guessing) played the resulting DVD on your home DVD set-top. Congratulations!

In the previous hour you created a basic DVD menu structure, built a chapter menu, and added video clips and a slideshow. In this hour you gave your DVD project a unique look and feel and then burned it to a recordable DVD disc.

Workshop

Review the questions and answers in this section to reinforce your MyDVD authoring skills. Also, take a few moments to take the short quiz and perform the exercises.

Q&A

Q I used the Build Motion Menu feature, but when I preview my project, not all the menus are in motion. Why?

A You have to use the Build Motion Menu button separately, on each menu, one at a time. Despite the extra manual labor, this is helpful. If you want to see only one menu in motion, this means you won't have an interminable wait while rendering your whole project.

Q I changed the text under the chapter button and that changed the title at the top of the chapter menu page. I want the title to say something different. But when I changed it, the text beneath the button changed to match that title. Is there some way around this?

A Sorry. There is no way around it. Submenu titles will always match submenu button text, and vice versa. MyDVD is a consumer-level product, and its designers did plenty to keep things simple. This was one of those design decisions that can confound more experienced users but does make sense to those just getting started.

Quiz

6

1. How do you use a photo or video and music from a personal collection to create a menu?
2. What's the correlation between the house and the little swirling arrow buttons in the submenus versus the Menu and Title buttons in the simulated remote control screen?
3. You see a MyDVD style you like on the Sonic Solutions Web site. How do you use it in a project?

Quiz Answers

1. In the Edit Style interface, click the Select Custom Video or Still Background button to select your image or video clip. Then click the Select Custom Music Track button to add music of your choice. If you use a lengthy video clip, you might want to save rendering time by reducing the loop time to a reasonable figure, such as 15–30 seconds.
2. The house icon Home menu button and your remote's Title button perform the same function: They return you to the project's main menu. The swirling arrow Return button and the remote's Menu button both take you back to the previous menu.
3. Download a copy and store it in a convenient file folder. With the MyDVD Edit Style interface open, click the Import Style button, navigate to the downloaded style file, select it, and click Open. It drops it into the Edit Style interface, and you can start using its background, buttons, and music right away.

Exercises

1. Use MyDVD to make a personalized DVD, which is what this section of the book is all about. You now have a facility with MyDVD and should be able to translate that to your own material.
2. Go through your collection of still images and graphics and select some to use as menu backgrounds. You might want to use a consumer-level graphics product such as Photoshop Elements to combine some graphics and images into customized menu backgrounds. Read the MyDVD help file for specifics on recommended image sizes and aspect ratios.
3. Download all the MyDVD styles from the Sonic Solutions Web site and install them in one folder. Take a look at them and consider using them in a personal project.
4. Create a DVD in MyDVD. Because MyDVD uses OpenDVD technology, you can reedit it. To do so, start MyDVD and, in the Welcome window, move the mouse pointer over DVD Video and click Edit an Existing OpenDVD Disc. If MyDVD is already running, select File, Edit DVD/VCD.

PART II
Creating Media

Hour

HOUR 7

Creating Still Images

To make high-quality DVDs, you need to start with high-quality media. Simply taking a bunch of poorly exposed snapshots or jerky, out-of-focus videos and slapping them on a DVD will not transform them into works of art.

In this four-hour section of the book, I'll give you tips that will help you create quality media.

This hour focuses on still images. I offer tips on digital camera selection with a hands-on test of two popular consumer digital cameras. I also present some picture taking dos and don'ts and cover three new digital image technological developments.

Finally, I tell you what to look for when buying an image scanner.

The highlights of this hour include the following:

- Digital or film cameras—what will work best for you
- Making high-quality photos—tips and tricks
- Evaluating two digital cameras—my hands-on experience
- Three rapidly changing digital image technologies
- Selecting a scanner for your video and DVD projects

Digital or Film Cameras—What Will Work for You

My quick answer is film for sure. Digital? Maybe.

I am not a big fan of digital still cameras. They cannot replace film cameras and have too many drawbacks, including the following:

- **Shutter lag time**—You must press the button and wait for up to two seconds to actually take the photo (see the sidebar titled "Consumer Digital Cameras Don't Do Action," later in this chapter).
- **Delay between shots**—This ranges from two to five seconds.
- **Battery consumption**—They devour batteries.
- **Slow auto focus.**
- **Poor flash metering**—They use "pre-flashes" instead of responding to light during the actual exposure.
- **Frequent color or white balance miscues.**
- **Digital focal multiplier**—Your 35mm SLR (single lens reflex) camera wide-angle lens won't work properly on a digital SLR (see the sidebar titled "Three Rapidly Changing Digital Image Technologies," later in this chapter).
- **Rapid obsolescence**—You can buy a digital camera one day and see a better, less expensive model advertised the next day.
- **Learning curve**—Yet another technology to learn with frequently complicated, arcane, and incomprehensible controls.
- **Poor low-light capabilities.**
- **Expensive printers and paper.**

Despite these drawbacks, millions of digital camera users can't all be wrong. Digital cameras have the following advantages:

- You get immediate feedback. If you don't like how the photo turned out, you can erase it and try again.
- You never have to buy film again.
- You don't have to pay to process your film.
- You can print only the photos you need when you need them.
- You can quickly and easily upload pictures to your PC.
- You don't need to use a scanner.

- Prices are dropping, and quality is increasing.
- You can give a digital camera to your kids without worrying about them wasting film.

Digital still cameras are not completely ready for prime time, but they can be tremendously useful for certain applications:

- Real estate agents emailing photos of homes to out-of-state clients
- Insurance adjusters photographing property damage
- Employee security badges, credit card photos, or driver's license photos
- Posting images to Web sites

Consumer Digital Cameras Don't Do Action

The first time you try a digital camera, I guarantee you'll wonder what's going wrong.

You'll look through the viewfinder and see nothing. You'll think, "Ooops, I need to turn the darned thing on." You'll press the On button or open the sliding lens cover, you'll wait, and the camera will finally finish its startup process and be ready to shoot.

You'll then compose a shot and press the shutter, but nothing will happen. You'll probably press the shutter again, a bit harder, and still nothing will happen. So, you'll hold it down longer—a second or two—and finally you'll hear a click and whir and an image will appear. But it won't be the image you thought you were going to get—it won't be that moment, frozen in time, that you visualized when you pressed the shutter.

That moment passed your digital camera by. Why? Electronics, surprisingly, can be slow.

Here's what occurs as you press the shutter: As with a film camera, a digital camera emits an infrared signal to set the focus, adjusts the autoexposure by changing the aperture (f-stop) and the shutter speed, and (if it's dark) sends out a small burst of light to determine how much flash to use. At this point, a film camera snaps the picture. But a digital camera has much more work to do.

That digital camera flushes the photosensitive computer chip's electric charge to prepare it to receive a new image.

Photons from the subject hit that chip. It converts them to electrons, changes them to digital data (typically at least two million chunks of color and brightness data), and moves them to an interim storage location. From shutter press to image capture, up to two seconds elapse.

If you're ready to take another picture, you have to wait from two to five seconds while your digital camera recycles. The camera has to compress that digital information and store it before it's ready to take another photo.

Shooting action photos is just about out of the question, and expecting portrait subjects to hold that smile for a second and a half is asking a lot. No longer is it, "Three, two, one, click." Now it's "Three, two, depress shutter, one, click."

> There is one way around this, though. You can spend a few thousand dollars for a professional digital camera, which uses one-click/one-shot, or *sequencing*, technology.
>
> But even then, depending on the camera, you might need to wait more than a second between photos. Also, you can't use the flash (it can't recycle fast enough), the photos might have lower resolutions than normal, and the color balance might be off.
>
> So, if you use a digital camera, you'll need to make some adjustments. See the section "Compensating for Lag Time," later in this hour.

Digital Camera Buying Tips

Other than convenience, I see no compelling reason to buy a digital camera for a DVD project. If you need to use archived photos in any project—family history DVDs, for instance—you'll need to buy a scanner anyway. Therefore, you can continue to rely on film.

But the demand for digital cameras continues to grow. Despite my reservations, you might want to buy one, or you might already own one and want to replace it—new technology is so tempting.

So, here are my digital camera buying tips:

- **Megapixels**—These are millions of picture elements or data points on the light-sensitive chip. The higher the number, the more you can enlarge the printed image and not lose details. Two megapixels is the minimum to make an average-quality 5"×7" photo printout, three megapixels is the minimum for an 8"×10" printout, and four megapixels is the minimum for an 11"×17" printout.
- **Storage capacity**—Larger megapixel images require more storage space. Note how much capacity comes with the camera and the cost for additional memory modules: CompactFlash, SmartMedia, Secure Media, or Memory Sticks. Don't buy cameras with floppy disk or CD storage because they're too slow.
- **Try before you buy**—The feel, size, and weight of the camera along with the location of its controls are important.
- **Optical zoom capability**—2X optical zoom capability is the minimum you should get, but 3X is much better. Ignore references to digital zoom; that just reduces the resolution of the image.
- **Rechargeable batteries and a charger**—These are a must. Buying them separately adds $30+ to the total price. NiMH (nickel metal hydride) rechargeable batteries are better than NiCad (nickel cadmium). Always keep a second set on the charger, and keep in mind that many rechargeable batteries tend to lose power over time.

- **Burst**—This is also called *sequence shooting mode*, and it compensates for shutter and shot-to-shot lag times.
- **Check out the software bundle**—Some cameras come with some excellent products, but most do not.
- **Color liquid crystal display (LCD) panel**—You use this to preview photos and determine whether exposure or color balance adjustments are necessary.
- **Macro function**—You use this to make extreme close-ups, from an inch or so away from the subject.
- **USB PC connectivity**—It's ubiquitous.

Evaluating Two Digital Cameras—My Hands-on Experience

Olympus loaned me two of its more popular digital cameras. Both offer DVD producers plenty of features.

I enjoyed testing the Olympus D-550 Zoom, which is a simple point-and-shoot digital camera. Shown in Figure 7.1, it has the familiar look and feel of many point-and-shoot 35mm film cameras. The D-550 retails for about $300 versus $200 for a similar 35mm film camera.

The D-550 functions much like the easy-to-use Stylus line: You slide open the lens cover, the lens extends, and the flash pops up. Then you simply point and shoot. The D-550 has three megapixels and a 2.8:1 optical zoom—plenty of muscle for this price range.

The LCD display allows easy access to your photos, and the menus and icons are large, easy to read, and reasonably self-explanatory. The sequence mode did not work any better or worse than others in this price range, meaning the results were unimpressive.

The camera ships with four nonrechargeable AA alkaline batteries, but an AC adapter/charger and rechargeable batteries are optional.

The software bundle is rudimentary but works well. Uploading images to your PC is remarkably easy: You simply use the USB cable to connect the camera to the PC, open the camera's door to switch it on, start Olympus's Camedia image editing software, and it automatically pops up thumbnails of every image in the camera (see Figure 7.2).

I heartily recommend the D-550.

The Olympus C-720 Ultra Zoom three-megapixel model, illustrated in Figure 7.3, is geared to prosumers who don't want to step all the way up to single-lens reflex (SLR) cameras but want some of their high-end features, manual override options, and powerful zoom lenses.

Selling for only $100 more than the D-550 ($400), the C-720 offers a substantial 8:1 zoom that fits the needs of most photo situations, an electronic viewfinder (as opposed to optical) to create accurate framing, and an extra powerful pop-up flash.

> The C-720 has several drawbacks that keep me from recommending it: At its highest zoom level, the autofocus rarely hits the sweet spot. You can't use the built-in flash for close-up macro mode shots because the long lens throws a shadow. When using that flash, the C-720 typically creates overexposed images. A manual exposure adjustment is available, but the LCD control typeface is so small I had to use reading glasses—something I don't normally keep handy.

FIGURE 7.1
Olympus D-550 Zoom digital camera has the familiar look and feel of the Stylus Zoom 140, a standard point-and-shoot 35mm film camera.

FIGURE 7.2
Olympus Camedia, bundled with the D-550 and C-720, has limited features but smoothly uploads your photos.

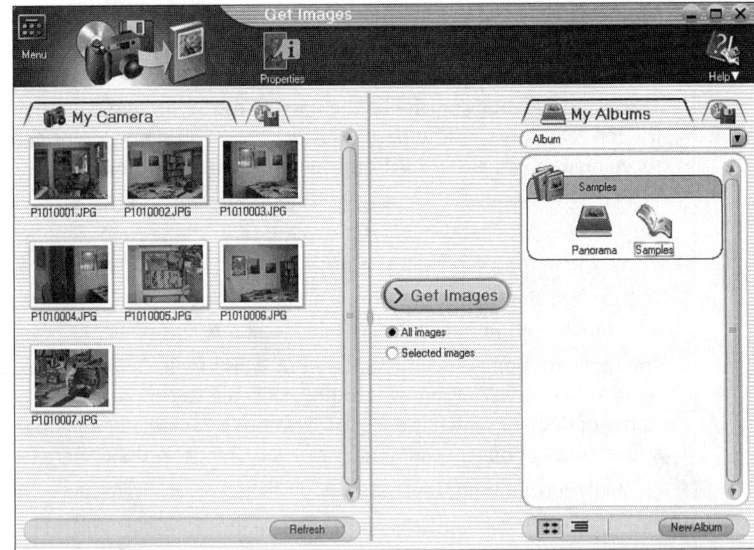

FIGURE 7.3
Olympus C-720 Ultra Zoom.

Saving Your Photos Without Using Third-Party Software

You don't need the Olympus bundled software, or any third-party software for that matter, to load images to your PC.

When you connect your camera to the PC, as shown in Figure 7.4, Windows gives you an option labeled Copy Pictures to a Folder on My Computer. Select that option to open the Scanner and Camera Wizard; then click Next.

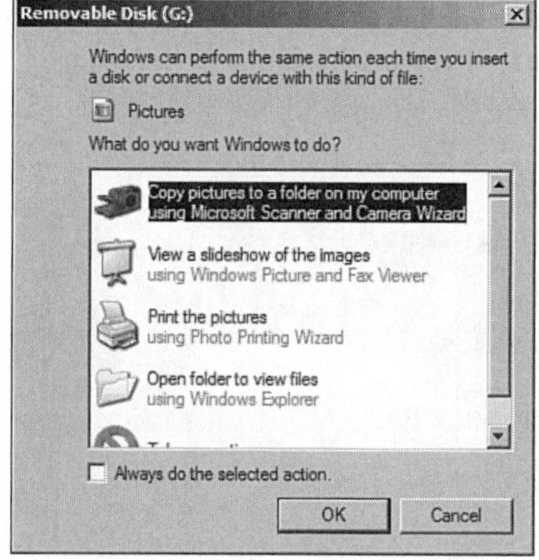

FIGURE 7.4
When you connect and turn on your digital camera, Windows pops up this Removable Disk screen (your camera is a storage device).

That opens the window shown in Figure 7.5. Select the photos to save, click Next, select a file folder, and save the photos.

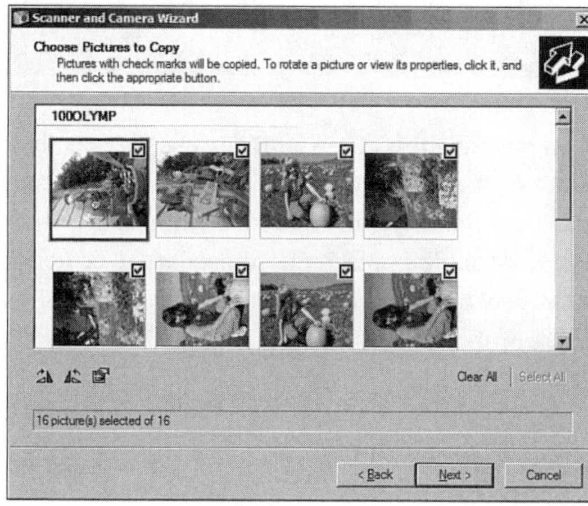

FIGURE 7.5
Use the Scanner and Camera Wizard to select "keeper" photos to store on your hard drive.

Making High-Quality Photos—Tips and Tricks

Whether digital or film, here are some standard tricks and tips that will help you improve your picture-taking results.

Putting an End to Blurry Images

The biggest problem in amateur photos is blurry pictures. The principal cause is camera movement, but there are several other reasons, including the following:

- **Camera movement**—Figure 7.6 shows a classic example of the results of camera movement. Instead of gently pressing the shutter, many amateur photographers abruptly push it, shaking the entire camera. Digital cameras exacerbate this because shutter lag time leads many digital camera users to press down even harder.

FIGURE 7.6
When everything in an image is blurry, you can bet camera movement is the culprit.

- **Autofocus on wrong subject**—Autofocus usually sets the focus based on whatever is in the center of the viewfinder. If you're framing a scene with something in the foreground, as in Figure 7.7, the autofocus might "see" the frame, not the subject. Adjust the camera angle to place the subject at the center of the viewfinder, depress the shutter halfway to set the autofocus, compose your shot, and then press the shutter the rest of the way.

FIGURE 7.7
The autofocus saw the cornstalk frame, not the pumpkin picker subjects.

Composing Your Shots

Composition is critical, and for most photos only a slight change in camera angle or location will make the difference between a mediocre snapshot and an effective, pleasing photo. Here are some tips:

- **Get close to your subject**—Instead of typical tourist shots of family members off in the distance standing directly in front of some fountain, frame the fountain to fill your viewfinder and then have your family stand close to the camera and a bit off to one side of the frame.

> If you put your family up close with the fountain some distance away and then focus on your family, will the fountain be out of focus? It depends. In daylight, the auto aperture (iris) will be very small, creating a deep depth of field. Foreground and background elements will all be in focus. In low-light settings, however, the aperture is wide open and the depth of field is very shallow. Therefore, the fountain will be out of focus. Using a narrow depth of field well can lead to dramatic images.

- **Add a foreground element**—Adding something between you and your subject gives depth to your images.
- **Use the rule of thirds**—As shown in Figure 7.8, divide your image into thirds and place the object of interest at one of the intersecting lines. That creates much more visual interest. One quick and easy way to adhere to this rule is to keep your subject off-center.

FIGURE 7.8
This image has elements from each of the three previous tips: it has a foreground element, the subject is off-center, and it uses the rule of thirds.

- **Shoot at oblique angles**—As shown in Figure 7.9, instead of shooting straight on, shoot a subject from a nonperpendicular angle.

Other Photo-Taking Tips

Fuzzy photos and poor composition are the most frequent culprits in the picture-taking snafu department. Here are a few other tips you can follow to improve your photo-taking skills:

- **Watch backlit scenes**—As shown in Figure 7.10, your camera's autoexposure sets itself for the light behind your subjects, meaning they'll be silhouettes. Either set the autoexposure on them first and then compose the shot, or use fill-in flash—or do both.

FIGURE 7.9
Use oblique angles to add interest.

FIGURE 7.10
When the sun's behind your subjects, silhouettes might be all you get.

- **Use fill-in flash**—Whenever you shoot outdoors, adding flash brings out the colors and details of your subject. Figure 7.11 shows how fill-in flash can overcome the silhouette effect of back-lit shots.

Figure 7.11
Fill-in flash can overcome backlit scenes (getting your subjects to stop squinting takes much more effort).

Flash has a very limited range, about 10–15 feet. Next time you're at a concert or nighttime sporting event, note all the fans with point-and-shoot cameras taking flash photos from 100 rows back. What they'll get is brightly illuminated backs of heads from a couple rows in front of them. Don't waste your time. The only way to use a flash is to get close to your subject.

- **Don't overexpose foreground objects**—As shown in Figure 7.12, when using flash, objects close to the camera will be over-illuminated. This is one time when adding a foreground element might not work.
- **Avoid stiff poses**—Encourage your subjects to do something, such as walking, talking, pointing—anything to add interest.
- **Keep the background simple**—Distractions draw the attention away from your subject.
- **Use lines to add interest**—S-curves and diagonal lines, such as those in Figure 7.13, add visual interest.

FIGURE 7.12
The flash tends to illuminate the closest object, which might not be your desired outcome.

FIGURE 7.13
Diagonal lines help draw attention to the subject (the backlit cloud of dust is a nice touch, too).

Compensating for Lag Time

Here are some tips to overcome lag time inherent to digital cameras:

- Turn on your camera before you need it but, if possible, keep the LCD viewfinder and flash turned off (they drain too much battery power).

- Get used to depressing the shutter release halfway to lock focus and exposure and then depressing the shutter all the way to take the picture.
- When shooting action, switch to the burst or sequence mode. However, those modes typically create less-than-perfect images.
- If you let your subjects know when you're going to take the photo by counting down from three, press the shutter on "one."
- Anticipate action by shooting sooner than normal.

Three Rapidly Changing Digital Image Technologies

Digital camera technology does not stand still. Here are three recent developments:

CMOS versus CCD—Most still cameras and video camcorders used CCDs (charge-coupled devices) to capture images. Kodak and Canon have challenged that dominance with two ultra-high resolution digital still cameras: the 14 megapixel Kodak DCS Pro 14n and the 11 megapixel Canon EOS-1DS both use CMOS (complementary metal-oxide semiconductor) chips.

CCD chips with the same resolution would be more costly and bulky. And CMOS chips use far less power than CCDs. Another advantage, the CMOS chip has the same frame size as 35 mm film, meaning there is no need for owners of SLR (single-lens reflex) cameras to buy new lenses.

- **Digital focal multiplier**—That CCDs are smaller than 35mm film creates a sometimes expensive inconvenience for owners of digital SLR cameras using their film SLR lenses. The smaller image capturing area effectively increases the focal length of any interchangeable lens used on a digital SLR camera. That boosts telephoto lens power—arguably a nifty benefit—but narrows the view of wide-angle lenses. Creating distortion-free wide-angle lenses is an expensive art—buying a new one just for a digital SLR camera can easily cost more than $1,000.

  Olympus, Kodak, and Fuji think they have an answer: the Four Thirds System (4/3 System), a standardized lens-mounting scheme for digital SLRs. If enough companies sign on, this will resolve the digital focal multiplier issue (you'll still have to buy new lenses, though). It will also lead to smaller, lighter lenses and ensure uniform lens mounts across all brand lines, something that does not exist for 35mm SLRs.
- **Foveon X3 image sensor**—CCDs and CMOS technology pale in comparison to the image clarity of Foveon X3 (`www.foveon.com`). These new chips capture three times the color resolution, feature a simpler design, and offer higher overall performance for digital still and video cameras.

  Standard digital camera chips use a mosaic pattern of pixels in groups of three red, green, and blue photodetectors. The resulting image, when viewed up close, looks like a checkerboard.

  As illustrated in Figure 7.14, Foveon embeds three layered photodetectors in silicon at every pixel location to capture all colors within each pixel. The result is sharper images with more accurate color reproduction.

> In November 2002, Sigma Corporation shipped the first Foveon X3-enabled camera, the Sigma SD9 ($1,800). Foveon expects that several other still and video camera manufacturers will soon follow suit.

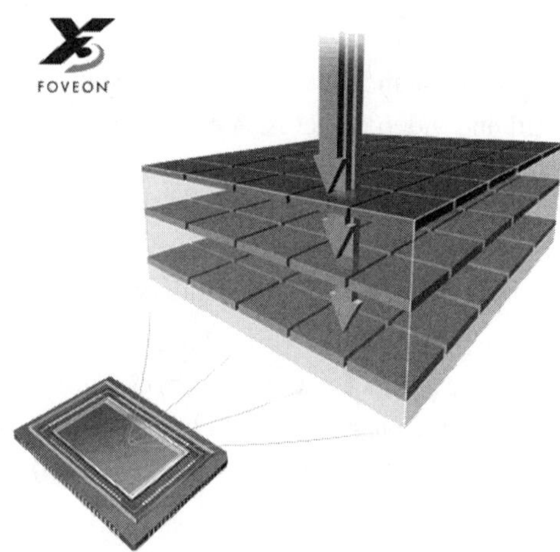

FIGURE 7.14
Foveon X3 Technology. Illustration ©2002, Foveon, Inc.

Selecting a Scanner for Your Video and DVD Projects

A scanner is a critical part of your DVD production toolset. You will frequently need to include nondigital photos, logos, graphics, or other printed material in your DVD projects. To do that, you need a scanner.

Now is a great time to buy either your first scanner or one to replace that old clunker in the corner. For $200 or less you can get plenty of horsepower. I discuss scanner usage tips in Hour 15, "Scanning and Formatting Images for DVD Authoring."

Scanner Buying Tips

Before heading off to the store or jumping online, take a look at the following shopping tips:

- **Dots per inch (DPI) and color depth**—1,200dpi is sufficient for video production work; 2,400dpi works well for high-end photo or prepress projects. 42- or 48-bit color depth is more than you'll ever use. Most image editing software scales down to 24 bits—8 bits per color (red, green, and blue).

DPI is supposed to be noted as horizontal by vertical, as in 1,200×2,400. The horizontal number is the true resolution and refers to the density of sensors in the image chip. The vertical number equals the steps per inch that the scanner motor moves the scanner head. Some manufacturers flip the numbers, so pay close attention.

- **CCD versus CIS**—Most scanners use CCDs, the same type of image sensor chips found in digital still and video cameras. A couple of companies, Canon and Mustek, rely on contact image sensor (CIS) chips. CIS chips use less power and are more compact, but they have trouble with books that don't lie absolutely flat on the scanner glass. Additionally, CIS scanners from Canon and Mustek are slower than the norm, so stick with the industry-standard CCD scanners.
- **Scanning speed**—This varies greatly and changes with each new model. I suggest checking online at either `http://www.pcworld.com` or `http://computers.cnet.com/` for current bench test results. At last word Epson, HP, and Visioneer have the best scanning speeds for low-resolution (300dpi) scans—typically about 20 seconds.

Generally, you don't need to scan at high resolution for display on TV; 300dpi usually is more than adequate (unless you have a very small photo you want to display full-screen). I cover scanning techniques for DVD and video projects in Hour 15.

- **Connection speed**—USB 2.0 scanners are slowly appearing on the scene. If you have USB 2.0 capability then you will see speed improvements, but only for higher-resolution images. Otherwise, USB 1.1 is adequate.
- **Transparencies**—Most consumer scanners do not handle slides or negatives as a standard feature. You'll need an optional tray, which typically costs about $25.
- **Onboard buttons**—Some scanners give you several controls on the scanner itself, which can be convenient and helpful.
- **Bundled software**—Most flatbed scanners come with the excellent ABBYY FineReader optical character recognition (OCR) software and a barebones image-editing package. Even though all scanners these days are TWAIN compliant, meaning products such as Microsoft Word and Photoshop can directly access and operate your scanner, the software bundle usually includes a rudimentary scanner control interface.

Summary

Still images play a vital role in DVD projects both as content and menu backgrounds. Digital cameras are a convenient way to create those images, but this new technology has too many drawbacks for it to replace film cameras. Consider buying a digital camera only as a supplement to your film camera. Following some basic photo shooting tips will ensure your images are of top quality.

Owning a scanner is a must. For $200, you can get an excellent model that does everything you need.

Workshop

Review the questions and answers in this section to reinforce your knowledge of still image creation tips and techniques. Also, take a few moments to take the short quiz and perform the exercises.

Q&A

Q I connect my digital camera to my PC, but nothing happens. What did I miss?

A Switch the camera to "on." How you do that varies depending on the model. With most consumer-level models, such as the Olympus D-550, you need to do the same thing that you do when you want to take a photo. In that case, slide open the lens cover. Higher-end models, such as the Olympus C-720, actually detect that you've plugged in the USB cable, turn off the photo mode, and automatically switch to photo transfer mode. When all else fails, check your batteries.

Q My flash photos are uneven. Foreground elements are too bright, and distant elements are too dark. What should I do?

A Flash is effective for only about 10–15 feet, which means foreground subjects get all the light and distant objects get none. Couple that with the autoexposure feature of most cameras that reacts to the average illumination of an entire scene and the close-in elements end up being overexposed and distant elements end up being too dark. When using a flash in a dark setting (as opposed to using fill-in flash), try to keep objects equidistant from the camera to ensure even lighting.

Quiz

1. How do you compensate for lag time?
2. What is the number-one cause of blurry photos?
3. What is the rule of thirds?

Quiz Answers

1. You can try relying on your camera's sequencing mode, but that usually isn't satisfactory. Otherwise, turn on the camera before you need it and get used to depressing the shutter halfway to lock in exposure, white balance, and focus. Finally, anticipate action by depressing the shutter at the penultimate moment.
2. Digital cameras exacerbate camera movement because users need to depress the shutter for so long. Keep a steady hand.
3. The rule of thirds is a standard photo composition tool. When looking through your viewfinder, think of it as being crisscrossed by perpendicular lines that divide it into nine blocks. Generally, placing the principle subject of your photo at one of the four intersections of those lines makes the resulting photo look more appealing.

Exercises

1. Using the photo-taking tips listed in this hour, grab your camera—film or digital—and take some pictures. Concentrate less on the subject matter and more on techniques. Look for interesting angles, s-curves, and foreground elements. Use the rule of thirds, and keep your camera steady.
2. Along the lines of exercise #1, come up with an easy-to-find subject you want to photograph. It could be park benches, car bumpers, or jelly donuts. Then grab your camera and go on a quest for that subject. This exercise forces you to come up with different ways to approach a subject. I did this with gravestones once, and it got pretty weird. But it opened my eyes to new possibilities.
3. *National Geographic* works with the best photographers in the world. Thumb through a few issues and take a critical look at the photographs. Note the oblique angles, placement of subjects, strong foreground elements, and action. You also notice might that most exterior photos have long shadows. The light is best early and late in the day.

HOUR 8

Acquiring Audio

Audio is critical to your DVD production. You'll want to use music to create a mood. Finding or fashioning that music is fairly easy these days. You can rip tunes from music CDs, license or buy compositions, or create your own professional-quality songs using inexpensive software.

Your narration is crucial. Building a personal voice recording area is the first step. Then, by following a few narration tips, you can add professional polish to your piece.

I'm not leaving out so-called "natural sound." It lends authenticity and believability to your video productions. I'll cover that topic in Hour 9, "Making Videos," and Hour 13, "Adding Audio, Tackling Text, and Improving Images."

The highlights of this hour include the following:

- Ripping music CDs
- Licensing music or buying royalty-free music
- Creating custom music with SmartSound Movie Maestro
- Recording high-quality narrations

Ripping Music CDs

The easiest source for DVD production music is next to your stereo: your personal music CD collection. All CD cuts are digital and easily ripped to your hard drive. Once there, you can use them in video productions or add them to DVD menus.

Those tunes on your CDs are all copyrighted. I am not an attorney and don't pretend to understand copyright law. That said, I'd suggest treading carefully when using someone else's music. Generally, if it's for personal use, it's considered fair use and there are no copyright issues. But just about any other use can step outside fair use. To be on the safe side, compose your own music or license or buy royalty-free music. I cover all three options in the following sections.

Task: Use Windows Media Player to Rip CD Cuts

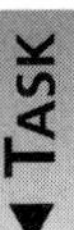

No, the book is not taking a violent twist. *Ripping* is just how some describe the process of transferring music from a CD to a PC.

Several products are available to do that. The one most likely to be at your fingertips is Windows Media Player. Here's how it works:

1. Open Windows Media Player. It's probably in the Start menu under Accessories. If not, its default location is `C:\Program Files\Windows Media Player\wmplayer.exe`. You can open My Computer, go to this location, and double-click `wmplayer.exe`. Doing so opens the interface shown in Figure 8.1.

You can open Media Player another way. Simply insert a music CD into your DVD or CD drive and, depending on your version of Windows, either Media player automatically starts playing that CD or you are given an option of playing the CD with one of several programs you have installed on your PC.

2. Click Copy from CD, as highlighted in Figure 8.1. That pops up a message asking you to insert an audio CD.
3. Insert a music CD. Either Windows will ask you which CD player you want to use (select Media Player) or Media Player will automatically start playing the CD. Click the Stop button, as shown in Figure 8.2.

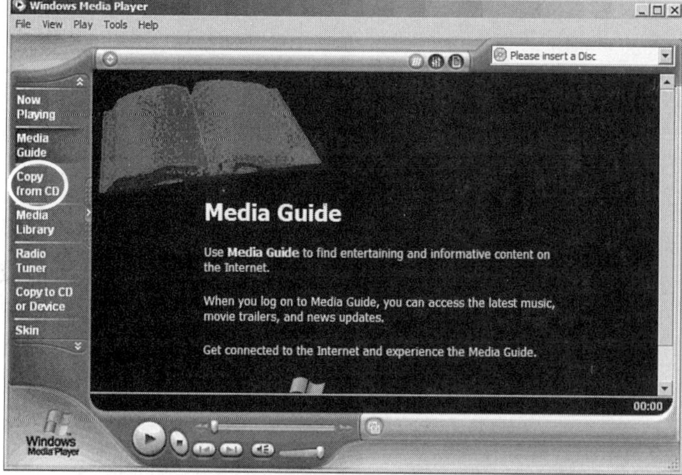

FIGURE 8.1
Use Windows Media Player to rip tracks from music CDs.

If you're connected to the Internet, Media Player can read the music CD's unique identifier number and retrieve the album information—CD title, artist, and track names—from the Web. Click Get Names, and then click Album Details for a full listing plus a review. Even my example, the little known but wonderful Comedian Harmonists CD, shows up.

FIGURE 8.2
Windows Media Player's Copy CD interface displays information about your CD retrieved from the Internet.

4. By default, all tracks are check-marked for copying to your hard drive. You can uncheck any you don't want to copy.

If you want to preview a track, click it to select it and then click the Play button.

5. The default copy location is `My Documents/My Music`. Media Player creates a folder for the artists and a subfolder for the selected album. If you want to change that location, select Tools, Options from the menu bar. Doing so opens the Options menu, shown in Figure 8.3. Select the Copy Music tab and change the directory.

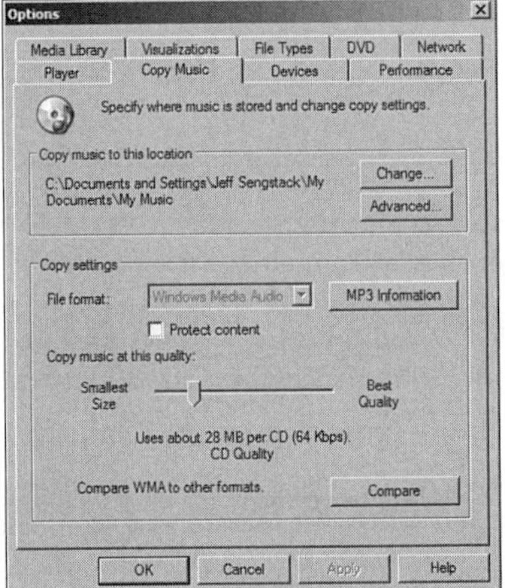

FIGURE 8.3
Use the Options interface to change the file folder storage location for your selected music tracks.

When selecting music for use in video or DVD productions, instrumentals generally work best. Vocals can step on your narration, natural sound, or voices in your videos.

6. When you're ready to copy the tracks, click the Copy Music button at the top of the interface. Media Player uses progress bars to let you know how things are proceeding.

Now that you have ripped a few tunes, you can use Media Player to burn a music CD. This is a great way to create personalized CDs of your favorite tunes.

8

> You can also use other CD/DVD file copying software to create music CDs, but some older products don't work with Windows Media Audio (WMA) files.

Task: Use Media Player to Create Custom Music CDs

Media Player has a built-in music CD burning module. Here's how it works:

1. Click Copy to CD or Device to open a new interface, as shown in Figure 8.4.

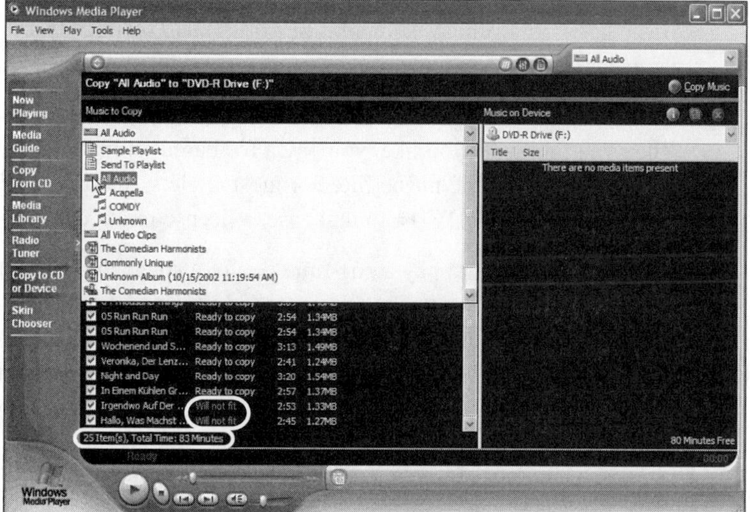

FIGURE 8.4
Use Windows Media Player to create personalized, customized music CDs.

2. Because you've probably stored your music selections in various locations, click the drop-down list above the music track listing and select All Audio. This places all your ripped tracks in the left window.
3. Place a recordable CD in your DVD recorder, and Media Player updates its interface by noting that your tracks are ready to copy.
4. Uncheck those tracks you don't want to include. The Total Time value, shown in Figure 8.4, changes as you update your selections.

> If your selected tracks will exceed the capacity of a CD (about 75 minutes), Media Player lets you know that by stating `Will not fit` for tracks at the end of the list (see Figure 8.4).

▼

5. When you're ready, click the Copy Music button in the upper-right corner. Media Player converts the WMA files to audio-CD-compatible files and burns a music CD. When completed, try the new disc in your home stereo.

Licensing Music or Buying Royalty-Free Music

For those who will venture beyond personal—noncommercial—DVD productions, you can avoid any copyright hassles by licensing or purchasing tunes. The Internet makes that remarkably easy.

Music licensing agencies abound. They offer song search capabilities geared to multiple parameters: music genre, mood, instrumentation, tempo, and so on. Most even have try-before-you-buy listening capabilities.

When you find what you like, you pay a fee based on planned usage and audience size and then download the music file. For most readers, that usage will be nonexclusive and for fewer than 5,000 DVDs. In that case, a license might cost \$75–\$300.

Another approach is to pay a one-time fee for a song and get the right to use it as often and in any way you want. That's a so-called *royalty-free* stock music service. In the pre-Internet days, stock music companies sent out boxes of music CDs to production houses. Most times they gathered dust until that one moment when a producer needed that special piece of music and had to slog through too many tracks to find it.

Now some stock music houses have put their products online with the same type of search and try-before-you-buy capabilities as the music licensing houses offer.

I contacted both types of companies, selected three to use as examples, and will go through the process you can use to buy music.

Licensing Music

Anyone with any Internet savvy and a search engine—`www.Google.com` is my favorite—can track down music licensing houses. The choices can overwhelm you, though. Here are two that take slightly different approaches.

Task: Use LicenseMusic.com to Audition and Download Tunes

TASK

LicenseMusic.com, based in Copenhagen, Denmark, represents more than 100 content providers with more than 100,000 songs. It characterizes itself as the "foremost Web-based music licensing company," and I see no reason to dispute that. Here's how to use its service:

1. Navigate your browser to `http://www.licensemusic.com/`. That takes you to the company's home page, shown in Figure 8.5. You can sign up if you want, and it's free.

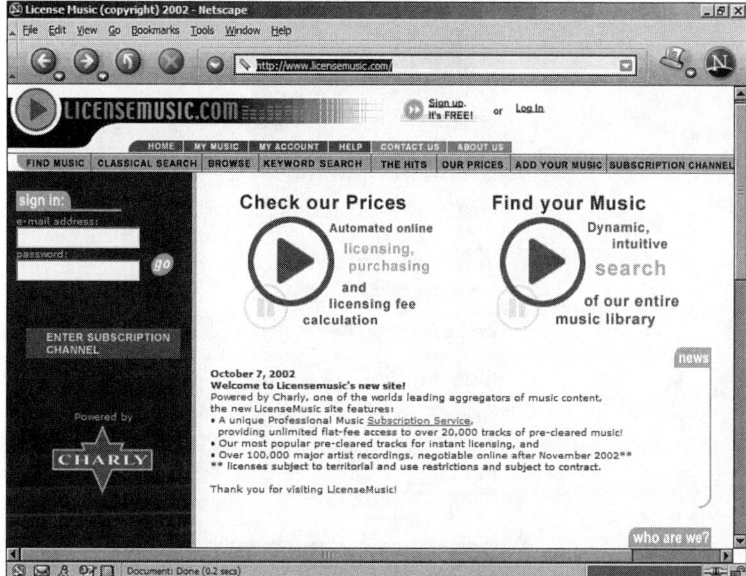

FIGURE 8.5
LicenseMusic.com offers a full range of musical production possibilities.

2. Click Search. As illustrated in Figure 8.6, doing so opens a collection of drop-down menus that let you narrow your search. You have a choice between two main music sources: Pre-cleared and Production Library. Pre-cleared tends to be more pop-oriented, whereas Production Library leans toward filling specific production needs. For this exercise, stick with the default setting: Music from Both Sources.
3. Go through the drop-down lists and narrow your search. As you add parameters, note that the Tracks Found number decreases. If you refine things too much, it can drop to zero (for instance, selecting Country Rock as the genre and Oboe as the instrument). You don't have to make a selection from each list.

> The Maximum Price menu displays from one to three dollar signs ($).The lowest-priced offerings for a limited distribution DVD typically license for about $75.

4. When you're done adjusting parameters, click Show Songs to display a list of songs. Clicking one, as shown in Figure 8.7, displays information about that tune and lets you listen to a 25-second excerpt.

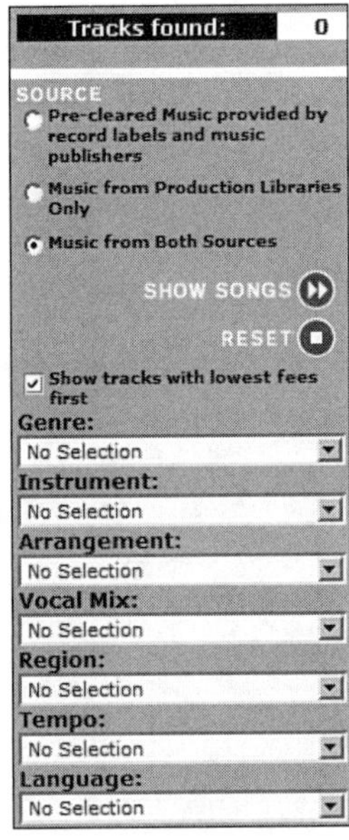

FIGURE 8.6
Use more than a dozen criteria to narrow your search for the best musical number to fit your DVD production.

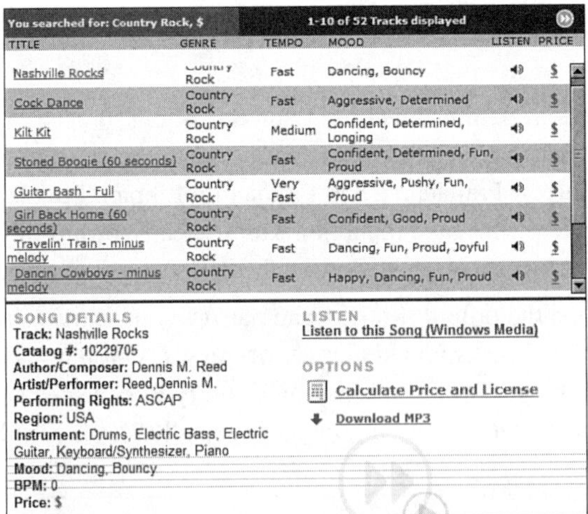

FIGURE 8.7
After you've narrowed your search, view and preview your music "hits."

5. You can download the selection (you have to be registered) at no charge on a try-before-you-buy basis. Right-click the Download MP3 link, select Save Target As, select a file folder location, and click Save.

> LicenseMusic.com watermarks its MP3 files. As you download, the company embeds inaudible yet readable digital code into those MP3 files that identify you as the buyer.
>
> If you later upload or transfer a LicenseMusic.com file to other users, the unique watermark can clearly identify the MP3 file's origins (you need the watermark manufacturer's software to do this). However, it does not prevent playback of that file.

6. To check the licensing fee, click Calculate Price and License. The page shown in Figure 8.8 opens. Your likely medium is Straight to Home Video/CD-ROM/Similar Media: Roll-Over (Limited # of Copies). Select it, and click Next.

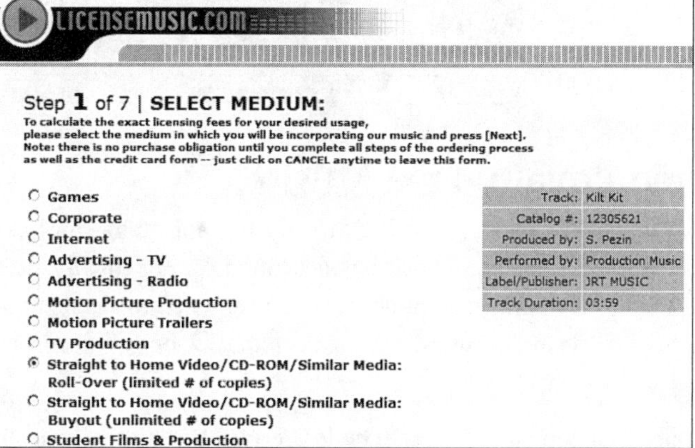

FIGURE 8.8
LicenseMusic.com bases its fees on your type of production and its audience size.

7. You'll see a couple other options about the number of DVD units; whether it'll be the theme song; whether your budget is less than $100,000; and whether you'll use more than a prescribed amount of time (usually 45 seconds). The bottom line is that your license will likely cost about $100.

I'd suggest checking out at least one other music licensing firm: Dittybase (`www.dittybase.com`), based in Victoria, British Columbia, Canada. It works similarly to MusicLicense.com in that it has a search engine and a try-before-you-buy option. Its selection is not as vast, but its service is more user friendly (see Figure 8.9).

It bases its rates solely on your usage as opposed to varying rates based on artist and publisher contracts. Typical business uses for a one-time, so-called "needledrop" cost about $80, and uses for fewer than 10,000 copies of a DVD for retail sale cost about $150.

FIGURE 8.9
Dittybase works much the same way as LicenseMusic.com but has a more user-friendly interface.

Using Royalty-Free Music

As with licensed music, so-called royalty-free music has hit the Internet. Only a few differences exist between the two business models. For royalty-free music, you pay a one-time flat fee for *unlimited*, nonexclusive use, whereas licensing agencies usually base their fees on a single use plus the expected sales or audience size and the type of production.

Royalty-free companies tend to be less broad-based than licensing firms. Frequently, they represent only a handful of artists or have limited genres. Theirs is a simpler, easier-to-manage business model, but they typically cost much less and the quality can be on par with licensed music.

One such royalty-free firm is Stock-Music.com. Figure 8.10 shows its home page. This small, Utah-based firm at last word had about 70 songs in seven categories. The company provided three of its songs for inclusion on this book's companion DVD.

To access those three MP3 files, go to the Stock-Music file folder. Each song comes in three lengths: full-length (typically 3 minutes), 30 seconds, and 15 seconds. The short versions are great to use with DVD menus that *loop*—repeat a brief audio clip over and over until the user clicks a button.

Readers of this book have permission to use these songs in any production, any number of times, at no charge. If purchased separately, these three songs would cost $60.

8

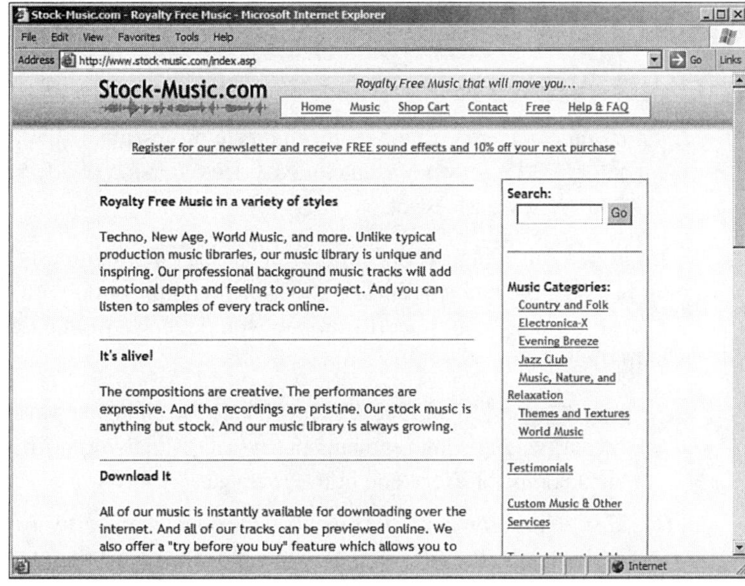

FIGURE 8.10 *Stock-Music.com is a small, royalty-free shop with several dozen tunes across a handful of genres.*

To try other songs from this company, visit `www.stock-music.com`. As shown in Figure 8.11, after selecting a genre, you can listen to brief excerpts and then purchase individual tunes or an entire collection.

FIGURE 8.11 *Stock-Music.com offers straightforward music preview pages.*

Creating Custom Music with SmartSound Movie Maestro

Even if you can't carry a tune, you can create professional quality music. On this book's companion CD/DVD, we have included a demo version of a fun and exciting product: SmartSound Movie Maestro.

With it, you can compose custom music that fits not only the style of your video production, but also its exact length. The $50 retail version comes with 26 songs. The following are what sets those songs apart from any you'd receive from a licensing agency or royalty-free firm:

- SmartSound Movie Maestro can change the character of each tune to fit your needs by rearranging song segments in several distinct ways. Effectively, you get about eight songs for every one in the package.
- Using those same song segments and some smart software, Movie Maestro gives each song a clear, decisive ending right when you want it. There's no need to fade it out at your video's close.

> Songs you create with the demo and retail versions of Movie Maestro are for noncommercial use only—meaning for home or school. If you want to use Movie Maestro for other purposes, such as business promotions, DVDs for retail sale, or in-house training DVDs, you must purchase music from Sonic Desktop's Audio Palette or Edge series.

Task: Use Movie Maestro to Add Music to a Video

TASK

The demo version of Movie Maestro, included on this book's DVD, has only three songs and expires after a month's use, but it works the same way as its retail big brother. Here's how to use it:

1. Install the Movie Maestro demo by locating its file folder on the DVD and double-clicking the following file: `MovieMaestroDemo.exe`.

> Users of Adobe Premiere and Pinnacle Studio 8 can access SmartSound Software's music composition core technology, called QuickTracks, through plug-ins or modules included with those two video-editing programs. Each product's SmartSound module offers slightly different sets of songs and functionality than the full retail Movie Maestro package does.

2. Open the program by double-clicking the SmartSound Movie Maestro icon on the desktop. After a brief musical introduction, the screen shown in Figure 8.12 pops up.

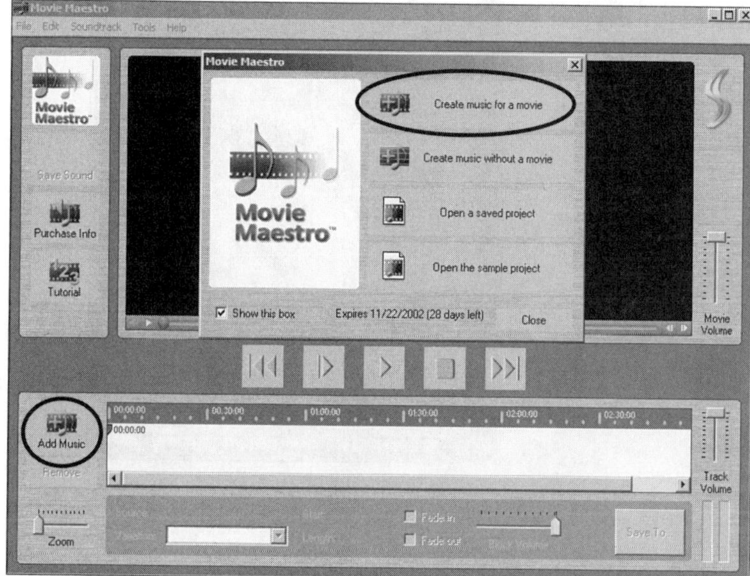

FIGURE 8.12
Create custom instrumental music in minutes with SmartSound Movie Maestro.

3. Click Create Music for a Movie, as shown in Figure 8.12. Doing so opens a file selection window with a default location of My Videos. If you have a video there, feel free to use it; otherwise, navigate to Movie Maestro's file folder (the default location is `C:\Program Files\Movie Maestro\Documentation`) and select `Sample Movie.mov`.

> Feel free to use Movie Maestro's built-in tutorial by clicking the Tutorial button on the left side of the screen. It, too, uses the `Sample Movie.mov` file to demonstrate the software's functionality.

4. Click the Add Music button to open the interface shown in Figure 8.13. This screen offers several musical styles as well as other means to make a musical selection.

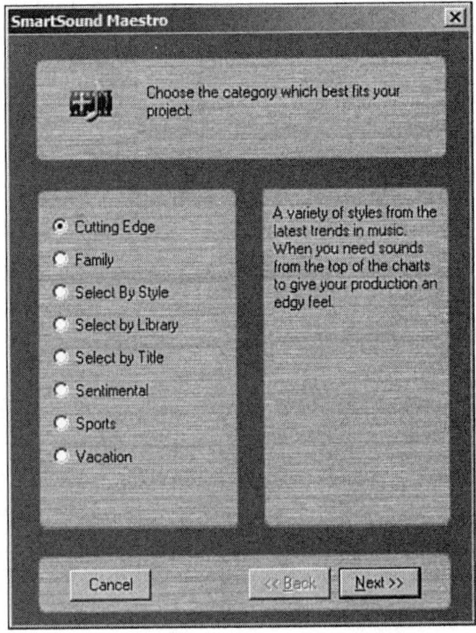

Figure 8.13
You can narrow your Movie Maestro music search using this interface. However, the demo version has only three songs, so don't expect too many choices.

It might appear that you have several choices in music, but this demo version has only three songs so the various options don't have much meaning. If you click through the categories, you'll find little choice is available. If this were the retail product, there would be several tunes for each category, with 26 songs in all.

5. Click through whichever categories you want, and select one of the three songs. Preview it by clicking the Play button at the bottom of the screen (see Figure 8.14). If it suits your needs, click Finish to automatically add it to the movie timeline and set its length to match the video.
6. Try your newly edited video, and be sure you drag the Play Indicator to the start of the video (see Figure 8.15). Then, click the Play button.

You can adjust the song's length by simply dragging and dropping its start and end points. As you move the cursor, the video moves as well, so you can fine-tune the music to start or finish with a specific shot in your video.

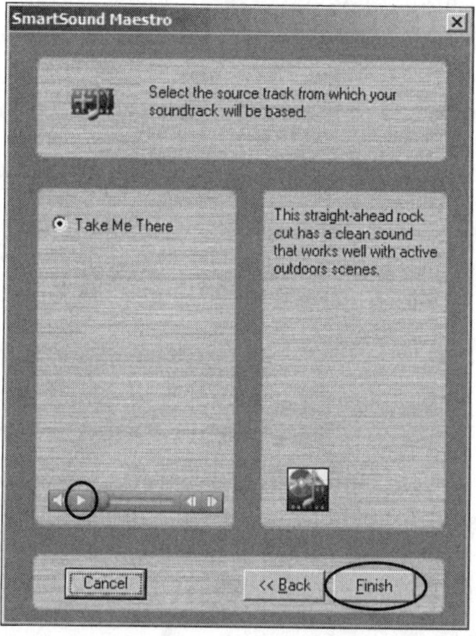

FIGURE 8.14
Try your selected tune by clicking the Play button.

FIGURE 8.15
Use this timeline interface to listen to how well your composition fits the video style.

7. As the music is playing, try the variations by clicking the drop-down menu shown in Figure 8.15. Select any variation you want, and the video clip jumps back to the beginning. Each variation has a distinctly different start.

8. After you've chosen a variation, save your musical selection by clicking the Stop button and then the Save Sound button in the upper-left corner of the screen. You have several save options: the sound file itself, as separate musical segments, or combined with the video. You also have three levels of audio: less than music CD quality (22KHz), music CD (44KHz), and digital video (48KHz).

If you choose to save the combined movie and soundtrack, the Save window uses the original filename as the default saved filename. *Be careful*: You don't want to overwrite your original video file. Give this file a new descriptive name, such as `oldfilename with soundtrack.avi`.

The Advanced Save button lets you convert the original video file into several other video formats. Some selections give you more options than you might have believed possible. For instance, select Export Movie to QuickTime Movie and then open the Use drop-down menu. None of these options is particularly suited to DVD production, but feel free to experiment.

If you like what Movie Maestro does, I recommend visiting Sonic Desktop at `http://www.smartsound.com`. There you can purchase the retail version and sample and order six CDs (60 songs) of additional noncommercial-use music, as well as several dozen CDs to create royalty-free music for commercial use.

If you want to take your music creation to a higher, more customized level, check out ACID Music 3.0 from Sonic Foundry (`http://www.sonicfoundry.com/products/acidfamily.asp`). ACID is an excellent product but much more complex than Movie Maestro. ACID lets you select from several hundred music loops of varying genres. You assign instruments to those loops, mix and layer them, and then apply sound effects if desired.

Recording High-Quality Narrations

Some of your DVD productions will need a narrator. You might not consider yourself to be the Walter Cronkite of DVD production, but if you create a quiet voice recording area, use a good handheld microphone, and follow my 10 narration recording tips, you might approximate that avuncular announcer.

Creating a Voice Recording Area

To create your voiceover narration, you'll need a quiet, sound-absorbing location. The easiest solution is simply to hang some thick blankets or fiberglass insulation on two joining corner walls (egg cartons, carpeting, and foam rubber do not work well). If you can create a four-sided cubicle, so much the better.

If you want to create a higher-quality acoustic environment, purchase foam sheets from the industry leader in sound absorption material, Auralex Acoustics (`http://www.auralex.com/`).

Figure 8.16 illustrates Auralex's $350 Max-Wall 420. This is not an isolation booth, but it *is* an effective way to cut down on noise echoing off nearby walls.

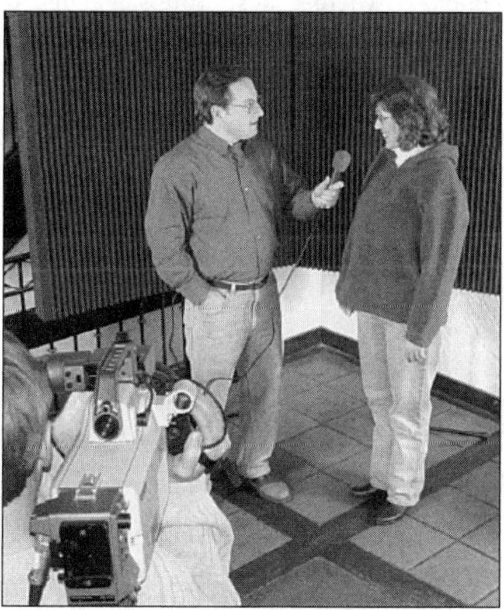

FIGURE 8.16
The Auralex Max-Wall 420 creates an excellent acoustic environment for your narrations.

When you use your voice recording area to record your narration, put your back to the sound absorbing material and point the microphone or camcorder toward you and the wall behind you. You should speak away from the sound absorbing material. It might seem counterintuitive, but the mic works sort of like a camera—it "sees" what's in front of it. In this case, it sees your mouth and the sound-absorbing material.

Using an External Microphone

Camcorders' built-in microphones are jacks of all trades and masters of none. Specifically, they don't work very effectively for narrations.

If you want to improve your video production, buy a good handheld, omnidirectional mic (they pick up sound from all directions). They're the rugged workhorses of the audio industry. Built with internal shock mounts to reduce handling noise, you can use these mics for interviews, place them on podiums to record speeches, and use them to create narrations.

A top-of-the-line, rugged, durable handheld costs $150–$250. I suggest buying a mic from Shure, the world's leading mic manufacturer (`www.shure.com`). Its industry-standard SM63 retails for $200.

Most camcorder mic inputs do not have enough amplification to hear standard, low-impedance (essentially unamplified) handheld mics. What's more, they use mini-plugs, whereas most professional mics use rugged, reliable XLR jacks. You might need a transformer with an XLR to mini-plug cable to increase the impedance and allow you to connect your mic to your camcorder. Such transformers are passive, meaning they do not require electricity. Shure has just such a transformer—the A96F—for $54 (see `http://www.shure.com/accessories/acc-problemsolvers.html`).

Voicing Solid Narrations

In my 15+ years in TV news and production, I've narrated thousands of stories and have worked with coaches and other reporters and anchors to refine my techniques. It boils down to fundamentals. If you follow these 10 tips, you should be able to create a comfortable, listenable voice-over:

- **Practice**—Record some narrations and then play them back and listen. Have others listen, as well. Most first-time narrators mumble, or *swallow*, words. Make sure you've made yourself clear.
- **Before recording your narration, read your copy out loud**—Your script should sound comfortable, conversational, and even informal.
- **Short sentences work best**—If you find yourself stumbling over certain phrases, rewrite them. Break long sentences into several shorter ones.
- **Stress important words and phrases**—As you review your copy, underline important words. When you record your voice-over, give them extra emphasis, meaning more volume and energy.
- **Avoid technical jargon**—That demands extra effort from your listeners, and you might lose them.

- **Mark pauses**—Mark logical breaks in your copy with short parallel lines. They'll remind you to pause at those points.
- **Avoid overly smooth and constant pacing**—That's characteristic of a scripted delivery. Keep your narration conversational.
- **Punch up your voice**—Add zest and enthusiasm to your narration. Pump up your projection, and speak as if the subject truly interests you.
- **Don't pop your *p*'s**—As you say *p* words, you project a small blast of wind at the mic. Use a windscreen, and don't speak directly into the mic.

Shure, Inc., is an excellent source for windscreens. Check out `http://music1online.com/windscreens.html` for a Shure product listing. Expect to pay about $6–$15 for most standard windscreens.

- **Wear earphones**—You'll discover how the mic hears you, and you'll see whether you pop any *p*'s or speak with too much *sibilance*—an overemphasis on the *s* sound.

Summary

Audio—music, narration, and natural sound—plays an important role in any DVD production (I explain how best to add audio to your projects in Hours 8 and 12). Finding or creating the right music to set the perfect mood is easier in this digital, Internet era. You can rip tunes from your personal CD collection, license songs from an online agency, buy royalty-free music for unlimited use, or create your own with software such as Movie Maestro. Creating a professional-sounding narration requires a quiet, sound-absorbing voice recording area; a good microphone; and a little practice using my 10 tips.

Workshop

This chapter doesn't lend itself to a typical Q&A and quiz format, so I'll dispense with those two features. Do tackle the exercises, though. They will help you enhance your productions with high-quality audio.

Exercises

1. Sift through your music CD collection looking for tunes that might fit your video and DVD productions. Instrumentals are your best choice. They also are a great source for *stings*—brief musical phrases you can use for emphasis in a production.
2. Create some personal compilation music CDs. They are great for long car rides when the radio reception, chatter, and ads can become aggravating.
3. Build a make-shift voice recording area using a couple of blankets. Set up your camcorder on a tripod or a sound-absorbing surface, such as a pillow on a chair. Record a narration, and then record the same narration elsewhere in that room. Listen to the two versions. My guess is that the pillow helped.

HOUR 9

Making Videos

If I haven't made it clear enough by now, I'll state it once again, unequivocally: Digital video (DV) is the way to go. True, I am not a big fan of digital *still* cameras, but I have no qualms about digital *video* camcorders. They offer strikingly better visual quality than consumer analog camcorders—VHS, 8mm, or Hi-8—and have unmatched conveniences and features.

I begin this hour by presenting several camcorder buying tips. I focus on features that figure prominently in consumer entry-level and mid-priced models selling for between $500 and $1,500.

While writing this book, I evaluated two entry- to mid-level camcorders—Panasonic's PV-DV102 and Sony's DCR-TRV25. I'll give you my take on both, and I'll take a brief look at a promising new camcorder medium: recordable DVDs. Will they replace camcorder videocassettes?

After purchasing your camcorder is a good time to ramp up your video-making skills. I'll tap my 15 years in the TV news and production business and pass along several video taking tips.

The highlights of this hour include the following:

- Making a case for digital video
- Choosing a digital camcorder
- Evaluating two popular consumer camcorders
- DVD camcorders—not there yet
- Video shooting tips

Making a Case for Digital Video

Digital video (DV) has significant advantages over analog. I'll explain a few here and touch on a few more in Hour 10, "Capturing Video—Transferring Videos to Your PC."

Digital Video Looks Better Than Analog

The reason digital video looks better than analog is that is has more lines of resolution.

This can be a bit confusing because TV sets, within each international standard, have the same number of lines of resolution: NTSC is 525 lines (although only about 500 are visible), and PAL is 650 lines. Anyone who has visited Europe and viewed TV there has probably noticed the higher-quality image.

But signals sent to those TVs might not have that full resolution, so the signal is interpolated to expand it to fill the screen.

VHS and 8mm offer 240 lines of information, and S-VHS and Hi-8 have 400 lines. Consumer DV tops them all with 500–530 lines. It therefore looks sharper than its predecessors.

Digital Video Retains Its Quality

DV is a data stream of digits: 0s and 1s. If you copy, transfer, edit, or add special effects to it, DV remains that collection of discrete digits. Analog, on the other hand, is a continuous waveform. Maintaining that exact shape edit after edit and through the airwaves is nearly impossible. Analog loses quality, whereas digital retains it.

Videotapes Lose Information with Time and Use

Videotape is a physical storage medium. Its magnetic coating can wear off because a spinning head reads the signal on a videocassette by coming in contact with its surface. This can cause some quality loss over time. True, you initially store DV on tape (usually), but you load it to your hard drive, edit it on your PC, and can archive it on DVDs. The opportunity for quality loss is virtually nil.

You can create DV backup copies of your analog videotapes. When you consider your purchase of a new DV camcorder, check to see that it has video-in and audio-in plugs. Then, you can connect your old analog camcorder to your DV camcorder and copy those analog tapes onto DV. One other benefit this has is that when the time comes to transfer that video to your PC, transferring digital video is easier than transferring analog video.

Videotapes Are Linear

Back in my TV news days, when editing news stories, each individual edit could take several minutes to complete. That's how long it would take to fast forward or rewind through a typical analog videotape to track down a specific clip. With DV on your PC, it takes only moments. Most DV starts its life on a mini-DV videocassette, but after it's transferred to a PC, rapid edits become routine.

Choosing a Digital Camcorder

Finding a digital camcorder that fits your project needs and budget is no small task. The possibilities can be overwhelming. So, your first task is to narrow down the field.

Doing Some Homework

Start by using the Internet to get an overview of what's available (later you'll visit a local electronics or camera store for a hands-on trial).

I suggest starting your online research at `www.CNET.com`. Do a search on "digital camcorders," under "All CNET." That'll open a results page. Click the Reviews options to open a series of pages with as many as 70 camcorders (see Figure 9.1).

The listing likely will start with models reviewed by CNET, with the highest-rated products displayed first. You can arrange the listing alphabetically by manufacturer or from lowest to highest price.

This is a fast-moving business so models, features, and prices will have changed by the time this book publishes, but Figure 9.2 shows some top-rated camcorders.

As you page through the listings, you'll note five main manufacturers. Topping the charts, in my view, are Sony, Canon, and Panasonic. The also-rans are JVC and Sharp.

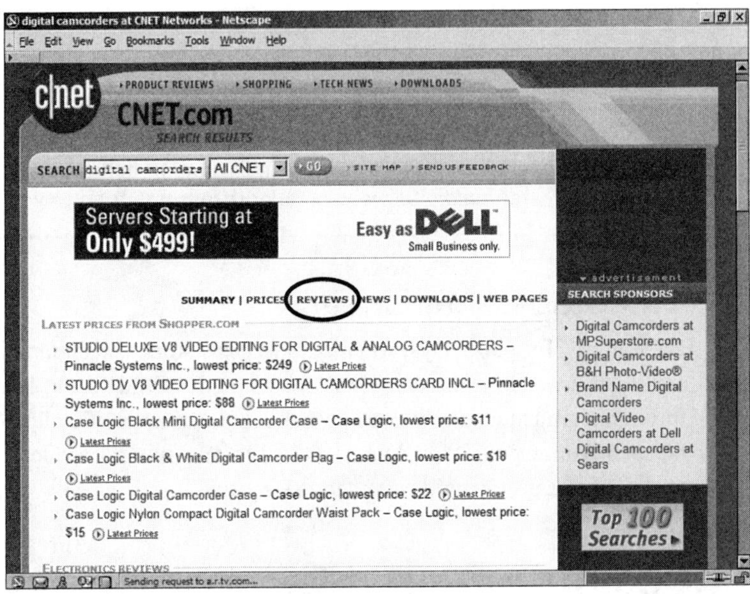

FIGURE 9.1
CNET.com is a good starting point for your DV camcorder quest.

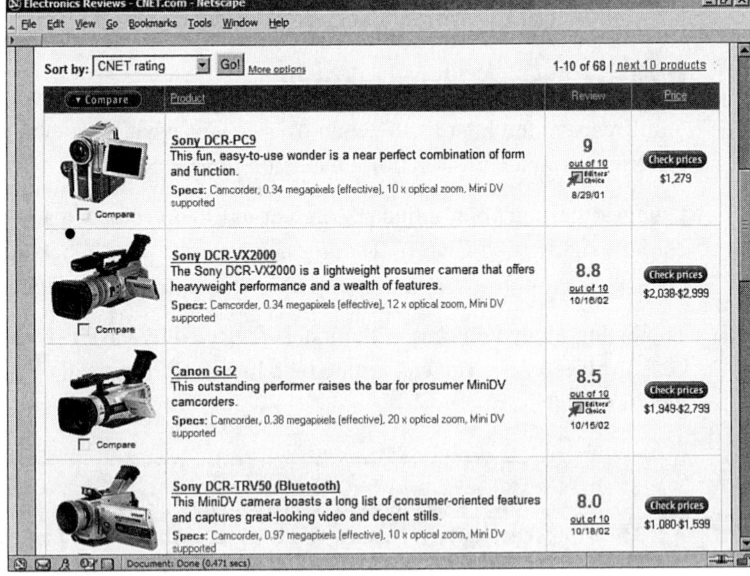

FIGURE 9.2
Most top-rated camcorders have a price tag to match. In this competitive business, you usually get what you pay for.

The three basic groups of camcorders are entry-level, mid-level, and prosumer. Entry-level camcorders range in price from $450 to $1,000. Figure 9.3 shows three popular, high-quality models near the top of this category's price range.

FIGURE 9.3
Three entry-level digital video camcorders. From left to right: Canon Elura 40MC, Panasonic PV-DV702, and Sony DCR-TRV27.

Canon Elura 40 MC

Panasonic PV-DV702

Sony DCR-TRV27

The Panasonic and Sony both have a traditional look and feel, whereas the Canon looks more like a pocket camera. I'll go over what features to look for in a moment. These three reasonably priced camcorders all have many options but are missing some higher-end characteristics, such as improved image quality, longer zoom, and smoother controls.

Mid-priced camcorders cost from $1,000 to $2,000. Figure 9.4 shows three excellent models in this category.

FIGURE 9.4
These three mid-priced DV camcorders cost between $1,000 and $2,000. From left to right: Canon Optura 200MC, Panasonic PV-DV952, and Sony DCR-TRV50.

Canon Optura 200MC

Panasonic PV-DV952

Sony DCR-TRV50

Finally, if you want the very best, you'll need to spend from $2,500 to $4,500. Figure 9.5 shows the three industry-leading models: two from Canon and one from Sony.

Single Chip Versus Three Chips

DV camcorders use the same basic image gathering technology used in digital still cameras: charge-coupled device chips (CCDs). But the following are two principal differences in how DV camcorders use CCDs:

- TV resolution is not as sharp as still photos, so DV camcorder CCDs typically have fewer pixels than those in digital still cameras.

- But color quality and low-light functionality are important issues. Therefore, the best prosumer camcorders have three CCDs—one for each primary color (red, green, and blue).

FIGURE 9.5
These are the best prosumer DV camcorders. From left to right: Canon XL1S, Canon GL2, and Sony DCR-VX2000.

Top-of-the-line camcorders also provide very accurate imaging and many manual override options, including focus, iris, shutter speed, and white balance.

> You should use a three-chip camera, for instance, if you will be projecting your videos on large screens for sales presentations or if you want to move into the professional video-production business. Showing up at a client's office with a palm-sized, single-chip camera will not win many converts.

Features to Consider

As you compare camcorders, there are some features worth considering and others that should have little or no bearing on your buying decision. Here are my deal makers or breakers:

- **Substantial optical zoom**—It should be at least 10X, but 16X or more is better.
- **Comfortable, accessible, logical controls.**
- **Easy-to-use viewfinder and easy-to-read LCD display**—These should also have sensible digital setup and controls.
- **Optical image stabilizing using prisms or some other means**—Instead of the less desirable electronic stabilization.
- **Input and output capabilities**—IEEE-1394 (the industry-standard means to transfer digital video) is a must, as is a means to record from and to a VCR or other camcorder (S-Video connectors are better than composite ones).

- **An external mic plug**—This is a necessity, as is a headphone plug.
- **High shutter speed settings**—These are necessary to capture crisp images of fast-moving subjects.

Camcorders have many superfluous features, including the following:

- **Digital zoom**—All you get are chunky pixels. Use your video editing software's zoom feature to handle this.
- **Titler, fade-in, fade-out, and digital special video effects**—Your video editing software will take care of these as well.
- **Still image capability**—Camcorders need CCDs rated at only about 500,000 pixels to make high-quality DVs. Most digital still cameras, however, have at least 3 million pixels—six times the camcorder resolution. So, when camcorder manufacturers use higher-resolution CCDs, their purpose is not to improve *video* quality, but to improve *still image* capture. But up to this point, no still image from a DV camcorder equals an image from a 3-megapixel still camera.
- **Presets for special lighting conditions**—These include backlit, low-light, portrait, sports, and extremely bright settings (surf and snow). You should use the manual controls to more accurately handle these situations.

> Visit a retail camera or electronics store. Camcorder buying is one of those things that might simply come down to "feel." Does the camcorder fit well in your hands, are the controls logical and accessible, do the menus make sense, and how do the images look?

Evaluating Two Popular Consumer Camcorders

There is no discounting the importance of a hands-on test when evaluating a camcorder for potential purchase. I took two such test drives while writing this book.

Panasonic and Sony each loaned me entry-level models for my evaluation. The Panasonic PV-DV102, shown in Figure 9.6, has an attractive price for such a full-featured camcorder, about $450. Sony's DCR-TRV25, also shown in Figure 9.6, costs about $300 more, but it offers plenty of extra value for the extra expense.

FIGURE 9.6
Two entry-level camcorders—Panasonic's PV-DV102 (left) and Sony's DCR-TRV25 (right)—give a lot of value for their prices.

Panasonic PV-DV102

The Panasonic PV-DV102 is a great first camcorder for budding video enthusiasts. It offers plenty of useful features at a bargain-basement price.

I like the look and feel of camcorders cut from a traditional mold. The PV-DV102 fits that bill. It's comfortable, it's easy to handle, and its camcorder controls are readily accessible.

Image quality is reasonably sharp and "warmer" than the Sony DCR-TRV25. The Sony does have better low-light quality, though. The Panasonic has a fast autofocus even when moving the camera quickly from a distant object to a macro close-up about an inch from the lens.

As do many camcorders, the PV-DV102 uses a knurled push-button knob to select menu functions. In this case, the knob is in a logical location and large enough for even the most ham-handed camera operator. It also has several manual override buttons, precluding the need to navigate through the menu for frequently used features such as manual focus and image stabilization. It has the nice extra feature of a built-in light, and the printed manual is well laid-out and very helpful, especially for first-time users.

I have a few minor gripes, however: The viewfinder is smallish and black and white (the larger LCD is color, but like all camcorder LCDs, it is a battery drainer). Additionally, to eject the DV cassette or access the IEEE-1394 connector, you have to open that LCD. Its electronic image stabilizer doesn't help much when zoomed all the way out to 10X.

Sony DCR-TRV25

The Sony DCR-TRV25 costs about $300 more than the Panasonic, but I don't think it has $300 worth of additional features.

> Earlier in this hour, I mentioned that the Sony DCR-TRV27 (as opposed to the TRV25) is an excellent camcorder at the higher end of the entry-level price scale. The only difference between it and its cheaper sibling, the TRV25, is that TRV27 sports a 3 1/2" LCD versus a 2 1/2" LCD on the TRV25.

The Sony DCR-TRV25 has the same traditional camcorder shape, the same-size LCD, and a slightly longer focal length lens (but the same 10:1 zoom) as the Panasonic PV-DV102.

It outshines the Panasonic in several ways. Its viewfinder is color, you don't need to remove the battery to charge it, it comes with a remote control, and it has a manual focus ring on the lens (so you don't have to fumble with a knurled knob).

It doesn't have a traditional onboard light, but it does offer a nightshoot mode using an onboard infrared beam. This is one of those attention-grabbing features you might never use.

The menu system is similar to the Panasonic, but the knurled knob is tiny. Finally, the manual is poorly laid-out and sometimes confusing.

DVD Camcorders—Not There Yet

As of early 2003, two manufacturers had shipped DVD camcorders: Hitachi and Panasonic. I've illustrated a sampling of their DVD camcorder lines in Figure 9.7. These camcorders use special mini-recordable DVD discs—DVD-RAM and DVD-R. For a few fundamental reasons, neither company's products outshine other comparably equipped DV cassette camcorders.

Hitachi DZMV270A

Panasonic VDR-M20

FIGURE 9.7
Hitachi DZMV270A (left) and Panasonic VDR-M20 DVD camcorders have some drawbacks that might relegate them to only videographers with limited video editing requirements.

The biggest drawback is video quality. Because DVD discs don't have enough room to store high-quality digital video, they rely on highly compressed MPEG-2 files. DVD camcorders can store 40 minutes of reasonably high-quality MPEG-2 video (or 2 hours of highly compressed, mediocre-quality video) on each two-sided recordable DVD-RAM disc. However, even the higher-quality MPEG-2 video cannot match DV.

DVD Camcorder Advantages

Despite the lower-quality video, DVD camcorders do have some clear advantages:

- **Your videos are already on a DVD**—You can take the mini-disc right from your camcorder and use it in your DVD set-top player, taking advantage of the rapid fast forward/rewind capabilities.
- **They offer substantial capacity for still images**—Up to 2,000 JPEG images per DVD.
- **They store video in MPEG-2 format using two quality levels**—Therefore, you don't need to encode them for your DVD productions. (MPEG has more disadvantages than advantages; see the following section for details.)
- **They search your recordable DVD media for blank spaces and will not record over other video.**

DVD Camcorder Disadvantages

In my view, the DVD camcorder disadvantages far outweigh the small advantages. Here are a few of those drawbacks:

- **They cost more versus camcorders with similar features**—The Hitachi and Panasonic models start at about $800.
- **You probably don't have a DVD-RAM drive**—You might not be able to use that rewritable format other than for in-camera editing and will have to use write-once DVD-R discs.
- **They use MPEG video compression instead of DV**—Depending on the quality of compression, MPEG typically has a lower quality than DV and sometimes has tearing or streaks during action sequences. It also does not edit as well.
- **The mini DVD-RAM and DVD-R discs needed for these camcorders cost more than standard DVD recordable media.**
- **The finalization process is lengthy**—It takes 5 minutes to finalize a DVD-R disc before you can remove it from the camcorder and use it in your PC or set-top DVD player.
- **Some offer only slow USB connections and not IEEE 1394.**

The bottom line is that DVD camcorders are a convenient means to create videos of acceptable quality for home use. If you want to do anything more than very basic editing or have projects requiring higher-quality video and higher-level editing techniques, DVD camcorders do not fit the bill.

Video Shooting Tips

Most of the still photography tips presented in Hour 7, "Creating Still Images," apply to shooting video as well. But those tips serve merely as a foundation.

Video camera work has two fundamental differences from still photography: action and sequential story-telling. With that in mind, here are my video-shooting tips.

Use Still-Camera Composition Tips

Adhere to the rule of thirds, add foreground elements, keep your subject off-center, and shoot at oblique angles.

Get Establishing and Closing Shots

An establishing shot sets a scene—lets viewers know where they are. Most videographers rely on wide shots or aerials, but consider using tight shots: a baseball slamming into a catcher's mitt, hands poised over a piano keyboard, or a scalpel with light glinting off its surface.

Your closing images are what your audience will take away from your video production. They wrap up your productions in satisfying ways. They can be as simple as someone shutting a door, capping a pen, petting a dog, turning out the lights, or releasing a butterfly from her cupped hands. As you start shooting your video, if you don't have a closing shot in mind, you should be constantly on the lookout for that one shot or sequence that will best wrap up your story.

Watch Your Lighting

Professional video photographers make liberal use of their light kits. You might not have that luxury of resources or time. Nevertheless, when shooting indoors, add as much light to your scene as you can (unless you want to set a low-light mood). Light adds depth to your subjects and fills shadows that look worse on TV than in still photo prints. When working outdoors, watch out for back-lit scenes (generally shoot with your back to the sun).

Watch your *white balance*. Different lights operate with different color temperatures. Most camcorders have auto-white balance that maintains a consistent overall look as you move from one lighting situation to another—from fluorescent to incandescent to sunlight for instance. Problems arise when you have two unbalanced light sources—a lighted room with a large window. In that situation, you might use a manual white balance or choose the outdoors preset white balance. If you balance for indoor lighting, the sunlit window will look blue.

Get a Good Mix of Shots

The following methods will help you get a good mix:

- **Use unusual angles**—Move your camcorder away from eye level. Get down on the floor when videotaping babies or puppies, and in other circumstances use ladders or rooftops.
- **Use wide and tight shots, matched action, and sequences**—Get wide and tight shots of the same scene because they add interest. Match action with multiple shots. That is, if you're shooting a wood carver, get a wide shot and then move in for a tight shot of his hands in the same position as in the wide shot.

Most novice videographers rely too heavily on their zoom lenses. Wide and tight shots work well, but instead of zooming in for the tight shot, physically move your camera in close. It's much more interesting to change positions rather than simply toggle that zoom button.

Shoot sequences mixing tight, wide, and matched action. They help tell your story, and they frequently can replace a narration. This works well with repetitive action—for instance, a golfer's face looking up a fairway, a view from behind as he turns to his golf bag, a tight shot as he selects a short iron, and so on as he lines up the shot and finally takes a swing. This takes some cooperation from that golfer as you set up each shot. A good finish is to shoot a wide shot from up the fairway and then a tight shot as the ball lands on the green (or in some unexpected, comical location).

Keep Your Shots Steady

Use a tripod. Failing that, lean against something or prop your camcorder on a stationary object. We all know photographers take the images we view on TV, but don't remind viewers of that. A shaky camera shatters that illusion.

Reasonably high-quality tripods—*sticks* in TV parlance—start at about $100. See Figure 9.8 for a top-of-the-line model built specifically for lightweight prosumer camcorders and one from Sony with a camera remote control built into its handle.

Amateur camcorder users often attempt to create action with zooms, pans, and tilts. Minimize those moves, and let the action take place in your viewfinder. If you need to pan or tilt to follow action, do that.

FIGURE 9.8
The Sachtler DA 75 L aluminum tripod (left) weighs only 2kg, and its DV 2 fluid head (middle) works well with lightweight camcorders. The Sony VCT-D680RM has remote camcorder controls built into its handle.

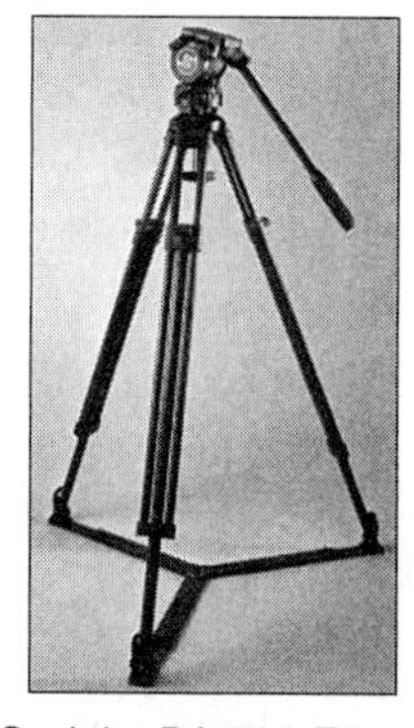

Sachtler DA 75 L Tripod

Sachtler DV 2 fluid head

Sony VCT-D680RM

Avoid fast pans, or *snap* zooms. These are MTV tricks and usually don't work well.

Don't let this no-fast-moves admonition force you to stop rolling while you zoom or pan. If you see something that warrants a quick close-up shot or you need to suddenly pan to grab some possibly fleeting footage, keep rolling. You always can edit around that sudden movement later.

If you stop recording to make the pan or zoom and adjust the focus, you can lose some or all of whatever it was you were trying so desperately to shoot, as well as any accompanying natural sound.

If you do zoom or pan, do it for a purpose: to reveal something, to follow someone's gaze from his eyes to the subject of interest, or to continue the flow of action. Follow a leaf as it floats downstream and then continue your camera motion past the leaf—panning and widening out to show something unexpected: a waterfall, a huge industrial complex, or a fisherman. During interviews, a slow, brief zoom in can add drama to an emotional moment. Again, do it sparingly.

Move with the Action—Trucking Shots

This might seem to cancel out the previous tip—keep your shots steady—but it's a different concept. If the situation allows it, get your camera off your shoulder and place it near a moving object, such as behind, in front, off to one side, or directly above it. Then follow it, using your arm and legs as shock absorbers. If you're not fast enough, hop on a moving vehicle (thus the name *trucking shot*). Try to move smoothly and keep the camera centered on the action. Because most DV camcorders have an LCD viewing screen, just tilt it so you can see what you're shooting.

Here's a corollary to the trucking shot: Put your camcorder *on* a moving object—a shopping cart, hospital gurney, bicycle, and so on—to create a different point-of-view. Here you don't move with the action; you become the action.

Get Plenty of Natural Sound

Remember that your camcorder has an onboard mic. Use it to get sound that will enhance your video project—crowd noise, machinery, music, or footsteps. Even if the accompanying shot is not the best, the sound might be all you need. When you edit your piece, you can always add the sound to another video clip.

Summary

Digital video has become the de-facto consumer/prosumer camcorder standard. It has many advantages over analog—VHS, 8mm, and Hi8. If you're in the market for your first camcorder, choose a DV camcorder. If you have an analog camcorder and your budget is tight, that can work for you, but it has drawbacks such as video quality loss during capture. I cover that issue in Hour 10.

Selecting a DV camcorder might simply come down to "feel," so be sure you try before you buy. Look for easily accessible controls as well as menus that make sense.

You probably want to go beyond creating simple, basic home videos, which is why the new DVD camcorders will not work for you. Their video quality simply is not good enough for anything other than barebones editing.

With your camcorder in hand, use my video shooting tips to give your productions some professional polish.

Workshop

Review the following questions and answers in this section to reinforce your DV camcorder and video shooting knowledge. Also, take a few moments to tackle the short quiz and the exercises.

Q&A

Q It seems my camera acts as some kind of human "ham-it-up" switch. I turn it on and people get goofy or stare into the lens. Is there some way to put a stop to this?

A Not entirely, but try a few techniques to minimize it. Let people know you are going to videotape them and that you will edit out all such behavior. Take a few wide shots and gradually move in close. This lets your subjects get used to the camera. If your camcorder has a tally light on it (the little, red LED near the lens that indicates when the tape is rolling), cover it up with tape. When interviewing people try to look *at* them rather than *through* the viewfinder. They'll talk to you instead of the camera.

Q When I visit my local camcorder retailer, they make a big deal about in-camera editing features and still-image quality. How important are those characteristics?

A They're not very important. It's a hassle to use those editing features and much easier and better to do it later with video-editing software. I've never used my camcorder for still images. When I go out on a shoot, I'm thinking video, not photos. Camcorders cannot match still cameras for image quality.

Quiz

1. Why is digital video the best choice?
2. How do you set a white balance in a room with a large picture window?
3. You want viewers to have a sense that they are watching events unfold as if they were there. How do you avoid shattering this illusion?

Quiz Answers

1. Better image quality, longer lasting storage medium, and no quality loss during transfer to your PC, to name a few reasons.
2. The daylight from the window will likely dominate the ambient lighting, so use the daylight preset white balance or press your camcorder's manual white balance close to the window.
3. Keep you camera steady and avoid unnecessary pans, zooms, and tilts. This is not a hard and fast rule, however. You can pan your camcorder to follow the action, and you can take it off your shoulder or tripod and use it for trucking shots.

Exercises

1. Practice, practice, practice. The only way to improve your camcorder skills is to get out there and tape something. Even when shooting the most mundane material, try to come up with different angles, points-of-view, or locations. Some might actually work for you.
2. Ask to go out on a story with a local TV station news team or out in the field with a video production crew. They usually accommodate budding video producers. Note how the photographers shoot a variety of subjects, choose camera angles, and set up lighting for interviews.
3. Critically view TV news magazine stories. Note the variety of styles. *60 Minutes* tends to do it straight with no abrupt camera moves. Other shows, aiming for younger demographics, tend to jazz things up. There's a fine line between effective camera work and camera actions that interfere with storytelling.

Hour 10

Capturing Video—Transferring Videos to Your PC

Here's help for those of you burdened with shoeboxes of old videocassettes and no easy way to show off your handiwork. In this hour, I explain how to transfer, or *capture*, those digital and analog videos onto your hard drive—the first step on the road to recording them to DVDs.

For digital videos (DVs), you'll need only a standard IEEE-1394 (FireWire) connection. On the software side, you can use the DV capture utility on the DVD included with this book or Microsoft's Movie Maker to perform those transfers. I'll ease you through that process for both products.

Analog video requires something less standardized than FireWire to transfer your tapes to your PC: a video capture card. I'll give you some hands-on instruction using a surprisingly inexpensive yet powerful little card. Then, I'll fill you in on some high-octane, more expensive, all-in-one hardware graphical powerhouses that convert your PC into a full-featured TV program and video recorder.

The highlights of this hour include the following:

- Transferring digital videos to your PC
- Trying two free video capture software products
- Capturing analog video easily and inexpensively
- Overview of four full-featured video capture hardware products

Transferring Digital Videos to Your PC

I'll start with digital video because it's much easier to work with than analog video.

Here's what you'll need:

- Digital video camcorder
- IEEE 1394 (FireWire) card
- Video capture software

At this point, I'm assuming you have a digital video camcorder. If you have only an analog camcorder or VCR, go to the next section of this hour, "Capturing Analog Video Easily and Inexpensively."

As mentioned before, DV is the best option. It simplifies everything, it has better image quality than all but the highest-level broadcast-quality analog camcorders, and prices are dropping rapidly. At some point, it just makes sense to make the move to DV.

To *capture*, or transfer, digital video to your PC, you need an IEEE-1394 or FireWire (an Apple Computer trademark) connection. Most newer PCs have IEEE-1394 connections either built in to the motherboard or on add-in hardware cards.

If you plan to work with a DV camcorder and your PC does not have an IEEE-1394 connector, you need to buy one.

Standard IEEE-1394 cards are inexpensive, typically costing less than $35. However, you might consider spending a few more dollars to buy a 1394/USB 2.0 combo card. The advantage to such cards is that they give you the best of both worlds. IEEE 1394 is a necessity for digital video, and USB 2.0 boosts the speed of that serial connection technology beyond even IEEE 1394. At last word, these cards ranged in price from $60 to $90. Figure 10.1 shows three brand-name products; the Adaptec offers the best software bundle.

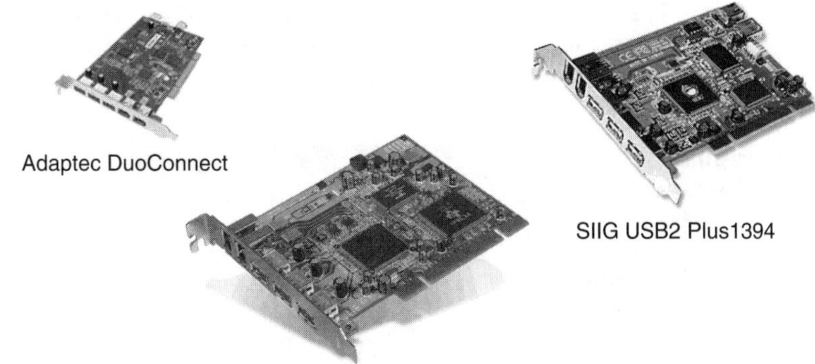

Figure 10.1
Combination IEEE-1394 (FireWire) and USB 2.0 add-in cards give your PC extra high-speed connectivity. Left to right: Adaptec DuoConnect, Belkin USB-FireWire 2, and SIIG USB2 Plus1394.

10

Trying Two Free Video Capture Software Products

Numerous software products are available that capture or transfer digital video to your PC. Most are modules built in to video editing or DVD authoring software. I'll cover video editing products in Part III, "Editing Media." For now, I'll take you through two free products already available to you.

Capturing Video Using DVIO

We've included a small (32KB) program on the book's DVD-ROM called `DVIO.EXE`. It's a surprisingly powerful DV input/output tool, and it's a great way to take your first steps into the realm of transferring video from your DV camcorder to your PC.

I came across DVIO when dealing with a balky, high-end video capture card while writing another book. The tech support folks suggested I try DVIO to see whether the problem was with my system or their card. DVIO worked fine, and I found out I had an uncooperative card. Read more about DVIO in the sidebar titled "DVIO Back Story," later in this chapter.

Task: Use the Included DVIO Freeware to Capture DV

TASK

DVIO is a remarkably simple tool to transfer DV to your PC's hard drive. Here's how you do that:

1. Copy `DVIO.EXE` from the book's DVD to any location on your hard drive.
2. Connect your camcorder to the IEEE-1394 card, and switch it to VCR (not Camera mode). In Windows XP, an icon pops up in the lower-right corner letting you know that Windows has detected the new hardware—your DV camcorder (see

Figure 10.2). Windows also displays a dialog box, shown in Figure 10.3, asking what you want to do with your camcorder. In this case, you don't want to do anything, so click Cancel.

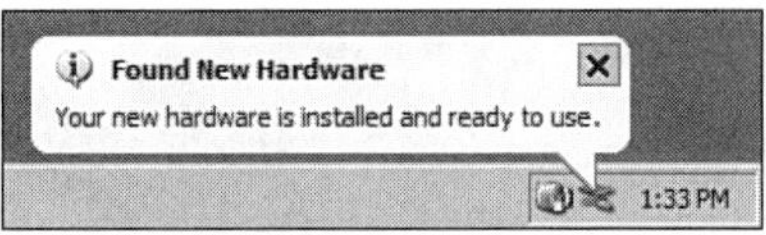

FIGURE 10.2
Windows lets you know it sees your DV camcorder when you connect it to your PC with an IEEE-1394 connection.

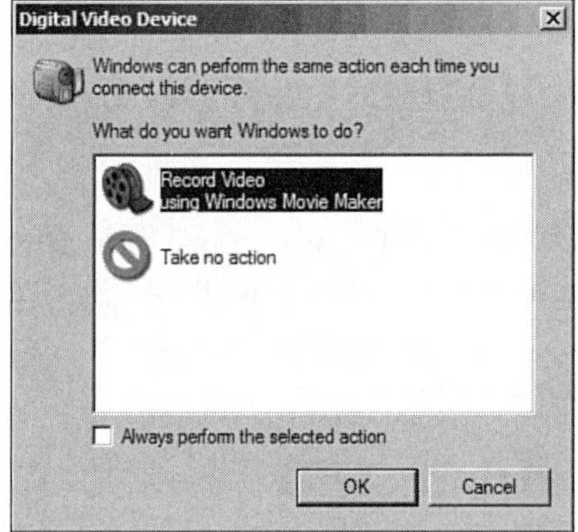

FIGURE 10.3
Windows also asks what you want to do with your newly connected DV camcorder.

3. Locate `DVIO.EXE` in My Computer, and double-click DVIO to open it. As you can see in Figure 10.4, the interface is very simple.

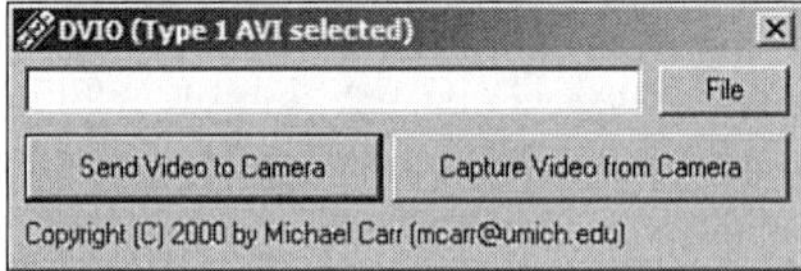

FIGURE 10.4
DVIO transfers digital video from your camcorder to your hard drive in two easy steps.

4. Click the filmstrip icon, shown in Figure 10.5, and select AVI Format, Type 2 (see the following sidebar titled "Type 1 Versus Type 2 AVI Files" for an explanation). Selecting the other option, Verify Overwriting, means DVIO will ask you whether you want to replace an existing video file with the one you are about to capture.
5. Click File, browse to a file folder where you want to store your video clip (My Documents/My Videos is a logical location), and type in a filename.

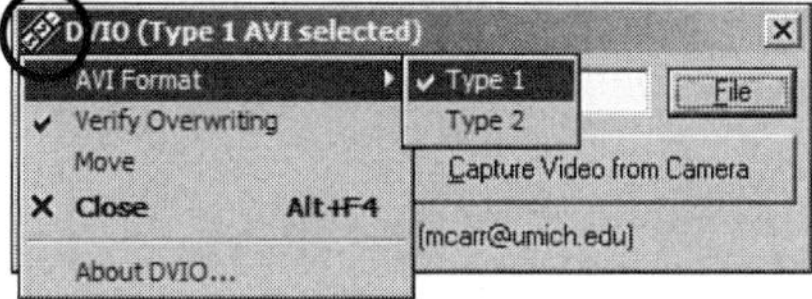

FIGURE 10.5
DVIO gives you only two options: AVI Format and Verify Overwriting.

6. Cue up your videotape to the start of whichever video clip you want to transfer.
7. In DVIO, click Capture Video from Camera, and then press Play on your camcorder. As shown in Figure 10.6, DVIO tracks the number of frames captured (at 30 frames per second) and the number of dropped frames (faster systems should see no dropped frames).

FIGURE 10.6
DVIO tracks recorded frames (at 30 frames per second) and dropped frames.

8. When you reach the end of the video clip, click Stop Capture from Camera.
9. Test that all went well by opening Windows Media Player; selecting File, Open; and then browsing to your newly created AVI file. Double-click it, and Media Player should play it back without a hitch.

Type 1 Versus Type 2 AVI Files

There is some debate whether you should capture video to Type 1 AVI files versus Type 2. Here's some background:

- DV is a stream of data. It is a well-documented standard and does not vary from manufacturer to manufacturer. What varies is the wrapper and some of the additional data that computer software adds to the DV stream. AVI (Microsoft Audio/Video Interleave) is one of those wrappers.
- Type 1 AVI files contain both audio and video mixed together into a single DV stream.
- Type 2 AVI splits out the audio into a separate stream while leaving it untouched or replacing it with garbage data in the original audio. Type 2 AVI files are therefore about 5% larger than Type 1.

- Some time ago Type 2 AVI files could not exceed 2GB. Now, so-called OpenDML (Open Digital Media Consortium) Type 2 seamlessly chain together files greater than 2GB by placing data at the end of each 2GB file that references the beginning of the next one.

Although Type 2 appears to be cobbled together, market-leading video editing products tend to handle it smoothly.

DVIO might be the only program that offers a choice of Type 1 over Type 2. My recommendation is that you stick with Type 2, but feel free to experiment. Capture a Type 1 video file greater than 2GB and try it in your video editing software to see how smoothly (or not) it works. Then, capture a 2GB or larger Type 2 file and try that as well.

The bottom line is that the AVI format, which was created in the mid-1980s by Electronic Arts for the Commodore Amiga and later adapted by Microsoft, might be on its last legs. Microsoft's rapidly improving Windows Media file formats might become the next standard DV wrapper.

DVIO Back Story

Michael Carr, a University of Michigan electronics engineering grad student, faced a dilemma. In 1999, he had just bought his first DV camcorder. However, he soon discovered that PC DV capture utilities left something to be desired for several reasons:

- They captured only to nonstandard or non-OpenDML (Open Digital Media Consortium) Type 2 AVI files, which had a 2GB limit—the equivalent of only a few minutes of video.
- Because DV was so new, most capture/export utilities were unreliable.
- Those DV utilities used a lot of CPU bandwidth largely because they included real-time onscreen previewing that couldn't be turned off.

As a full-time graduate student, it was cheaper for Carr to write a DV capture utility without real-time preview than it was to upgrade his computer to something that could handle it. Researching a means to that end, he discovered that Microsoft's DirectX has built-in capture capability.

All he then had to do was create a graphical user interface (GUI) to access that capability, and, voilà, `DVIO.EXE` was born.

Although Carr lost the original DVIO source code in a hard drive crash, the compiled version is readily available for no charge at his Web site: `http://www.carr-engineering.com/dvio.htm`.

There has been so much demand for his product that he wrote a commercial version from scratch called DVIO Pro. Several television stations use DVIO Pro to export a sequence of DV clips seamlessly to a transmitter. A university uses DVIO Pro to automatically capture television news broadcasting on a nightly basis and then feed the captured video into another software program that analyzes human speech patterns.

Microsoft Me and XP's Movie Maker

DVIO reduces video capture to its most fundamental level. On the other hand, most retail video capture, video editing, and DVD authoring applications feature multiple video capture modes—DV and analog, manual and automated—as well as a host of video file formats. Some offer scene change detection, creating separate files every time you push the camcorder's record button, or encode the video to MPEG files.

To introduce you to these capabilities, I'll take you step-by-step through Microsoft's Movie Maker.

This is your best choice for stepping up your video capture skills a notch. Why? It's free and resides on your hard drive awaiting your commands. However, it has few bells and whistles and one major video capture flaw: It lacks MPEG encoding tools (it also leaves something to be desired in its collection of video editing options).

Task: Use Microsoft Movie Maker to Capture Digital Video

Movie Maker offers a basic video capture module. It features multiple video file formats, real-time video display, and camcorder device controls. Here's how you use it to transfer DV to your hard drive:

1. Find Movie Maker and open it. If it's not in `Start>Accessories`, its default hard drive location is `C:\Program Files\Movie Maker\moviemk.exe`. The opening screen looks similar to Figure 10.7.

> Movie Maker's opening screen is its video editing interface. I will not include Movie Maker in my discussion of video editing software because it simply doesn't have that much to offer. It does, however, serve as an excellent introduction to higher-level video capture.

2. Switch to Video Capture mode by clicking the Record button. That opens the Record interface shown in Figure 10.8.
3. Make sure your DV camcorder is attached to your PC; then, switch the camcorder to the VCR setting. If everything is working smoothly, the Movie Maker Record interface should automatically update with the name of your Camcorder (see Figure 10.8).

> If you have more than one capture device (while writing this chapter, I had an analog card installed along with my IEEE-1394 card), click Change Device to select the correct video source.

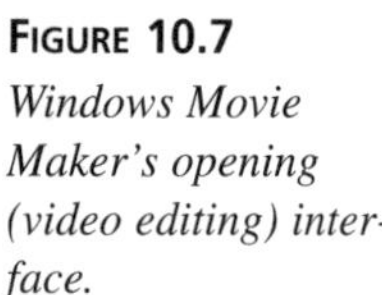

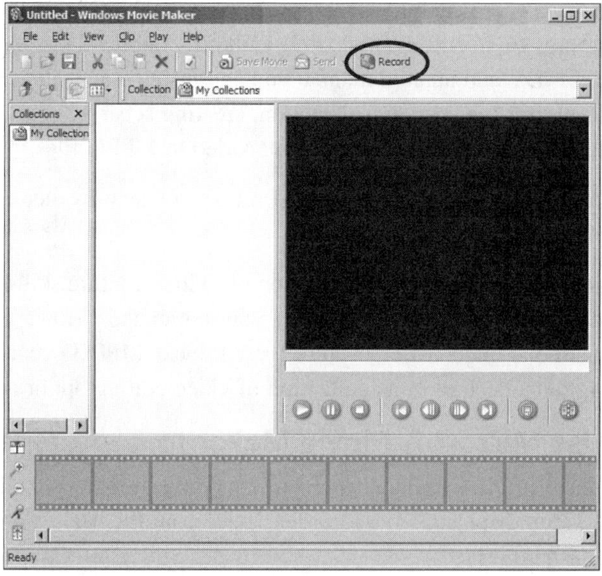

FIGURE 10.7
Windows Movie Maker's opening (video editing) interface.

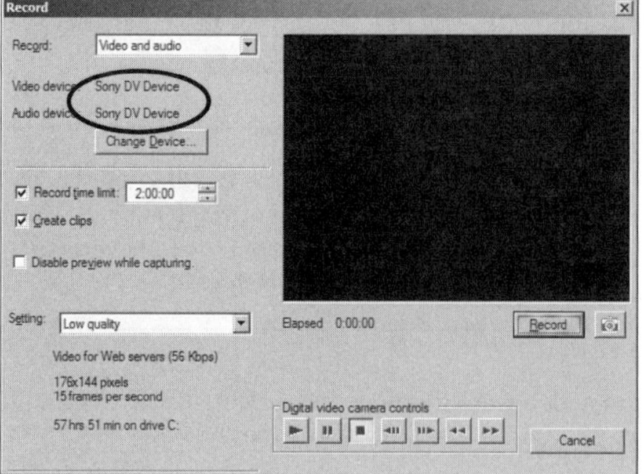

FIGURE 10.8
Movie Maker's video capture screen notes when you've connected your DV camcorder.

4. What sets Movie Maker, and most other DV capture software, apart from DVIO is device control. Note the set of VCR-like controls shown in Figure 10.9. Use them to cue up your tape to the start of a clip you want to transfer, and then pause your camcorder there.

If this is the first time you've used your PC to control your camcorder, you might have the same thoughts I had: "Wow. This is great!" It really is. You can't do this with most consumer analog camcorders because they don't have remote device control. DV camcorders, however, do.

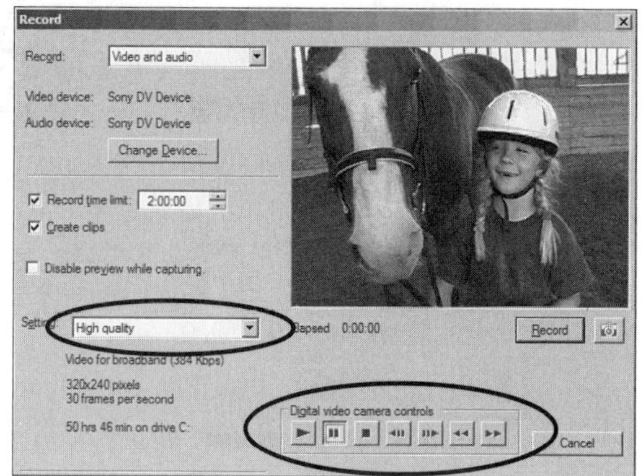

FIGURE 10.9
A feature of Movie Maker and other video capturing modules is device control.

10

5. Before performing the video transfer, specify the file type by clicking the Setting drop-down menu in the lower-left corner of the interface (refer to Figure 10.9). There are three quality settings—Low, Medium, and High—and all are Windows Media Video (WMV) files. The differences are in frame size and frame rate. In this case, you don't want to create a Windows Media file, so just select Other.

When capturing video, you should capture at the highest quality you can. Later, when you export your finished product to DVD, videotape, or the Web, you can adjust the quality setting to suit the medium. If you know what the final video quality will be, feel free to choose a file type and a quality that match the output.

6. Clicking Other opens a hidden drop-down menu. Click it and, as you can see in Figure 10.10, your options suddenly increase dramatically. These, too, are WMV files of varying qualities. Despite all these choices, only one size really fits all: DV-AVI (25Mbps). That's the standard DV25 that will match your DV camcorder's output. There will be no discernible difference between the source video and the file created by Movie Maker. Select DV-AVI (25Mbps).

FIGURE 10.10
Movie Maker offers many video file format options, but DV-AVI is the one that duplicates the quality of the original digital video.

> The number 25 in DV25 stands for 25 mega*bits* per second (as opposed to mega*bytes* per second). DV25 is an industry-standard data stream geared to consumers and prosumers and is one of a half dozen digital video flavors. As mentioned in the "Type 1 Versus Type 2 AVI Files" sidebar earlier this hour, DV AVI is Microsoft's wrapper that, in this case, Movie Maker attaches to DV25 to ensure that the captured file is compatible with other Windows-based video products.

7. It's time to capture your clip. Click Record, and Movie Maker automatically puts your camcorder in Play and captures the digital video stream.
8. When you've captured all you need, click Stop. That opens the Save Windows Media File window in its default My Documents/My Videos file folder. You can select a different folder or keep it here. Name your file (note that it will have an AVI file extension), and click Save. Movie Maker takes a few moments to save the file and then opens the editing interface that greeted you when you first started Movie Maker.
9. Test your handiwork by using the VCR buttons shown in Figure 10.11 to play your captured video clip.

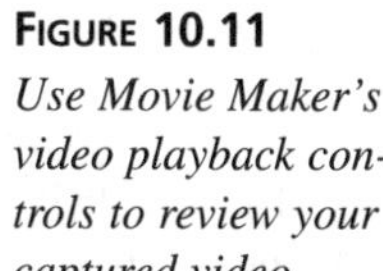

FIGURE 10.11
Use Movie Maker's video playback controls to review your captured video.

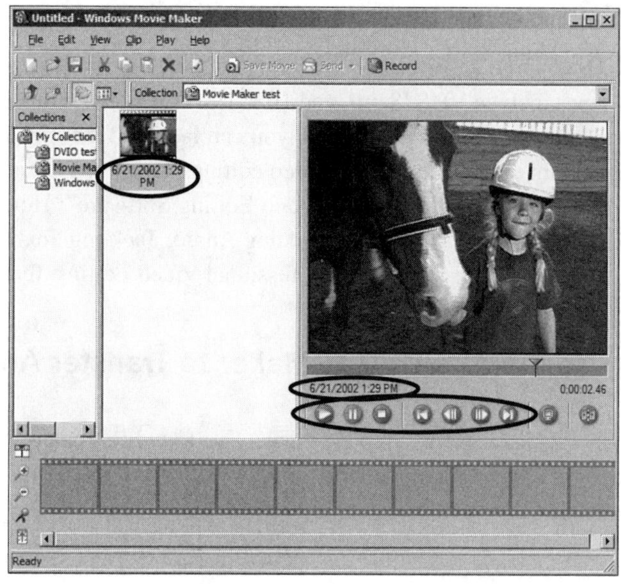

Note the dates shown in Figure 10.11. Those do *not* correspond to the date on which I *captured* this video file; rather they represent the date I *shot* the original video footage. This is yet another cool characteristic of DV. It records all sorts of data—date, time, and camera characteristics—along with the video and audio.

Capturing Analog Video Easily and Inexpensively

Capturing analog video takes some specialized hardware and is not nearly as user friendly as DV. That hardware ranges in price from $50 to more than $1,000.

I'll give you a hands-on look at the $50 card and present a brief overview of four $200+ cards that offer a full range of video capture and TV program recording and time shifting. Three of those latter products also have high-end 3D graphics functions.

I'll cover some $500+ cards (these tend to be geared specifically to high-end video editing) in Hour 14, "Applying Professional Video Editing Techniques with Adobe Premiere."

Taking AVerDVD EZMaker for a Test Drive

If you have analog video tapes and your video editing needs are simple, this $50 card from AVerMedia is a slick solution.

Analog tapes include VHS, 8mm, and Hi8. The object, from this book's perspective, is to get them onto your PC's hard drive so you can later add them to a DVD. Along the way, you might want to edit them using video editing software. That is discussed in Hour 11, "Crafting Your Story and Selecting Video Editing Software"; Hour 12, "Editing Video with Pinnacle Studio 8"; Hour 13, "Adding Audio, Tackling Text, and Improving Images"; and Hour 14, "Applying Professional Video Editing Techniques with Adobe Premiere."

Task: Use the AVerDVD EZMaker to Transfer Analog Video to Your PC

▼ TASK

AVerDVD EZMaker is the simplest means I've found to transfer analog video to your PC. Since we have not included this product with the book, I will use this task to simply demonstrate it. Here's how it works:

1. Install the card and software. The tiny card, pictured next to its retail box in Figure 10.12, slips into an open PCI slot. If you're averse to opening your PC, this is another reason to use DV. Windows prompts you to load the software, which is a painless process in this case.

FIGURE 10.12
The tiny AVerDVD EZMaker PCI card captures analog video.

The AVerDVD EZMaker card performs a relatively simple function: It separates an analog NTSC or PAL TV signal into its luminance and color components (the so-called YUV format) and then converts that YUV signal into uncompressed digital data. Next, the software bundled with this card converts that digital data stream into an MPEG file. NTSC is the standard TV signal in the United States, Japan, and Korea. PAL, on the other hand, is used in Australia, China, South America, and most of Europe.

2. Hook up your camcorder to the PC. You have two choices: Composite or S-Video. Composite uses a simple, so-called RCA cable (not supplied with this card but readily available at Radio Shack). S-Video is a cleaner, higher-quality signal and

▼

requires a slightly more expensive but also readily available cable (one might have shipped with your analog camcorder). You patch in audio using RCA cables connected to an RCA-to-phono jack (provided with card) and connect that to your PC sound card's line-in plug.

Another advantage of DV is that you don't have to mess with Composite video versus S-Video versus Component (an even higher-quality signal available on more expensive video capture cards). Plus, you don't need separate audio cables because everything moves through one pipe—the IEEE-1394 connection.

3. AVerDVD EZMaker's bundled software, mediostream's neoDVD Standard, takes over the chores from here. The opening screen, shown in Figure 10.13, lets you choose from two primary functions: video capture or DVD authoring. In this case, select the first option: Capture.

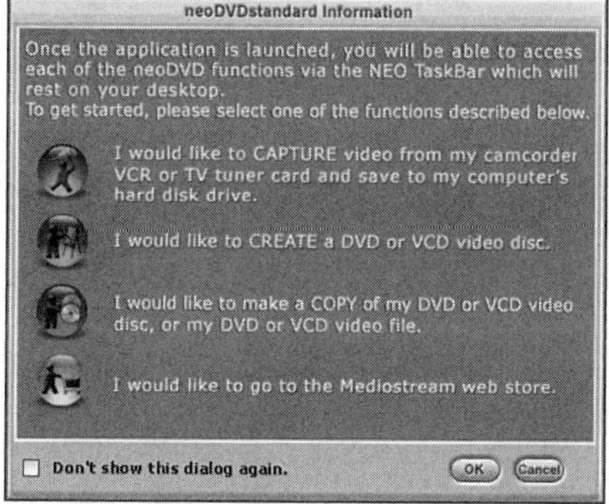

FIGURE 10.13
neoDVD comes bundled with AVerDVD EZMaker and handles the video encoding chores.

As is the case with most video capture software, neoDVD offers other functionality, including rudimentary video editing and DVD authoring. MyDVD, the consumer-level DVD authoring product provided on this book's DVD does a better job at both. Because AVerDVD EZMaker ships with neoDVD, I use it to explain this analog video capture process. You can use virtually any other video capture software with the AVerDVD EZMaker card.

4. The capture screen shown in Figure 10.14 opens with several options highlighted. Here, you select the video source (in this case, AVerMedia Capture), whether to use NTSC or PAL, and the audio input (in this case, my sound card's line-in plug).

5. Select the output file folder and name, choose between MPEG-1 or MPEG-2, and select the quality level (good, better, or best) for either MPEG format.
6. Cue up your camcorder to the start of your clip and either put it in pause or stop it.

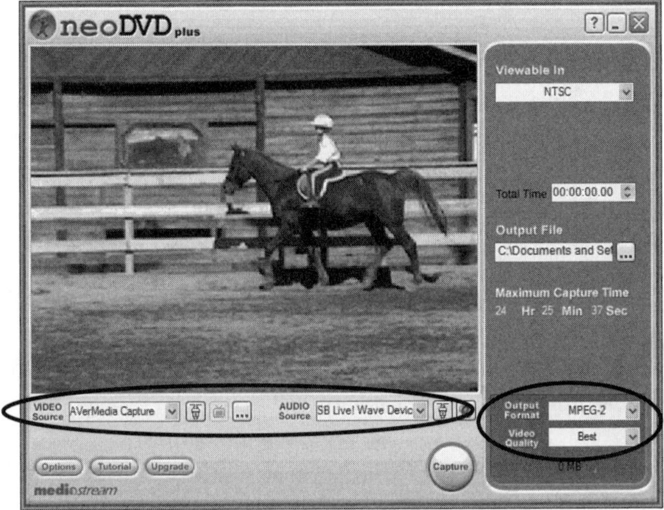

FIGURE 10.14
neoDVD's video capture screen gives you several options.

7. Click Capture and press the Play button on your camcorder. The video display shrinks, as shown in Figure 10.15, and it appears to have several dropped frames. But those two display changes free up processor cycles for MPEG encoding. On my 1.8GHz Pentium 4, neoDVD encoded to MPEG in real-time with excellent quality—surprisingly good performance for what amounts to free software.

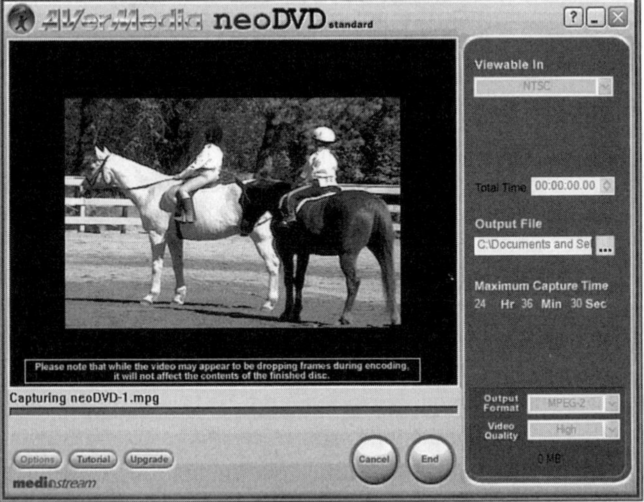

FIGURE 10.15
neoDVD captures the analog video signal and encodes it to MPEG-1 or MPEG-2 in real-time at high quality.

neoDVD encodes analog video only to MPEG files. That's great if you simply want to transfer your archived analog videos to DVDs or do only simple edits. But if you want to do higher-level video editing on those videos, MPEG is *not* the best choice—DV is. When considering mid-range and professional video editing products, you'll note that some capture analog video and convert it to DV.

8. When you reach the end of your clip, click End. neoDVD automatically stores your new file on the hard drive. You can use neoDVD's DVD creation and video editing module to view that MPEG video file.

The AVerDVD EZMaker card has a very narrow purpose: transferring your analog videos to your hard drive and storing them as MPEG files. You might want to keep the card in your computer only long enough to transfer your archived tapes and then remove it.

Overview of Three Full-featured Video Capture Cards

Instead of using the single-function AVerDVD EZMaker card, you might want to install a multipurpose card that handles all your video and graphics duties. Three cards worth considering are the ATI All-In-Wonder 9700 PRO, Matrox Marvel G450 eTV, and NVIDIA Personal Cinema (see Figure 10.16).

It used to be that these Swiss army knives of the PC video world featured yesterday's graphic chip sets. But PC gamers demanded the latest and greatest and ATI, and NVIDIA took note (the Matrox card is two years old). Both the ATI and the NVIDIA feature incredible 3D graphic performance.

Each has a remote TV tuner and offers several time-shifting functions, ranging from using automated programming updates to track and record your favorite TV shows to pause, fast forward, and rewind while watching in real-time.

Using the ATI, you can connect both your TV and PC monitor to the card. Then, your family can use its wireless RF remote to watch TV in the living room, taking advantage of all its functionality, while you work on your PC in your home office.

FIGURE 10.16
The top three, all-in-one video cards feature analog video capture, a TV tuner, and 3D graphics. From left to right: ATI All-In-Wonder 9700 PRO, Matrox Marvel G450 eTV, and NVidia Personal Cinema.

Each video card comes with a software bundle. At last word, ATI's looked most appealing because it ships with Pinnacle Systems' Studio 8, my favorite mid-priced video editing product.

Prices range from $200 to $450.

Pinnacle PCTV Deluxe—An *Almost* All-in-one Solution

If you simply want to add a full-featured, time-shifting TV tuner with video capture capabilities, the Pinnacle Systems PCTV Deluxe shown in Figure 10.17 is a good choice. It's easy to set up because it's an external box that connects to your PC's USB input, and it offers very high-quality hardware MPEG encoding. But it does not replace your PC's video card, as do the three all-in-one cards.

FIGURE 10.17
Pinnacle Systems' PCTV Deluxe.

Summary

Before you can put your videos on DVDs, you need to get those videos into your PC. If you work only with digital video DV, you're in luck. All you need is an IEEE-1394 (FireWire) connection and DVIO (provided on this book's CD) or Microsoft Movie Maker (provided for free with Windows Me and XP) to create AVI files. Those AVI files, for all practical purposes, are exact copies of your original DV footage.

If you work with analog video, you need an analog capture card. Converting that analog video into DV AVI files requires a mid- to high-level video editing product. MPEG is your analog capture file format of choice for simple, video-to-DVD projects.

Finally, you can use a low-cost analog capture card for a project and then remove it from your PC to free up room for other cards. Or, you can use all-in-one video capture, 3D graphics cards.

Workshop

Review the questions and answers in this section to reinforce your video capture knowledge. Also, take a few moments to take the short quiz and perform the exercises.

Q&A

Q I used Microsoft Movie Maker and AVerDVD EZMaker's neoDVD to capture digital video to a DV AVI file. But when I try to use either of those products to capture analog video, I can't find an option to create an AVI file. What's the problem?

A Movie Maker and neoDVD simply do not have that capability. It doesn't take much software coding to add an AVI wrapper to DV, but there's much more to converting analog video to AVI. You'll need a higher-end video editing product to do that—I cover them in Hour 11. neoDVD does at least offer MPEG encoding, but Movie Maker does not.

Q When I open Movie Maker or neoDVD and select the AVerDVD EZMaker analog video capture card, I don't see my video. Why?

A Do some troubleshooting. Is your camcorder connected to the PC? Is the battery charged? Have you switched the camcorder to VCR (as opposed to camera)? If all else fails, go to My Computer, click View System Information under the System Task section, select the Hardware tab, and click the Device Manager button. Look for any yellow (caution) check marks next to Sound, Video and Game Controllers. If so, the software might not have installed properly, or there might be a hardware conflict. In either case, a call to tech support is in order.

Quiz

1. What is the difference between AVI and MPEG?
2. You simply want to take all your videos and store them on DVDs. What's the best capture format to do that?
3. When you attach a DV camcorder to your PC, your video capture software displays camcorder controls—play, fast forward, pause, and so on. No controls show up when you attach an analog camcorder. Why?

Quiz Answers

1. AVI (Audio/Video Interleave) is a Microsoft digital video format. At its root is DV, but it adds a wrapper to that DV to make it compatible with Windows-based video editing and playback products. MPEG is a standardized video compression scheme, usable on multiple platforms (Macs and PCs, for instance) and is the standard video format for DVDs.
2. MPEG-2 offers the best compression in terms of quality per MB and compatibility with DVDs. If you later choose to do some simple editing (removing some poorly shot scenes, for instance) before putting the MPEG-2 files on a DVD, some video editing software will let you do that with no loss in quality. On the other hand, if you plan to do extensive editing using special effects and scene transitions (dissolves, wipes, 3D moves, and so on), then you should capture your videos as DV AVI files.
3. DV camcorders have what's called device control. All but the most expensive broadcast quality analog camcorders do not.

Exercises

1. This is a good time to take those many dusty analog or digital videotapes in your closet and transfer them to your hard drive. For $50, you can't go wrong buying the AVerDVD EZMaker and using it to encode your videos—analog or digital—to MPEG-2 files. If you run out of hard drive space, simply burn the files onto some DVDs using the DVD creation software bundled with your DVD recorder.
2. Experiment with the various file format options in capture programs such as neoDVD or Movie Maker. How does MPEG-1 compare visually and aurally to MPEG-2? What about the various WMV files? You might decide that a lower-quality (and lower file size) format suits your purposes just fine.

Part III
Editing Media

Hour

HOUR 11

Crafting Your Story and Selecting Video Editing Software

You'll use DVDs to tell stories—from "Our Family Vacation" to "Our Company's Product Line." Because most DVD projects are nonlinear and interactive, each DVD can have several self-contained story segments.

In most cases, you'll create those segments using video editing software. I cover that process in Hour 12, "Editing Video." But before you can tackle the technical side of story creation, you'll need to handle the organizing and scriptwriting.

First, get your assets in order—your pictures, videos, music, narration, graphics, and files. This chapter covers filenaming conventions and storage procedures. You should be selective when you capture and name video clips, so I'll offer some tips to help you focus those efforts.

Then, you'll craft your story. I'll offer more than a dozen story writing tips gleaned from my 15 years in the TV news and video production business.

Finally, I'll present an overview of the top four entry- to mid-level video editing products and note my selection as the overall winner of this category.

The highlights of this hour include the following:

- Organizing your assets
- Creating your story
- Evaluating the top four entry- to mid-level video editing software titles

Organizing Your Assets

Your first video projects might be no more than sequences of video scenes, which will present only minor creative challenges. So, putting them on a DVD will take only a few mouse clicks.

But I don't think that's why you bought this book. Rather, you want to produce involved and engaging DVDs. Your best productions will use all manner of media, and as your projects increase in complexity, keeping track of those assets becomes cumbersome. So, you need to establish an asset management process.

Using File and Folder Naming Conventions

Consider a DVD promoting an upscale housing development. It likely has photos of model homes, individual properties, and interior decorating choices. Viewers might be able to access architectural drawings, video walk-throughs of several houses, and taped comments from the developer.

This project is not atypical for a DVD, and tracking all those assets could become overwhelming.

Therefore, you need to organize your project's file folders and use descriptive and logical naming conventions.

Task: Organize File Folders

▼ TASK

Let's start with file folders, which some video editing products refer to as *bins*. Think of them as baskets full of photos, video clips, music, and so on. My recommendation is to create a main file folder for each project, with subfolders for each set of assets. Here's a way to do that:

1. Open My Computer (or whatever you've named it) by double-clicking its desktop icon. Depending on the display options you've selected, it should look similar to Figure 11.1. As mentioned earlier, I prefer the Details view. To change to this view, click the Views icon shown in Figure 11.1 and select Details.

FIGURE 11.1
Use My Computer to create new file folders for your media assets.

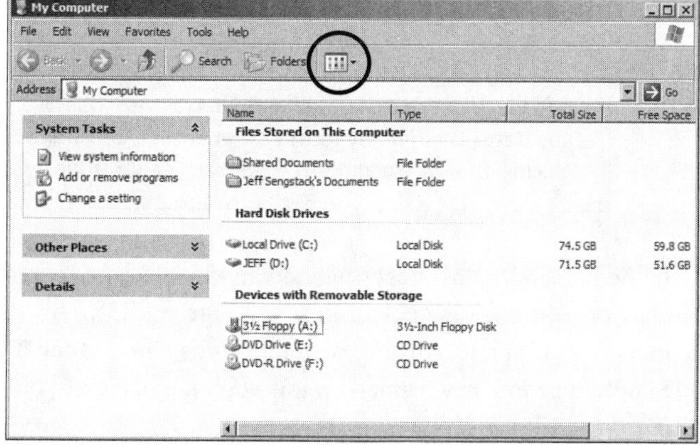

The My Documents folder (or whatever you might have renamed it) and its subfolders can seem like the logical place to put your assets. Open it to see why. As Figure 11.2 shows, by default My Documents has multiple media subfolders: My Music, My Pictures, and My Videos. Resist the temptation to use these file folders for your projects. It's best to have one folder for each project with subfolders for separate asset types. Instead, if you choose to use Windows' My Videos folder, you'll have to put all your videos for all your projects in only one place, defeating the purpose of storing your assets on a project-by-project basis.

FIGURE 11.2
The My Documents folder and its media subfolders can appear to be a logical place to store your media assets, but resist the temptation.

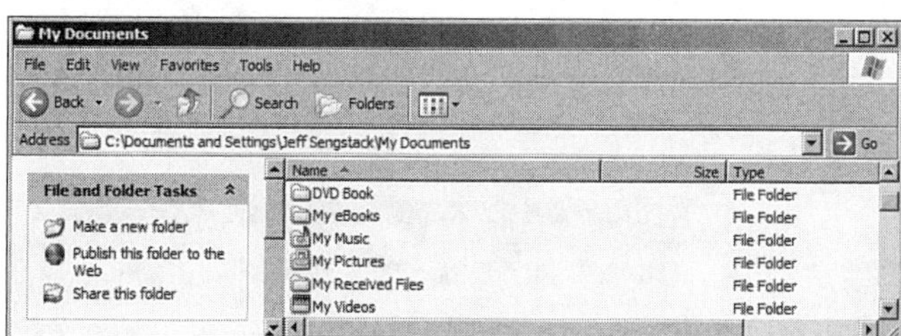

2. If you have two drives, open the non-system drive (typically the `D:` drive) by double-clicking its icon. You'll store your project assets here.

A good rule of thumb to follow is to install programs on your operating system (OS) drive (wherever Windows is located) and put data files—documents, videos, photos, and music—on a different hard drive. Do this for two reasons: First, it's easier to perform regular backups if your data is in one place. Second, if your OS drive crashes, you have not irretrievably lost any data. The only fly in this ointment is that Microsoft blithely insists on placing the My Documents folder on the OS drive.

3. In the File and Folder Tasks window in the upper-left corner of My Computer, click the Make a New Folder link (see Figure 11.3). That places a folder at the bottom of your directory listing with the words `New Folder` highlighted in blue. Simply type in a new name—in this case, the name of your project. For this book's purposes, I'll use `New Project`.

FIGURE 11.3
The File and Folder Tasks, Make a New Folder link is the fastest and easiest method to add a folder to My Computer.

If you're a long-time Windows user, the recently added My Computer, Make a New File Folder shortcut might not be familiar to you. You might instead prefer the former file folder creation/naming methodology; feel free to continue using it. As shown in Figure 11.4, select File, New, Folder and type in the folder name, replacing the blue highlighted `New Folder`.

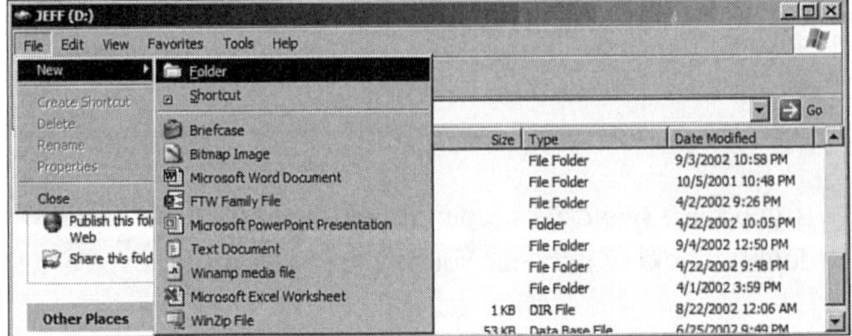

FIGURE 11.4
The old-fashioned way to add a new folder to My Computer.

4. Double-click your newly named New Project file folder to open it. Because it's empty, no files or folders appear in this window.

5. Make new subfolders for each media type by clicking the Make New Folder option, naming your folder, and pressing Enter. To make more folders, click somewhere in the My Computer window to stop highlighting that newly created folder and click the Make a New Folder option again. Create a subfolder for each type of media: Videos, Photos, Graphics, Music, or Documents. When completed, your New Project folder might look similar to Figure 11.5.

> Even though it can seem cumbersome, I suggest naming each media folder something such as `New Project Videos`, `New Project Photos`, and so on. If you have several projects, this will help you track your assets on a project-by-project basis.

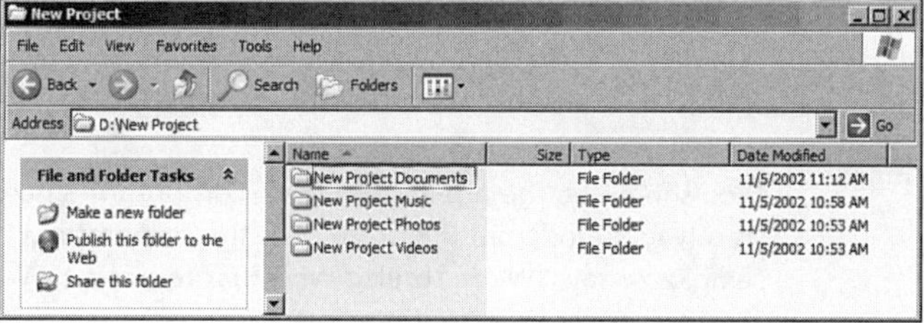

FIGURE 11.5
Use a consistent and descriptive file folder naming convention to avoid confusion during video editing.

> If you have a lot of one type of asset, you might want to create sub-subfolders, particularly for your New Project Videos folder. You might have interview segments with several different people, scenes from various soccer games, or multiple subjects within the same client folder. These sub-subfolders will help you track these different items.

Filenaming Conventions

Your file folders, or bins, organize your assets into logical, easy-to-track locations. Next, you need to use descriptive and consistent file—or asset—naming conventions to name your files.

Think through how you're going to name your video clips, photos, and other assets. You might end up with dozens of files, and if you don't give them descriptive names, your editing and DVD authoring process can become slow.

Naming Video Clips

You probably have already captured some video clips—perhaps several tapes, one tape at a time, one file per tape. Such large files are cumbersome when it comes time to edit them, so you should cut them into smaller, digestible chunks. One way to do this is as you capture them.

Some video editing products let you capture videos one clip or scene at a time. As you do that, you should use naming conventions that fit the nature of the clips.

You might have interview segments (called *sound bites* in TV news parlance), various specific scenes (a home run, sunset, wedding reception toast, and so on), and excellent instances of natural sound (applause, bird calls, or a burbling brook).

> Sound bite selection is an art. A sound bite's primary role is to present emotions, feelings, and opinions—not facts. That's your job as the writer/producer because you can tell the story better than the people you interview.
>
> You should be the one to say, "XYZ Corporation's new product line has received kudos from the press." Let the corporate CEO or assembly line employee say, "We're thrilled with that reception. We really think it's the best thing since sliced bread."
>
> In general, keep sound bites short; let them be punctuation marks, not paragraphs.

You might use a naming convention for sound bites such as `Bite-1`, `Bite-2`, and so on. However, adding a brief descriptive comment, such as `Bite 1 Laughs`, will help. For example, with natural sound, you could use the name `Nat 1 Hawk screech`.

With all other scenes (that is, besides natural sound and sound bites), you can drop the prefixes and just give them consistent, yet descriptive, names: `Soccer Goal-3`, `Crowd React-2 Applause`, and `Hawk Soaring-4`.

> As you capture video, reject superfluous scenes and sound bites. If you've shot more than one take of a scene, find the one that works best. If you've videotaped a softball game, select all the home runs, great catches, and enthusiastic crowd reactions, but skip most of the routine outs. The rule of thumb in the video or film production world is that you will shoot a lot more raw footage than you'll put in your final production—probably at least five times what you'll need.

Creating Your Story

A lot more is involved in crafting stories than you might think. The best TV news and video production writers use clever techniques to grab and hold your interest. They create a story flow that pulls you along and makes you want to pay attention.

You, too, can use those tips and tools to your advantage. You might not write network-quality stories immediately, but as you gain experience and revisit your old stories, you'll continually increase the quality of your work.

Overall Story Creation Tips

In my 15 years in TV news and video production, I've received advice from some real pros, including NBC reporter Bob Dotson and noted TV news media consultant Mackie Morris. Here are some of the tips I've gleaned from those contacts:

- **Use people to tell your story**—Your pictures and words might be fascinating, but if you add a compelling or colorful character, your story will have much more impact. People add interest.
- **Tell as much of the story as you can without narration**—Tell it visually by letting the images move your story. Unlike a radio sports announcer, you don't have to do a play-by-play.
- **Give viewers a reason to remember your story**—What do you want them to take away from your piece? If your viewers can feel something about the story, or its subject, they will remember it.
- **Listen to other TV writers**—Bob Dotson is my favorite feature story reporter. And Andy Rooney (yes, the *60 Minutes* Andy Rooney) is the dean of TV writers.

General Writing Tips

As you craft your videos, keep the following standard writing concepts in mind:

- **Keep your writing simple**—Short sentences work best. Write one thought to a sentence. Limit your story to only a few main ideas. Give the audience the best possible chance to understand the story.
- **There's no need to tell viewers what they already know or what the visuals tell them**—For instance, you don't need to tell viewers, "The children were ecstatic," when your video shows kids jumping for joy.
- **Write for the ear**—People will listen to your copy, not read it. Always read your copy aloud. How does it sound? Does it have a logical flow? Unlike the printed word, your viewers will not be able to reread a section that did not make sense.

- **Allow for moments of silence**—Stop writing occasionally and let 2 seconds or more of compelling action occur without voiceover. NBC-TV reporter Bob Dotson once told me, "For a writer, nothing is more difficult to write than silence. For viewers, sometimes nothing is more eloquent."
- **Use natural sound**—Let your videos "breathe" by adding brief moments of "wild" sound: a gurgling brook, the crack of a baseball bat, or assembly line equipment.
- **Add surprises to your piece**—They can be brief sound bites, exciting natural sound, or a clever turn of phrase. These moments build viewer interest.
- **Keep your interview sound bites brief**—Use them to add emphasis to your piece. Refrain from using sound bites as a substitute for your storytelling.

> When interviewing people, try not to ask questions. Merely make observations, as in, "That must have made you feel great." That loosens people up, letting them reveal their emotional, human side to you.

Specific Writing Tips

The following tips apply more to the mechanics of writing:

- **Write factually and accurately**—A standard newsroom admonition is, "When in doubt, leave it out." The corollary is, "Get the facts."
- **Write in the present or present perfect tense**—Immediacy is more interesting. When you use past tense, include a time reference—yesterday, five days ago, last year—to avoid confusion.
- **Write in the active voice**—Instead of writing, "This widget was made by our company," write, "Our company made this widget." This is a very important concept that takes some explanation. I'll tackle this in the next section, "Writing in the Active Voice."
- **Use of the "rule of threes"(as opposed to the photo composition "rule of thirds")**—When writing, try to group items by threes, such as red, white, and blue; left, right, and center; over, under, and through. Saying things in groups of three always sounds more interesting.
- **Avoid numbers**—Listeners have trouble remembering numbers. If you do use numbers, simplify them. Instead of writing, "We reduced expenses from $240,000 to $180,000," write, "We cut expenses by 25%."
- **Avoid pronouns**—Many times listeners have difficulty connecting a pronoun with the correct individual. Instead of saying, "Bob kidnapped Bill. Police later found him," say, "Bob kidnapped Bill. Police later found Bill."

Writing in the Active Voice

Active voice is a simple concept but difficult to put into use. Old habits die hard. Writing in the *active voice* means placing the receiver of the verb's action after the verb.

> The principal proponent of active voice writing is Mackie Morris. He's the former chairman of the Broadcast News Department at the University of Missouri School of Journalism. He later worked as a vice president and lead consultant for Frank N. Magid Associates, a major media consulting firm. These days, he remains very busy conducting writing workshops. I gleaned the following active voice tips from one of those seminars and through follow-up contact with Morris.

Consider the following *passively* voiced sentence:

John Doe was arrested by police.

Doe is the receiver of the action and is ahead of the verb. Change that to *active* voice by moving the receiver of the action after the verb, like so:

Police arrested John Doe.

Here's another sentence that uses passive voice:

A bill was passed by the Senate.

The bill is the receiver of the action. You would write it in active voice like so:

The Senate passed a bill.

Passive voice complicates and lengthens writing, whereas active voice makes your copy tighter and easier to understand.

Shifting from passive voice to active voice isn't easy. One major hurdle is recognizing passive voice writing. The biggest giveaway is the use of the "to be" verb, such as in these examples:

- The students *were* praised by the teacher.
- The unruly customer *was* told to leave by the maitre d'.
- The forest *was* destroyed by fire.

All these sentences have instances of the "to be" verb. When you move the receiver of the action after the verb, the "to be" verb magically disappears:

- The teacher praised the students.
- The maitre d' told the unruly customer to leave.
- Fire destroyed the forest.

That one fundamental technique—moving the receiver of the action after the verb—makes your sentences simpler, shorter, and more logical.

Besides simply switching the sentence around, you can fix passive sentences in three other ways:

- **Identify the missing actor and insert it into the sentence**—Change "The airplane was landed during the storm" to "A passenger landed the airplane during the storm."
- **Change the verb**—Instead of writing, "The bell was sounded at noon," write, "The bell rang at noon." (Or tolled, pealed, chimed—using active voice fosters the use of more descriptive words.)
- **Drop the "to be" verb**—Change "The spotlight was focused on downtown" to "The spotlight focused on downtown."

Not all "to be" verb phrases are passive, though. "The man was driving south" contains a verb phrase and a "to be" helper, but the man is performing the action, not receiving it. Therefore, the sentence is active.

That's not to say that you'll write exclusively in the active voice. You should write, "He was born in 1984," or "She was injured in the accident," because that's what people say.

Writing in the active voice forces you to get out of your writing rut. Instead of saying the same old things in the same old "to be" passive way, you will select new active verbs and constructions. You'll therefore write more conversationally and with a fresher and more interesting style.

Evaluating Four Video Editing Software Titles

Beginning in Hour 12, I'll take you through some standard video editing steps using video editing software. I'll use the trial version of Pinnacle Studio 8 included with this book for that.

I selected Studio 8 after evaluating it and three other entry- to mid-level video editing software products. Studio 8 is far and away the best of the field. I'll give you my quick take on all four in a moment.

Video Editing Software Update

I've worked with myriad video editing products—from slow, analog tape-only systems to expensive, standalone computerized hardware. What I find exciting these days is that powerful PCs, coupled with ever-improving video editing software, mean high-end video production tools now are available for anyone.

Software video editors are called *nonlinear editors (NLEs)*. They look and feel very different from standard, analog, linear videotape editing systems.

On tape systems you need to lay down edits consecutively and contiguously. If you decide to expand a story already edited on tape by inserting a sound bite in the middle, you simply cannot drop that bite into the piece and slide everything after it farther into the story. You need to edit in that sound bite *over* your existing edits and *reedit* everything after it.

NLEs, as you'll see in Hour 12, have no problems with any last-minute additions or deletions.

That said, you get what you pay for. I edit with Adobe Premiere, a $550 product. I'll demonstrate it in Hour 14, "Applying Professional Video Editing Techniques with Adobe Premiere." For this hour and the next, I revisited entry- to mid-level editing products that I haven't tested for more than 2 years. I discovered that these $100+ titles still have limited functionality, that creating high-end video editing tools is no simple task, and that not much has changed in 2 years.

These four NLEs have several things in common:

- Multiple video capture options
- A simple storyboard approach to editing, in which you drag and drop video clips to a film strip interface
- Myriad between-scene transitions and video special effects
- Several audio tracks for natural sound, music, and narration
- DVD authoring tools

Pinnacle Studio 8

Pinnacle Studio 8 leads this market niche by offering plenty of options while keeping things simplified. As shown in Figure 11.6, Studio 8 uses a timeline to supplement the storyboard functionality (I explain timeline editing in the next hour).

FIGURE 11.6
Studio 8 has a classic, straightforward, nonlinear editor layout.

It ships with the industry-leading text creation tool TitleDeko, plus SmartSound QuickTracks (refer to Hour 8, "Acquiring Audio") and some cool transitions lifted from Pinnacle's higher-end editing products. On the other hand, its DVD authoring module, shown in Figure 11.7, is rudimentary and virtually undocumented.

Studio 8 is the best means for a budding video producer to move into nonlinear editing or enhance her editing skills. It does, however, take a little extra effort to exploit all its bells and whistles.

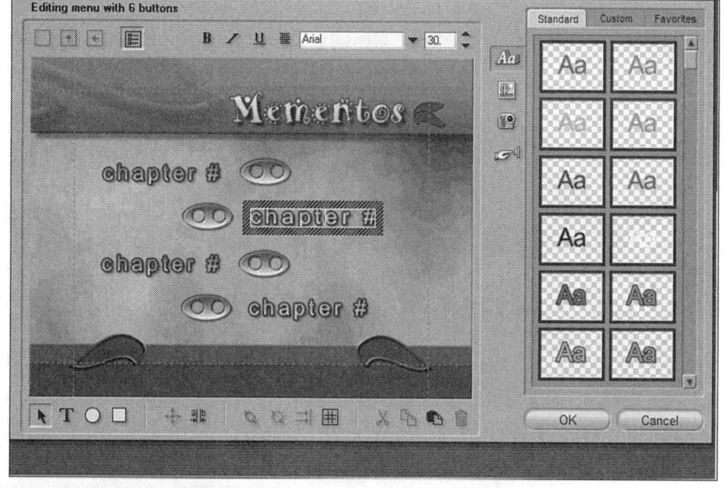

FIGURE 11.7
Studio 8's DVD authoring module is unimpressive.

Ulead VideoStudio 6

Ulead VideoStudio 6 gives Pinnacle Studio 8 a run for its money in some areas. As shown in Figure 11.8, it also offers a timeline, a full range of transitions, and a detailed text tool.

FIGURE 11.8
VideoStudio 6 has a full collection of features, but their execution leaves something to be desired.

As illustrated in Figure 11.9, it has an extra video track that enables you to create a very simplified picture in picture. However, the VideoStudio 6 interface has many quirky usability issues, and its DVD authoring module is barely functional.

Roxio VideoWave 5 Power Edition

This used to be MGI VideoWave. Roxio bought out this product line and has done nothing to bring it up to today's standards. The interface, shown in Figure 11.10, is nonintuitive and old-fashioned: It hasn't changed since I reviewed a previous version for *PC World* more than 2 years ago. For instance, you still have to change screens to do each function.

FIGURE 11.9
VideoStudio 6's picture-in-picture feature has minimal flexibility.

VideoWave 5 features a minimalist storyboard and some basic clip trimming tools. Its strength is its huge library of special effects, image editing, and digital animation tools. Its DVD authoring module, shown in Figure 11.11, is better than others in this category, but it is still far from MyDVD or other Sonic Solutions products.

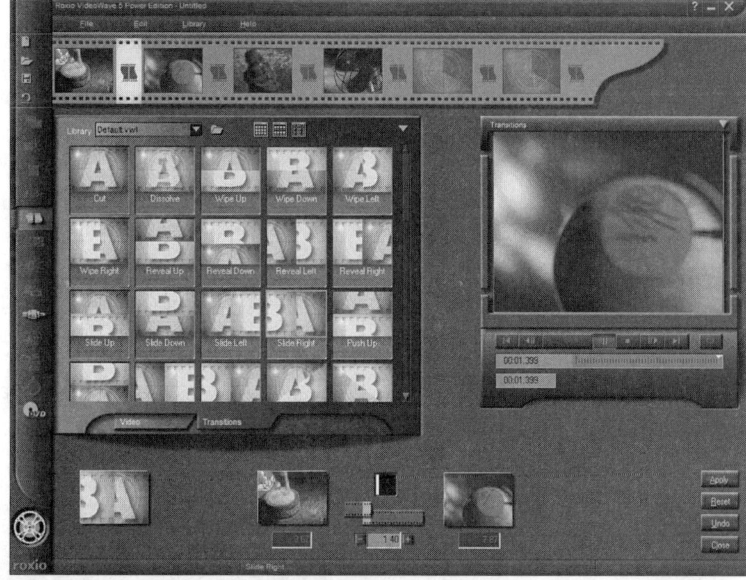

FIGURE 11.10
VideoWave 5 retains its out-of-date and awkward interface.

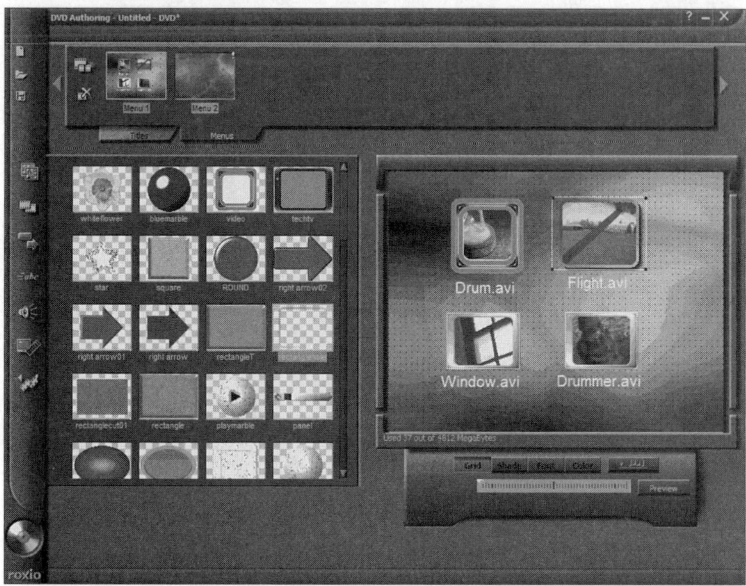

FIGURE 11.11
VideoWave 5 does offer the best DVD authoring module of these four NLEs, but it's still a far cry from MyDVD.

11

CyberLink PowerDirector Pro 2.5

CyberLink continues to release new versions of PowerDirector Pro, but this nonlinear editor still brings up the rear of this four-horse race. Although it does offer a plethora of special effects, as shown in Figure 11.12, it uses only a very simplified storyboard and a bizarre dial interface to change functions—title, transitions, audio.

The DVD authoring wizard, shown in Figure 11.13, is easy to use but has few menu design options—it has no text, no custom buttons, and no placement of icons on the page.

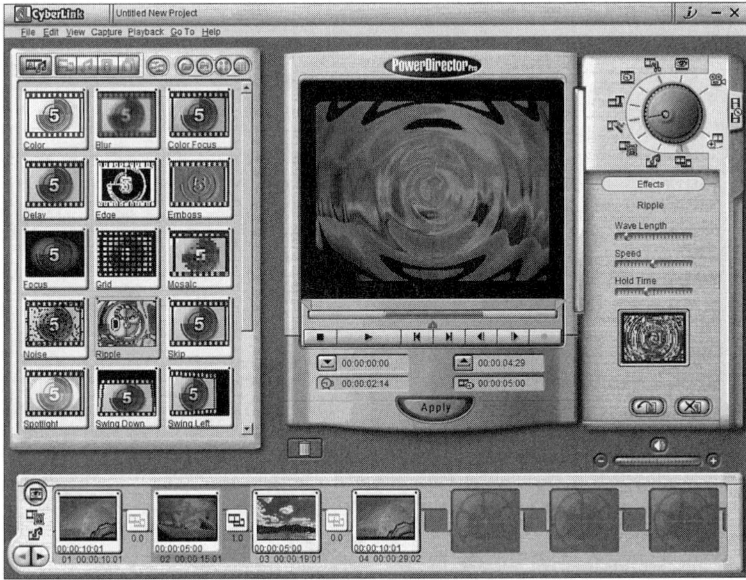

FIGURE 11.12
PowerDirector 2.5 has many special effects, but its unusual interface limits functionality.

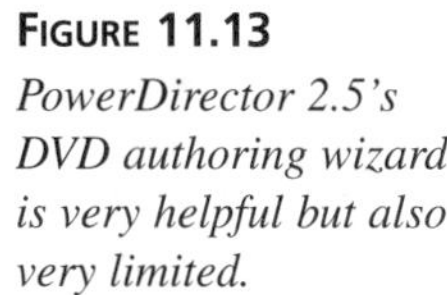

FIGURE 11.13
PowerDirector 2.5's DVD authoring wizard is very helpful but also very limited.

Summary

DVDs are great story-telling media. Their nonlinear interactivity gives you plenty of options, but you'll mainly use the video segments you create for your DVDs.

Tracking your media assets through file folder organization and filenaming conventions makes that creative process run more smoothly. In addition, narrowing your assets down to a manageable collection of best-of files will foster tighter and higher-quality stories.

By following this hour's writing tips, gleaned from experts in the TV news and video production business, your video productions should have an even greater impact.

Finally, you need to choose a nonlinear video editing product that works for you. My suggestion is to use Pinnacle Studio 8 (we've included a trial version with this book). I think it's the best of the entry- to mid-level NLEs.

Workshop

Review the questions and answers in this section to reinforce your story crafting skills. Also, take a few moments to take the quiz and do the exercises.

Q&A

Q You mentioned the "rule of thirds" and the "rule of threes." What's the difference?

A The "rule of thirds" is the photo composition technique that divides your image into thirds, horizontally and vertically. The "rule of threes" is a writing technique of listing items in threes, such as Citius, Altius, and Fortius or faster, higher, and stronger.

Q I interviewed a colorful character who says things in ways I wouldn't dare try. I want to use him as a narrator, but you said to keep my sound bites short. What should I do?

A Break my rule. When you find someone who can say it better, funnier, or quirkier than you, by all means use that person. However, you don't want a "talking head"—a person speaking on-camera—for the *entire* piece (even if he's wonderfully eccentric). Cover at least some of the interview with images that reinforce the words.

Quiz

1. What does "write for the ear" mean?
2. How is a nonlinear editor different from a tape editor?
3. How do you add a new file folder to your hard drive?

Quiz Answers

1. People will listen to your videos and DVD projects—they won't read them. With printed text, readers can go back or skip around. In contrast, an audio script has to flow smoothly, have logical transitions, and moments of natural sound or silence to let the viewers absorb what they've heard.
2. With tape, the only way to change a project is to go through the time-consuming and laborious editing process, which sometimes takes several minutes per edit. With a computer NLE, you can just drag and drop your edits in no time at all.
3. Open My Computer, double-click a hard drive, click Make a New Folder (in the File and Folder Tasks window), and type in the folder name.

Exercises

1. Visit a local video production studio and sit in on a video editing session. They might use a PC-based system with Premiere or a Mac with an NLE, such as Final Cut Pro. Or, they might have some higher-end, dedicated video editing hardware. Note the steps and work flow they follow to create a finished project.
2. Take a critical look at the local newspaper and try to find passive voice writing. Most newspapers do not make a point of using the active voice because they know readers can always look back a paragraph or two in case they missed something.

   Write a script. Look at your video clips, select some sound bites, and note instances of natural sound. Then write a script that features the best images, bites, and sound. Read it aloud.

HOUR 12

Editing Video with Pinnacle Studio 8

The foundation of a great DVD is its video content. How you edit that content can make it or break it.

In this hour, I present the first installment of a three-hour crash course in video editing fundamentals. Using the trial version of Pinnacle Studio 8 included with this book, I walk you through some hands-on editing basics.

You'll work with the storyboard and timeline, trim clips, do cuts-only editing, make transitions, and apply special effects.

The highlights of this hour include the following:

- Getting acquainted with Pinnacle Studio 8
- Using the storyboard and timeline for cuts-only editing
- Adding transitions
- Using special effects

Getting Acquainted with Pinnacle Studio 8

We've provided a trial version of Pinnacle Studio 8 on this book's companion DVD. There are a few differences between it and the full retail version, including these:

- The trial version expires 30 days after installation.
- It displays only in 1024×768 resolution. The retail version also offers 800×600 (see the following note).
- It has only three SmartSound QuickTracks tunes versus 25 in the retail version.
- You cannot import, edit, or encode MPEG-2 videos (it does support the lower-resolution MPEG-1 format).
- You cannot create Super Video CDs (SVCDs) or DVD discs, but you can burn VCDs. Video CDs are compatible with most set-top DVD players and use MPEG-1 videos.

The lack of MPEG-2 support in the trial version of Studio 8 is inconsequential as far as working through the topics in this book. You can use one or more of the three Sonic Solutions DVD authoring products included on the companion DVD to encode videos to MPEG-2 and to create DVDs.

Install Studio 8 by locating and opening the Studio 8 file folder on the companion DVD and double-clicking `STUDIO8MAIN.EXE`. This starts a product demo. When you're ready, exit the demo and click the Install Studio 8 link.

It's not my intention to give you a complete tutorial on Studio 8. A few options and features are not covered here or in the next hour. If you want to delve more deeply into this excellent product, I suggest consulting Studio 8's help section.

When you start Studio 8, it asks you to register. When you've completed registering and closed out of that window, Studio 8 opens to its default user interface screen, shown in Figure 12.1.

Checking Out Studio 8's User Interface

Studio 8 opens with a sample video project ready to edit. The thumbnail images represent clips from the photo shoot for the retail box and starting splash screen.

FIGURE 12.1
Studio 8's user interface opens with a collection of sample video clips.

You might notice that the Studio 8 screenshots used in this book do not necessarily match what you see as you work with the trial version of Studio 8. The reason is resolution: An 800×600 screen resolution displays more clearly in print than a 1024×786 resolution. So, I used the retail version of Studio 8 and set it to 800×600 before making my screen grabs. Because your Studio 8 view is 1024×768, you'll have more screen real estate and your user interface likely will feel less cluttered.

The Album

Take a look around the interface, starting with the collection of tabs shown in Figure 12.2. These give you access to various pages in Studio 8's album.

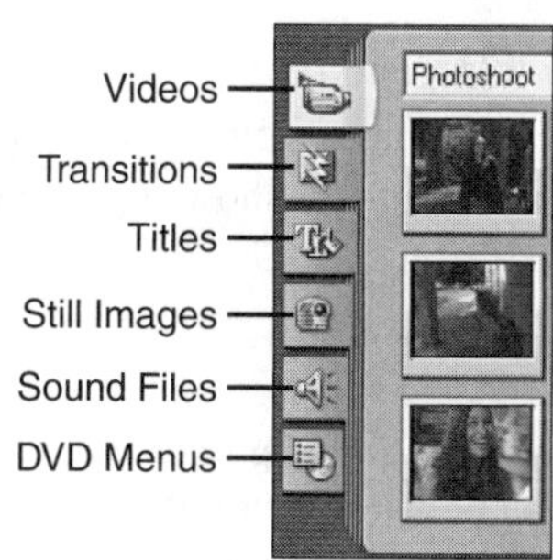

FIGURE 12.2
The tabs open pages in Studio 8's "Album."

I cover each of these features in more detail later in this hour and Hour 13, "Adding Audio, Tackling Text, and Improving Images." For now, just click through the following tabs to get an idea of what's available:

- **Videos**—The top tab displays thumbnail images of video clips you've selected for a video project. This is where you'll access your own video clips a bit later.

If you click a thumbnail, it pops up in the TV monitor (Pinnacle calls it the player) on the upper-right side of the screen. Use the VCR controls at the bottom of the monitor window to play the clip. When it ends, it moves seamlessly to the next clip in the thumbnail window. Note that the Play button has become a Pause button. Click Pause to stop the video playback.

- **Transitions**—This tab displays a collection of transitions you can use to move from one scene to another. Studio 8 offers 142 transitions plus demos of dozens more.

Feel free to check out some transitions. As shown in Figure 12.3, when you click a transition, Studio 8 gives a little demo of how it works in the TV monitor. I selected Push Up. In this case, the "B" scene (the next scene) pushes the "A" scene up and off the top of the screen.

FIGURE 12.3
Studio 8 displays a preview of a transition in the monitor, or player screen.

- **Titles**—This is Studio 8's text editing interface. Click this tab to see a collection of *templates*, which are 36 helpful tools that simplify adding text to your video. As shown in Figure 12.4, when you click one, it shows up in the TV monitor window. Text editing is discussed in Hour 13.

At any time during this get-acquainted process you can take a tour of Studio 8. Select Help, Guided Tour to view a more-or-less noninteractive overview of Studio 8's features.

FIGURE 12.4
The title page offers 36 templates that you can alter to fit your project.

- **Still Images**—Studio 8 displays two that come with the program. You access your stills through this tab.
- **Sound Files**—Studio 8 notes that this file folder is for sound effects, but you can add narration or music files here for use in your project.
- **DVD Menus**—I will not cover DVD authoring using Studio 8 because it simply does not come close to the authoring quality of MyDVD or the rest of the Sonic Solutions product line. I cover DVD authoring and production in Part IV, "Authoring DVDs."

The Movie Window

Now check out the movie window, which is the filmstrip-like interface at the bottom of the screen. This is where you piece together or edit your video. You'll start doing that in a few minutes. For now, though, note the three icons in the upper-right corner of the filmstrip window (see Figure 12.5).

FIGURE 12.5
These icons access three movie window interfaces: storyboard, timeline, and text.

12

The default view is the filmstrip-like storyboard. For first-time video editors, this can be an intuitive way to approach editing.

Roll your cursor over the middle icon, and note that the pop-up tip tells you it's for the timeline view. Click it to switch the editing section to the view shown in Figure 12.6. I'll explain the various elements along the left side in a few minutes.

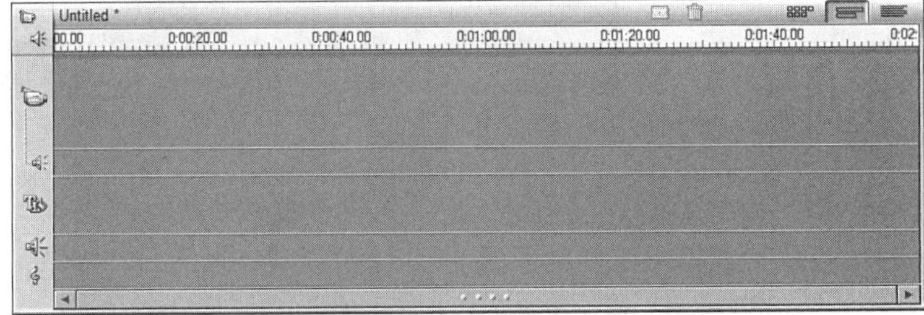

FIGURE 12.6
This timeline editing approach offers more editing options and flexibility than the storyboard.

Finally, open the text window by clicking its icon on the far right. You might never use this editing feature because it just displays time codes for each edit.

Taking a Quick Look at Studio 8's Video Capture Utility

Click the Capture tab; your screen should look similar to Figure 12.7. If you're transferring DV (as opposed to analog video), you can use the VCR-style controls on the camcorder to control your videocassette.

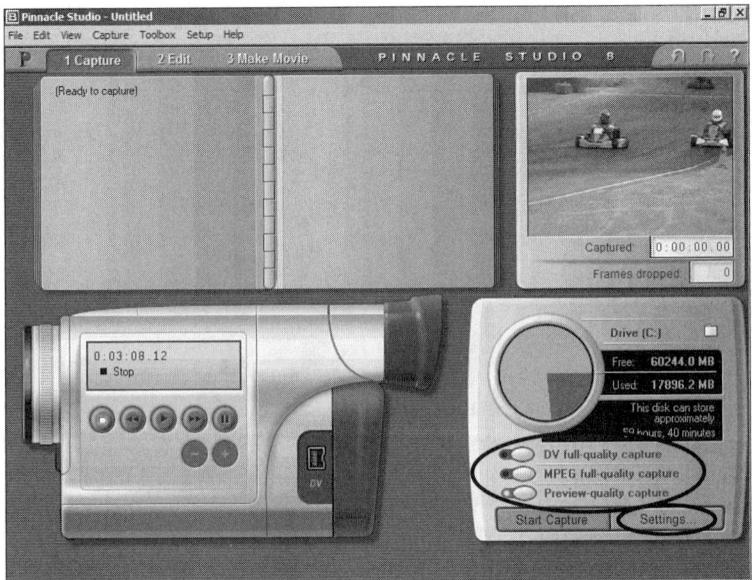

FIGURE 12.7
Studio 8's capture module uses familiar controls to simplify this process.

In Hour 10, "Capturing Video—Transferring Videos to Your PC," I walked you through two capturing processes, although neither offered much in the way of options.

Pinnacle is a much more powerful product than DVIO and Microsoft Movie Maker and has several useful capture features.

Notice the three capture settings in the lower-right corner: DV, MPEG, and Preview. Studio 8 lets you capture in a low-resolution (preview) format that saves disk space and speeds up editing. When you've finished editing and want to create your video, Studio 8 recaptures, in full DV or MPEG, only the portions of your original video included in your edited project.

> You can find out more about the capture settings by clicking the Settings button shown in Figure 12.7. That opens the Setup Options interface shown in Figure 12.8. Use the Presets drop-down menu to fine-tune your capture quality settings. However, the trial version offers only MPEG-1, versus multiple MPEG options in the retail version.

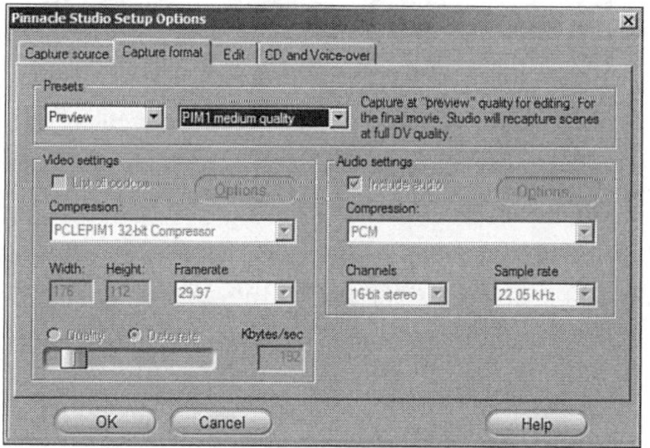

FIGURE 12.8
Capture settings let you choose your video quality.

Studio 8 has a helpful scene detection feature. As it captures video, it creates separate clips at natural breaks in the original video. You can fine-tune that by selecting Settings, Capture Source.

Using the Storyboard and Timeline for Cuts-only Editing

It's time to start editing a video. At its most basic level, the process involves dragging and dropping video clips in a desired order onto the storyboard. Then, you can trim clips to remove unwanted material. *Cuts-only* means all edits are straight cuts, with no transitions such as dissolves or fancy spinning boxes.

Task: Edit a Video Using the Storyboard

TASK

You'll use the sample video provided with Studio 8 to create your first video. Here's how:

1. Click the Edit tab at the top of the screen to return to the storyboard interface and the Photoshoot sample video thumbnail images.

You can use your own video instead of Photoshoot. To do that, click the file folder icon shown in Figure 12.9, locate your video, and click Open. Studio 8 performs an automatic scene detection and displays thumbnail images of each scene.

FIGURE 12.9
Use the highlighted file folder icon to access your own videos for editing.

2. Drag and drop the first scene to the first placeholder in the storyboard by clicking the scene, holding down the mouse button, and dragging the scene to the filmstrip. Repeat that for six additional scenes. When you're done, your screen should look similar to Figure 12.10.
3. Play this rough-cut video by clicking the Go to Beginning (Home) button and then clicking Play (refer to Figure 12.10). You can scrub through this collection of clips by dragging the slider shown in Figure 12.10.
4. Double-click the first clip in the storyboard to open the Video Toolbox. You'll use this interface, shown in Figure 12.11, to trim individual clips.
5. Drag the in and out points to change when the clip begins and ends. Note that the duration of the clip changes as you move the edit points. Three seconds is a comfortable length for a video scene.

FIGURE 12.10
Drag and drop video clips to the storyboard to begin editing a video.

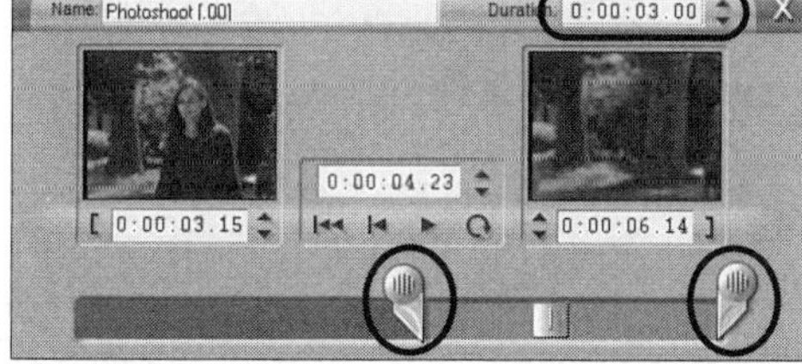

FIGURE 12.11
Move the in and out points to trim clips in the Video Toolbox.

12

> The example in Figure 12.11 shows a standard editing technique. The in and out points let the model start onscreen and then walk out of the scene at the end, creating a more comfortable transition to the next scene. If you let her stay in the scene, the view in the next clip would jump from in front of her to behind her—an awkward and abrupt edit.

6. Trim subsequent scenes by double-clicking them in the storyboard filmstrip and using the Video Toolbox to make adjustments.
7. If you don't like a scene (the third clip doesn't work for me), remove it from the storyboard by right-clicking it and selecting Delete, or by selecting it and pressing the Delete key, or by clicking the Trash icon above the timescale.

Removing a clip from your storyboard does not remove that clip from the Video Album. You can still use that removed clip elsewhere in your video.

8. Rearrange your video clip sequence by dragging one scene to a new location between two other scenes and dropping it there.
9. Insert a scene by dragging it from the Video Album to the storyboard and placing it between two clips at the beginning of your video or the end. As shown in Figure 12.12, a line with a little plus sign icon (+) appears at the insertion point.

FIGURE 12.12
Insert a video clip anywhere in your story by dragging and dropping that clip to the storyboard.

The removal, rearrangement, and insertion processes demonstrate just how much easier and faster nonlinear editors are than traditional, linear, videotape systems.

10. Save your project by selecting File, Save Project; then select a file folder location and name.

You can view how your edits have worked to this point by clicking the Go to Beginning (Home) button in the Player/Monitor window and then clicking Play.

Touring the Timeline

Your piece could probably use some fine-tuning. The timeline is the best means to that end.

At first, the timeline might not be as intuitive as the storyboard, but I think you will come to rely on it. It's the standard interface used by countless video editors.

Switch to the timeline view by clicking the Timeline icon at the upper-right of the storyboard filmstrip. This opens the timeline interface shown in Figure 12.13.

The timeline displays more than simple thumbnail images. It also shows the overall length of your video across the top and the relative length of each clip (based on the box size associated with each clip's thumbnail image). Plus, as noted in Figure 12.14, it displays the audio tracks for your original video clips and gives you places to add text, a narration, sound effects, or music.

FIGURE 12.13
The timeline interface is the standard nonlinear editing tool.

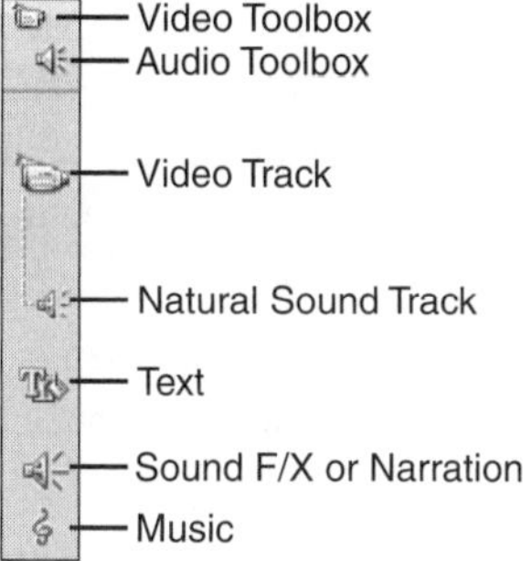

FIGURE 12.14
The timeline interface displays additional tracks—three for audio and one for placing text over your videos—that are not readily accessible in the storyboard view.

Drag the scrubber through your project (refer to Figure 12.13). Note that each clip's box on the timeline turns blue as you scrub through it.

Your edited video probably does not fill the full width of the timeline. Some short duration clips might be nothing more than thin, vertical lines. As shown in Figure 12.15, you can expand the project view by moving your cursor over the timeline's Timescale (it turns into a clock cursor with double arrows). Then, right-click and select Entire Movie.

FIGURE 12.15
Use a right-click menu to expand the Timescale and get a better view of your clips.

You can use keyboard shortcuts to expand or contract the workspace. The equal sign (=) key expands the view (meaning each clip takes up more space on the timeline), and the minus sign (-) key shrinks the project view. You can also specify time increments using the right-click menu. Finally, you can drag the clock cursor right or left along the timescale to expand or contract the view.

Task: Trim Clips on the Timeline

▼ TASK

To get a feel for how the timeline works, here's a quick run-through on how to insert, rearrange, trim, and slice clips:

1. You can insert, remove, and rearrange clips on the timeline just as you did in the storyboard. Insert a clip by dragging it from the Video Album to the timeline. The bars on either side of the clip will change from red to green when you've arrived at a legal insert point. Rearrange your clips by clicking one in the timeline and dragging it to a new location. Finally, remove a clip by selecting it, right-clicking, and selecting Delete.
2. Trimming a clip takes more steps. Select a clip and then expand that clip width by clicking the equal sign (=) several times. This helps you select more precise in and out points.
3. Place the cursor at the beginning or end of the clip. As shown in Figure 12.16, the cursor changes into an arrow pointing into the clip.

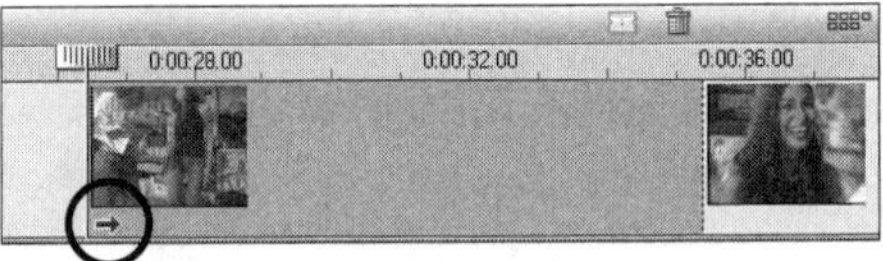

FIGURE 12.16
Drag the edge of a clip in the direction of the arrow cursor to trim it.

Make sure you actually select a clip by clicking it; otherwise, you won't see the clip-trimming arrow cursor. Also be sure to select only one clip (you can select more than one by holding down the Ctrl key while clicking other clips). By selecting two clips, you change the in point of one clip and the out point of the other.

4. Drag the edge of the clip in the direction of the arrow to shorten it. Watch the Player TV Monitor to select the trim point that works well. Release the mouse button, and the clip or the adjacent clip(s) will slide over to cover the empty space created.

▼

Just as you did in the storyboard, you also can trim clips by double-clicking a clip to open the Video Toolbox and using it.

5. You can split a clip in two as a means to insert another clip between the two pieces. Select a clip and move the scrubber to the point at which you want to cut the clip. Click the Razor Blade icon, shown in Figure 12.17. You now have two separate clips, between which you can drag another clip.

FIGURE 12.17
Use the Razor Blade icon to slice a clip in two.

You can always undo a bad edit. If you split a clip in the wrong place or inserted a clip in the wrong location on the timeline, click the Undo button or press Ctrl+Z (see Figure 12.18). You can undo as many steps as you like, but you cannot selectively undo one thing you did several steps back without undoing everything that came after it. The Redo button, next to the Undo button, restores whatever you just undid. Get it?

FIGURE 12.18
Use the Undo tool to move back an editing step.

12

Adding Transitions

Watch TV news stories, and you'll notice that most edits are cuts—quick jumps from one scene to the next. No transitions, such as dissolves, wipes, or pushes, are used. The reason is that viewers expect cut edits. They know they're watching news, not a feature film, so they expect the stories to move quickly from one scene to the next.

You might want to soften those scene changes by using transitions.

Studio 8's transitions look very cool. You'll probably be tempted to use a different one between each scene of your video, but resist that urge. Transitions tend to be distracting, so use them judiciously.

Task: Add Transitions to Your Edited Video

Adding transitions works similarly to adding video clips to your project—you use a simple drag-and-drop method. Here's how to do it:

1. Expand the timeline so two adjacent clips take up most of the space. As you can see in Figure 12.19, that makes seeing how to place a transition on the timeline easier.

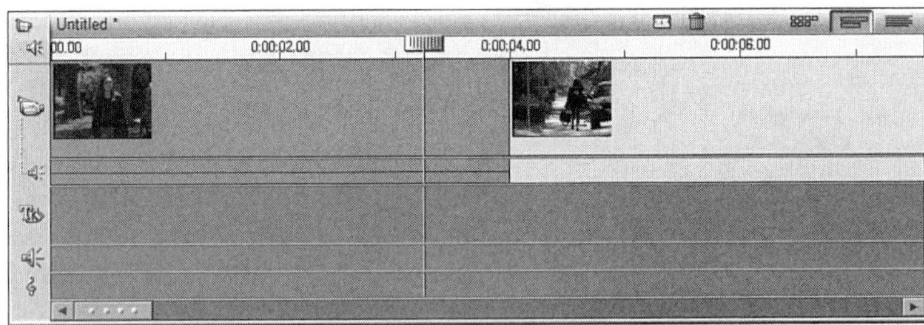

FIGURE 12.19
Expanding the timeline makes placing the transition between two clips easier.

2. Click the Transitions tab (the lightning bolt) in the album (see Figure 12.20).
3. Switch from Standard Transitions to the Alpha Magic transitions collection by clicking the drop-down menu and selecting Alpha Magic.
4. Drag and drop the transition shown in Figure 12.20 to the edit line between the two clips. You'll know that you've positioned it correctly when a thin, green line appears at the edit point as well as another line two seconds into the "B" clip.

FIGURE 12.20
Click the Transitions tab to access Studio 8's many transitions.

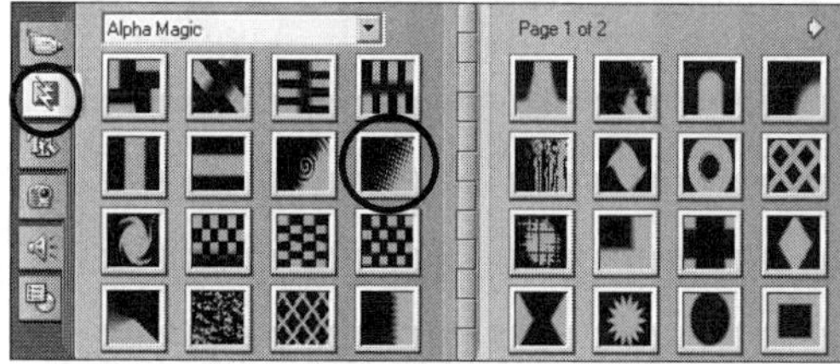

Adding transitions to the storyboard is just as easy: Simply drag and drop a transition to the space between two images on the filmstrip.

Take a close look at the audio track beneath the transition in the timeline (see Figure 12.21). The two blue lines show an audio cross fade. When you add a transition, Studio 8 automatically fades down the "A" clip audio and fades up the "B" clip audio to avoid an abrupt sound change during what is a gradual visual change.

FIGURE 12.21
Adding a transition automatically creates an audio crossfade.

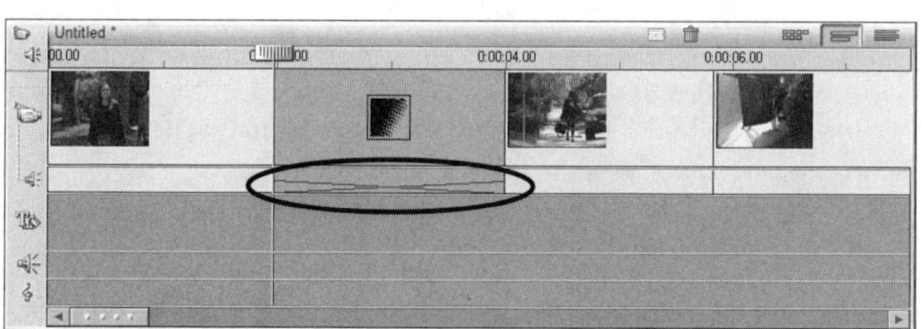

5. Double-click the transition on the timeline to open the Clip Properties tool, shown in Figure 12.22. Here you can change the length of the transition or, if available, have it move in reverse (in this case, from left to right).

FIGURE 12.22
The transition Clip Properties toolbox lets you fine-tune your transition.

You can make your video start with a fade-up from black and end with a fade-down to black. As I've done in Figure 12.23, simply add dissolve transitions to the start and finish of your video.

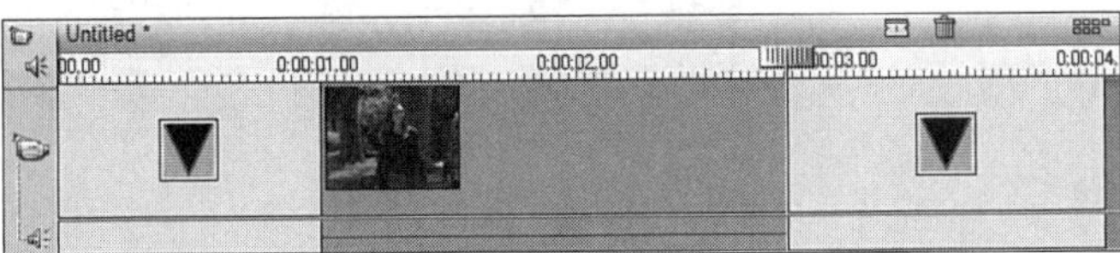

FIGURE 12.23
Use a dissolve transition at the start and end of your video to fade up from black and fade down to black.

Studio 8's Consumer-oriented Transition Behavior

Studio 8 is a consumer-level product and has a default transition behavior geared to that market. It's counterintuitive to me because I'm used to professional-level editors such as Adobe Premiere. But this might be a non-issue for you depending on how you create videos.

In Studio 8, every time you add a transition to your video, it shortens the overall length of your project by the length of the transition. This is also true for the editor in MyDVD 5 covered in Hours 4–6, "Authoring Your First DVD Project Using MyDVD: Part I–III."

Transitions overlap adjacent clips by blending the end of the "A" clip with the start of the "B" clip. Studio 8 accomplishes this by sliding the "B" clip (along with all subsequent clips) to the left by the length of the transition (the default time is 2 seconds).

Higher-end NLEs, such as Adobe Premiere, use extra footage from the original video clips—tailroom and headroom—to let that overlap happen without shifting the "B" clip (and all subsequent clips) to the left on the timeline.

This is important for professional editors because they frequently edit in the narration first and then choose video clips to match the audio. If each transition shortened the video, that narration would fall out of sync with the images.

Studio 8's programmers assume you will lay down your video clips, insert transitions, add titles and music, and then perhaps add a narration.

The bottom line is that you should either narrate your video as the last step or give clips extra tailroom and headroom at edits where you plan to insert a transition.

Using Special Effects

Studio 8 offers several special effects. They either affect the appearance of a clip or its playback speed.

Visual Effects

Double-click a clip to open the Video Clip Properties toolbox. Clicking the Sun/Moon icon opens the Visual Effects toolbox (see Figure 12.24).

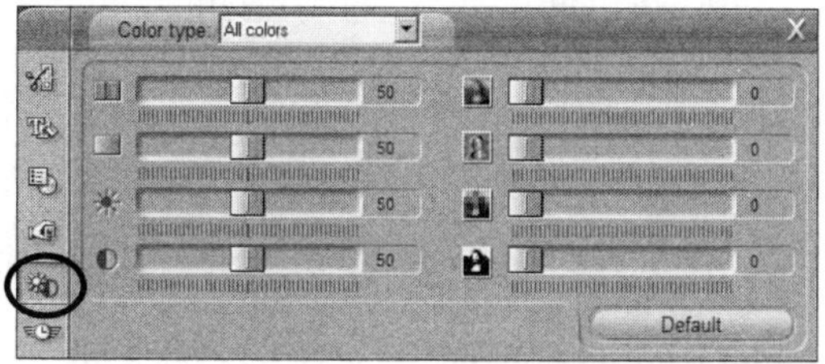

FIGURE 12.24
Use the Visual Effects toolbox to change the appearance of a clip.

Eight visual effect options are available, plus three overall effect controls. The drop-down list lets you change your clip into a black-and-white clip, sepia-toned clip (which gives it that old-fashioned flavor), or any single color.

The four slider bars on the left let you adjust the Hue (overall color), Saturation (color intensity), Brightness, and Contrast settings.

The four slider bars on the right let you blur your image, give it a 3D embossed feel, turn it into a mosaic of colored squares, or posterize it by blending similar colors.

> The Player Monitor TV screen gives you real-time feedback as you use the Visual Effect toolbox to make changes to a video clip. Experiment by combining some effects. For instance, use Blur and Emboss to create a softened 3D look.

Playback Speed

Click the winged clock icon shown in Figure 12.25 to access the Playback Speed toolbox.

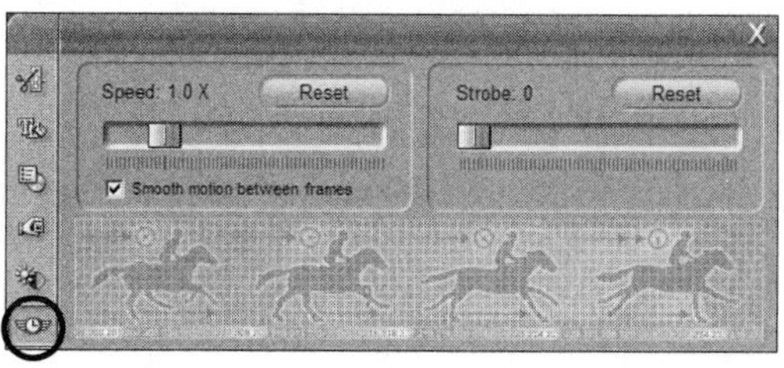

FIGURE 12.25
The Playback Speed toolbox lets you change a clip's speed or add a stop-action, or strobe, effect.

The Speed slider lets you set the clip speed from one-tenth normal to ten times normal. The Strobe effect creates something like a stop-action video; the number displayed—from 0 to 50—indicates how many frames will go by between scene changes. Video runs at 30 frames per second, so a setting of 15 displays two frames of video per second (each frame stays onscreen for one half of a second). A setting of 30 displays one frame per second.

> As you move the Speed slider, you might notice that the selected clip's length changes. Speeding up a clip shortens it, whereas slowing down a clip lengthens it. After changing a clip's speed, you might want to change how you trimmed it to compensate for the new length.

Summary

The better your video story-telling, the better your DVDs. Applying effective editing techniques to your videos can make all the difference. In this hour, I covered some editing basics: the storyboard and timeline, cuts-only edits and transitions, and special effects and varied clip speeds.

At this early stage, keeping it simple is your best bet. Avoid injudicious use of transitions because they can distract viewers.

Workshop

Review the questions and answers in this section to reinforce the video editing concepts covered in this hour. Also, take a few moments to take the short quiz and do the exercises.

Q&A

Q It's so easy to use the storyboard. Why should I use the timeline?

A You don't have to use the timeline. The storyboard might be as far as you need to go. It's an easy way to assemble video clips in a logical order. From the storyboard, you can access the Clip Properties toolbox to trim individual clips and easily add transitions. Only if you want to have more control over clip timing and placement of sound effects, narration, and music should you move on to using the timeline.

Q I really love all the cool transitions. Why shouldn't I use them for all my edits?

A This is a standard beginning editor mistake. Now that nonlinear editors make adding transitions so easy, folks go nuts over them. But check out the pros by watching feature films, TV programs, and music videos with a critical eye. They rarely use transitions, and almost all the edits are cuts. Plus, when they use transitions, they fit the circumstances or show style. Campy programs, such as the old *Batman* TV series, were big on obvious transitions. And *Star Wars* films copy the old movie theater science-fiction serial wipes and pushes. But few videos call for that editing approach.

Quiz

1. You've dragged and dropped a clip onto the timeline. But it's too long. How do you trim it?
2. You want a slow fade to black at the end of your production. How do you do that?
3. You want to create a clip for a family tree project that has a sepia tone and is a bit out of focus. This is a nice background for text. How do you create this special visual effect?

Quiz Answers

1. Use one of two methods. Click the clip to select it and drag in its edges. Or double-click the clip to open the Clip Properties toolbox and use the sliders to make your trim.
2. Drag the dissolve transition to the end of the last clip in your piece; then, double-click the transition to open the toolbox. Change the duration in the time window in the upper-right corner. Or, only when using the timeline, select the transition and drag its right edge to the right to lengthen it.
3. Double-click the clip to open the toolbox, and then click the Sun/Moon icon to open the Visual Effects toolbox. Select Sepia from the Color Type drop-down menu and move the focus slider until you've achieved the desired effect.

Exercises

1. Create some cuts-only videos using your captured footage. Drag clips to the Studio 8 storyboard and trim things down. Take a close look at your edited work. Are the sequences logical? Do they tell a story? Are the clips too long? Find areas to cut or rearrange.

2. Look at your edited cuts-only videos and see whether some scenes seem to cry out for a transition. For instance, is there movement from left to right? If so, use a transition that pushes or wipes from left to right. Do you have a view through a window? Use the Blinds wipe (page two of the Standard Transitions). Or a '60s party? Use the Pinwheel wipe (page three).
3. Be a critical TV and movie viewer. Watch edits. Under what circumstances do they use transitions? What types of transitions do they use? How long do individual clips last?

HOUR 13

Adding Audio, Tackling Text, and Improving Images

This hour takes you through the second lesson of my three-chapter video and image editing crash course. I present a rapid-fire collection of editing examples and tips focusing on audio, text, and still images.

Studio 8 features a powerful text (title) creation tool, which I demonstrate how to use.

I show you how Studio 8 handles freeze frames and use Adobe Photoshop Elements to show some basic image editing techniques.

Finally, using Studio 8, you'll learn how to add a narration, lay in music, and use an audio mixer to create TV news-style fade-ups and fade-unders.

The highlights of this hour include the following:

- Adding titles to your videos
- Grabbing freeze frames

- Editing images with Adobe Photoshop Elements
- Adding narrations, music, and sound effects to videos using Studio 8
- Using Studio 8's audio mixer

Adding Titles to Your Videos

Text enhances your video. You can add a location *super* (superimposed text) to state where the video took place or insert a date, the name of whomever is speaking on camera, statistics, an opening title, or closing credits.

Studio 8's Title Editor—a derivative of Pinnacle's well-regarded, professional video text product TitleDeko—is an outstanding feature. No other NLE in this price range matches its high quality.

To explain it in detail would consume an entire hour of this book, so I'll just present some highlights.

Task: Explore Studio 8's Title Editor

TASK

To demonstrate what the Title Editor can do, let's step through a few brief exercises. I'll start with selecting a background:

1. Open the Title Editor by double-clicking anywhere in the Title Track in the timeline. Figure 13.1 shows the screen that appears, which have several features, including Title Types, Templates, Backgrounds, and Object Tools.
2. The text background tools are in the upper-right corner, and Transparent is the default setting (see the following note for an explanation). Select Still Image (the small cactus icon) and use the slider on the right to check out all the available graphics. Note the checkerboard pattern in several of the graphics (see Figure 13.2). This indicates that that portion of the background is transparent, meaning it lets the video clip show through.

Normally, when you use text in a video, it appears over a video clip. So, you want to give your text a *transparent* background to let that video appear behind the text. Sometimes, though, you'll want text to appear over a still image or graphic background; that's where Studio 8 goes the extra mile—it gives you all sorts of text background options.

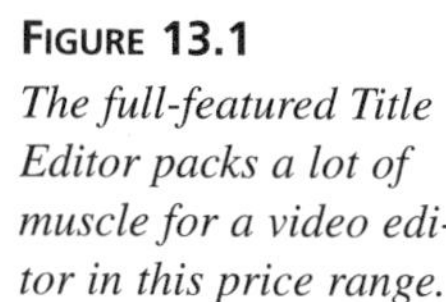

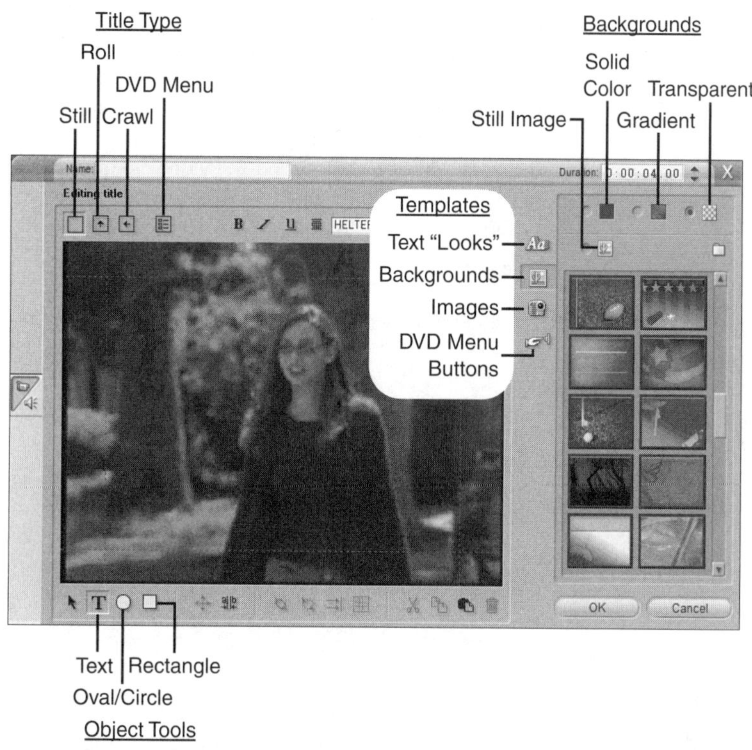

FIGURE 13.1
The full-featured Title Editor packs a lot of muscle for a video editor in this price range.

FIGURE 13.2
Studio 8 offers dozens of background graphics. Checkerboard portions are transparent, letting the video show through.

3. Select Solid Color to open the color selector interface. As shown in Figure 13.3, you can select a color using the crosshair tool and adjust the color intensity using the slider on the right side. Click OK.

You can adjust any color's opacity, meaning you can give it some transparency to let the video clip show through. In this way you can place a transparent color on top of a video clip (or still image) to give it a tint and then apply text over the video and the tint. That's a great way to emphasize the text without obliterating the video. To give a selected color some transparency, in the Color selection screen, move the opacity slider and then return to the Title Editor window to check the results (see Figure 13.3).

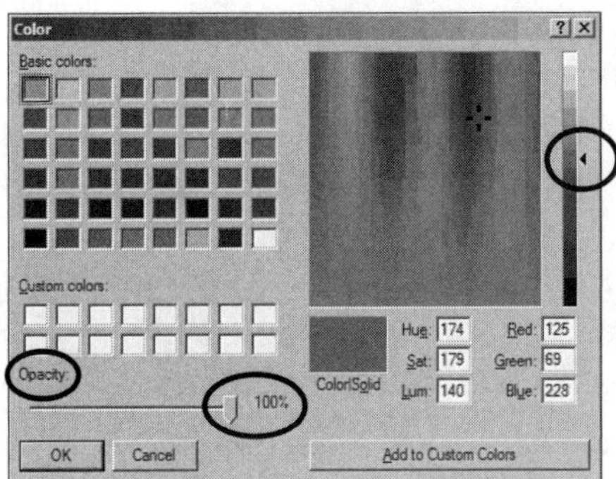

FIGURE 13.3
Use the Color Selection tool to create a solid or gradient background.

4. Select Gradient and note that you can change four colors. Click one of the corner boxes to open the color selector tool. You'll use this gradient background for your text.
5. Now, to create some text, make sure the Text Tool icon (T) is selected. Click somewhere in the upper-left of the Text Editor window and drag and drop to create a text bounding box in the Title Editor window (see Figure 13.4).
6. Click the text Looks button (the Aa icon) shown in Figure 13.4. Scroll through the type styles and select one that appeals to you. Figure 13.4 shows my selection.
7. Position your cursor in the text box you've created and start typing. As with word processors, you can change the typeface, font size, and other characteristics using the tools in the upper-right of the title window. Use the Justify tool at the bottom to arrange your words (refer to Figure 13.4).

The intersecting dashed red lines in the Title Editor window delineate the NTSC safe zone. Most TV sets display less than a full-screen view by truncating the edges of the original video. By keeping your text within the red rectangle, you'll guarantee all your words will appear on all TV screens.

FIGURE 13.4
Use the Title Editor and its template Looks collection to create text for your video project.

> You also can manipulate the text box by changing its size, moving it, and rotating it. Move your cursor around in the text box to see how it changes depending on the function it can perform in each location.

8. Select the text by clicking and dragging your mouse over it. Then, click the various looks to see all the possibilities. When you're done, select a simple type look (the simpler looks are found toward the top of the Looks collection). This will give you more options in step 8.
9. Click the Custom tab, shown in Figure 13.5. Its interface lets you fine-tune your typeface.

> In the Custom interface, the Face (typeface) option lets you change the type color, make it a gradient, or make it transparent. You also can give it an out-of-focus look. The Edge option has similar features and lets you change the edge width and add a glow. The Shadow option works the same way and lets you set the direction the shadow falls by using the dial in the lower-right corner.

10. When you've created a title you like, you can use a different background or select a transparent one. Then, you just click OK to put the title on the timeline. You can change its length by either dragging the start or end or dragging it to another location on the timeline. You can also view it by playing your video at that point.

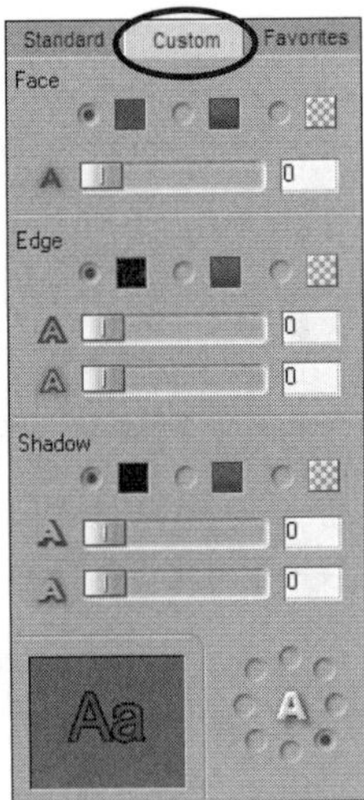

FIGURE 13.5
The Custom interface gives you detailed control over the look of your text.

Using the Shape Tools

You can make ellipses and rectangles using the Title Editor. These are great for title backgrounds or for highlighting a portion of your video.

To create them, open the editor again by double-clicking the title track. Click the Ellipse Object tool (see Figure 13.6). Then, click in the Title Editor window and drag to create a shape. As with text, you can change a shape's characteristics, including color (with gradients), border, and shadow, by opening the Custom interface.

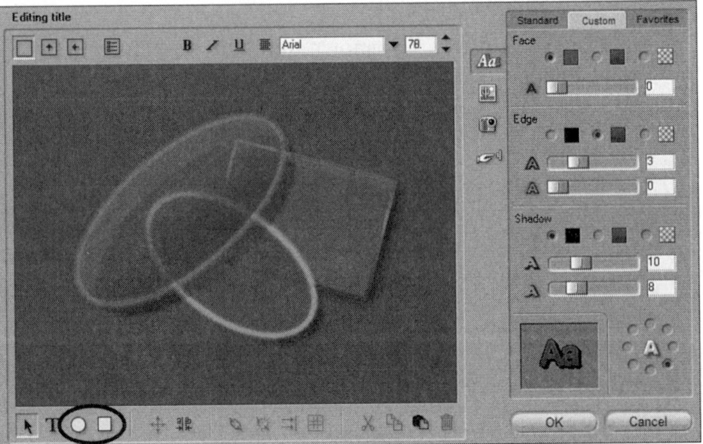

FIGURE 13.6
Use the Shape Object tools to create ellipses and rectangles.

I created three shapes, rotated them a bit, and used colors and gradients (with some transparency thrown in) to give the feeling of depth. You can move these shapes from front to back or vice versa by selecting one and using the menu shown in Figure 13.7.

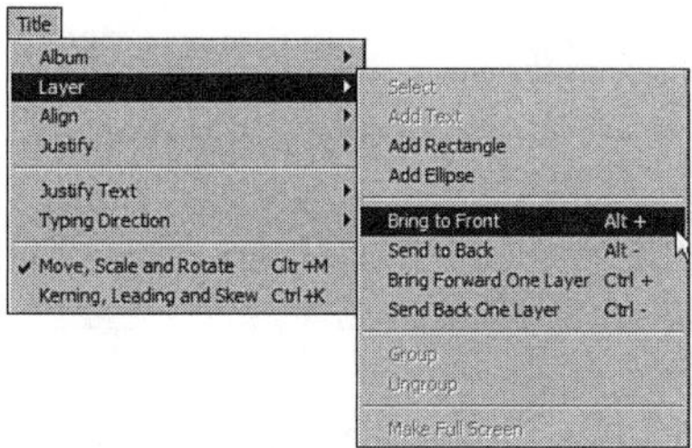

FIGURE 13.7
Move shapes from front to back, or reverse, using this menu.

When using transparent colors with an object or text, be sure you reduce the opacity of its shadows as well. To maintain a realistic look, set all the shadows in the same title screen in the same direction. The distance can vary, giving the impression that some objects or text are farther above the background than others.

Grabbing Freeze Frames

You might occasionally want to use still images grabbed from your original videos in your final edited production. Sometimes it's good to end your piece with a freeze frame and then fade to black. Or, you can use a still image as a background for a title.

Task: Use the Frame Grabber Tool

You can grab freeze frames using your camcorder or from video you've already transferred to your hard drive. I'll show you the latter. Here's how it works:

1. Drag the clip with the image you want to grab to the timeline.
2. In the main menu, select Toolbox, Grab Video Frame to open the interface shown in Figure 13.8.

13

FIGURE 13.8
Use the Frame Grabber to create a freeze frame from your video.

3. Move the timeline slider to the frame you want; then click the large Grab button.
4. Click the Add to Movie button to drop the frame onto the video track in the timeline just ahead of the currently selected clip. The default length is 4 seconds, but because it's a still image, you can set the length to whatever you want by dragging the start or end point.

You can tell the grabbed image is not video because it doesn't have an audio track.

5. Save the image to your hard drive by clicking the Save to Disk button.

To use the grabbed image as a title background, save it to disk and click the Still Images (also called Photos and Frame Grabs) page of the Studio Album. Your new still image should be the first on display. Drag it to the title track, and double-click it to open the Title Editor.

Editing Images with Adobe Photoshop Elements

If you're a graphics professional, you use Photoshop. It's a given. If you're an entry- to mid-level DVD producer, you might already—or someday will—use Photoshop. If Photoshop is not on your PC, try its younger, leaner sibling: Photoshop Elements.

Photoshop Elements offers standard photo touch-up tools that are found in many consumer-level photo products. Even some bundled with digital cameras have features on par with Photoshop Elements.

But Photoshop Elements goes beyond simple touch-up to professional graphics creation that can come in handy as you increase your DVD menu creation skills.

I will give you only a brief overview of what this full-featured product has to offer. I've included a trial version on this book's companion DVD; feel free to install it and follow along.

Photoshop Elements' opening interface, shown in Figure 13.9, eases users into this powerful program.

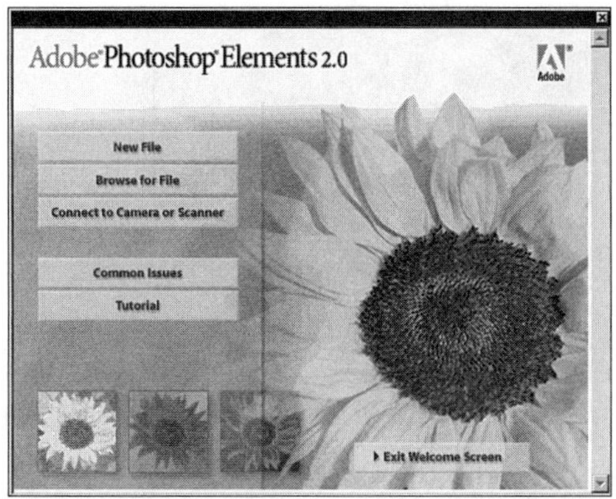

FIGURE 13.9
Photoshop Elements offers extra help not found in Adobe's professional product lines.

Fixing photos is its forte. Access its file browser by selecting File, Browse in the main menu. As shown in Figure 13.10, the file browser is an excellent way to preview photos while also viewing image data.

FIGURE 13.10
Photoshop Elements' file browser improves image access.

I chose an old photo full of imperfections. To make repairs, I selected Filter, Dust & Scratches in the main menu (see Figure 13.11).

I also used the Clone Stamp, which is accessible in the toolbox and shown in Figure 13.12. It grabs image data from a selected area and then lets you apply it over a flawed region in the photo to repair it.

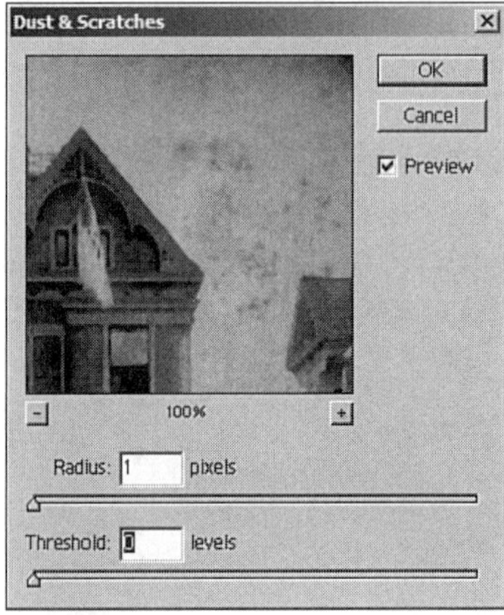

FIGURE 13.11
Photo touch-up is a powerful feature.

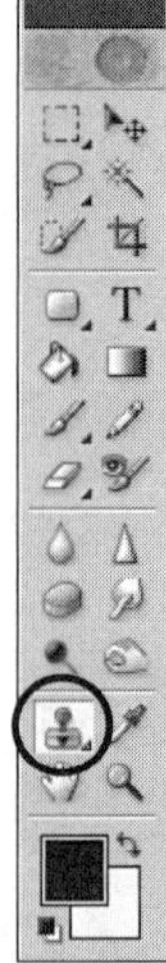

FIGURE 13.12
The Clone Stamp lets you fix photos using other portions of the same image.

Other touch-up tools include

- **Sponge**—This tool subtly changes the color saturation.
- **Focus**—This tool sharpens or blurs an image.
- **Dodge and Burn**—These darkroom-style tools change the exposure in specific areas of an image.
- **Red Eye Brush**—This tool repairs that vampire glow caused by camera flashes reflecting off the back surface of your subjects' eyes.

- **Color and Contrast.**
- **Straighten and Crop.**

Photoshop Elements has several drawing tools, which you can use to create buttons for your DVD menus. As shown in Figure 13.13, I used the rounded rectangle tool, applied color, applied a bevel from the Style drop-down menu, and added text. The options are endless and readily accessible. You can create multiple identically shaped buttons in layers.

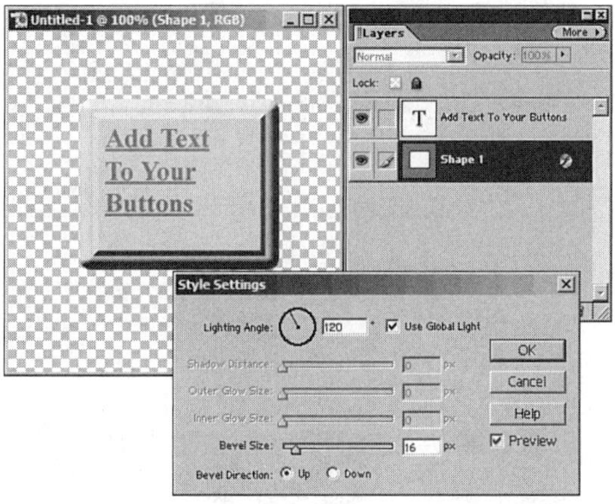

FIGURE 13.13
Photoshop Elements provides several tools to create DVD menu buttons.

> One characteristic of DVD menu buttons is how their appearances change when you use the remote control to highlight them or click them. Most DVD authoring products have built-in state changes for buttons. Creating identically shaped buttons in layers in Photoshop Elements lets you determine how a button's appearance will change depending on the viewer's actions.

One other handy tool is the video frame importer shown in Figure 13.14. You can use it to capture an image from your video.

After saving the freeze frame, use Photoshop Element's tools to change its characteristics: alter the color; replace or remove features; or apply any of the many filters, some of which are shown in Figure 13.15. After you've altered it, the frame can make an excellent-looking DVD menu or background for a title.

FIGURE 13.14
Use the video frame importer to grab freeze frames that you can alter and use as DVD menus or backgrounds for titles in your video.

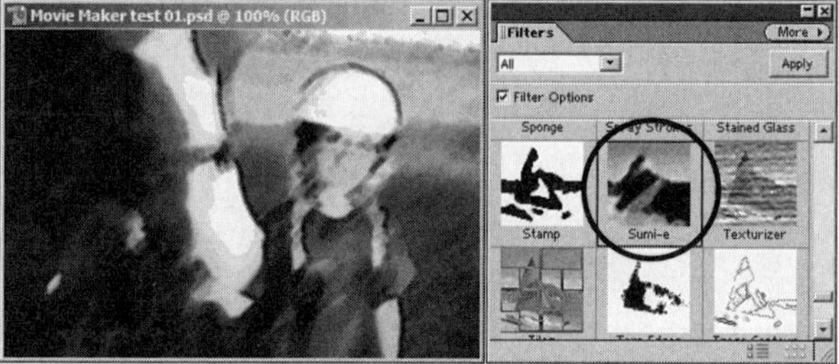

FIGURE 13.15
Elements offers dozens of filters and effects that you easily can apply to images, graphics, or text. In this case, I used Sumi-e to create an abstract oil painting look.

Adding Narrations, Music, and Sound Effects to Your Videos

Most professional video editors build a video in chunks. They might begin by laying down part of the narration or music, covering it with video to match the announcer's words or the music's beat, adding a snippet of video with audio, and then laying down another narration segment. That's not the approach used by most entry-level editors, however. Typically, adding audio—a narration or music—is the *last* step.

Studio 8's designers created this NLE with that fact in mind, which is why they included a built-in real-time narration tool (something nearly unheard of in professional-level NLEs). As mentioned in Hour 12, "Editing Video," this is why they have no qualms about Studio 8's transition process that shortens the overall length of the video with each added transition. They fully expect you will edit your video and then use the built-in narration tool to voice your piece as you watch it play.

Voicing a Narration

Depending on your audience, you might want to write a script or simply improvise. In either case, you should view your edited project and take some notes before building your voice-over.

Task: Use Studio 8's Voice-over Tool to Narrate Your Video

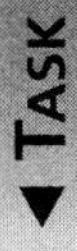

Studio 8 makes recording your voice directly onto your edited video surprisingly easy. Here's how you do it:

1. Open Studio 8 to an edited project by selecting File, Open Project and then locating an edited video and clicking Open.
2. As shown in Figure 13.16, click the Audio Toolbox icon in the upper-left corner of the storyboard or timeline (whichever you're using).

After you start adding audio to your projects, you should forego the storyboard and use the timeline exclusively. The storyboard simply does not give you the access you'll need to the various audio tracks.

FIGURE 13.16
Click the Audio Toolbox icon to access the Voice-over tool.

3. In the Audio toolbox, click the Microphone icon, shown in Figure 13.17, to open the Voice-over tool. Speak into your PC's microphone and check your Volume Unit (VU) level on the right.

Try to keep the VU meter in the yellow zone. Recording in the red leads to *over modulation*, a rough, scratchy, muffled voice-over. Also, avoid popping your *Ps* by moving the mic a little off to one side, away from the puffs of air you make when you say *P* words.

4. Move your video to its beginning by clicking the Go to Beginning (Home) button on the Player Monitor TV screen (refer to Figure 13.17). Or, you can move the slider to the location in the video where you want to start your narration.

13

FIGURE 13.17
The Voice-over tool lets you create real-time narrations.

You can record narration snippets by starting and stopping in one location and then moving to another spot in your video and recording a narration there.

5. When you're ready to record, click the Record button; watch the countdown, shown in Figure 13.18; and begin your narration when the Record light comes on. When completed, click Stop.

FIGURE 13.18
Watch the countdown to begin recording your narration.

6. Listen to your recording by playing the video from where you started your narration.

If you don't like what you hear, you can delete your narration segment and start over. However, the only way to do this is in the timeline. As shown in Figure 13.19, your narration snippets appear in the sound-effect/voice-over track. To delete a narration segment, click it to select it and press Delete.

You can slide your voice-over segment to another timeline location. Roll your cursor over the narration clip and, when the cursor changes to a hand, click and drag your clip. Note the cursor also turns into a blue speaker that lets you grab the blue volume rubberband and drag the clip volume up or down. I'll cover more on that later in the task titled "Using the Blue Audio Rubberband."

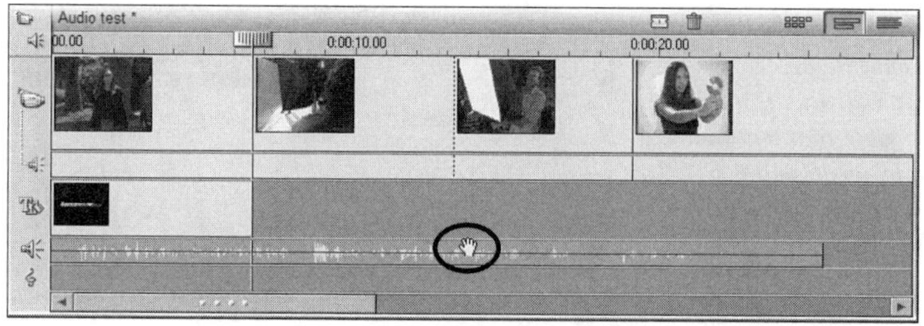

Figure 13.19
The only way to delete or move your narration segments is in the timeline.

Making Music

Studio 8 gives you two ways to make music: You can rip a tune directly from an audio CD or use SmartSound QuickTracks. SmartSound is covered in Hour 8, "Acquiring Audio," so I'll touch on it only briefly here.

Task: Rip Tracks from CDs

Using Studio 8'sCD Audio tool, you can rip music from a CD and add it to your video. Here's how:

1. Open the Audio toolbox by clicking its icon in the upper-left corner of the timeline.
2. Open the CD Audio tool by clicking the CD icon, shown in Figure 13.20.

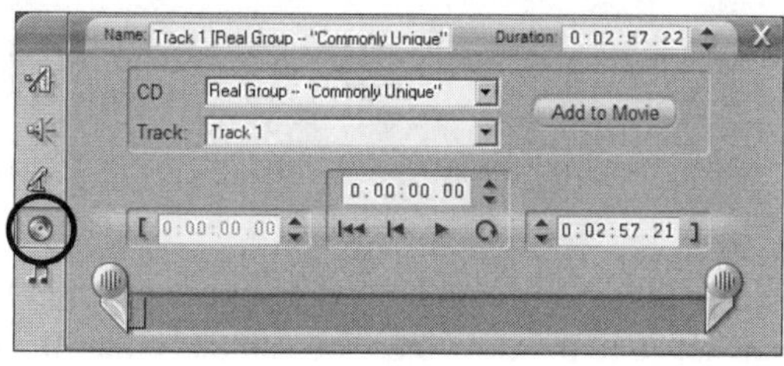

Figure 13.20
Use the CD Audio tool to rip songs from CDs.

3. Place a music CD in your DVD drive. Studio 8 might ask you for the CD's name, or it might find it automatically.
4. Select the track you want to use, and click Add to Movie.

The music selection appears in the background music track (the one with the treble clef icon). You can drag it to other locations in your piece and trim it by dragging in the ends.

SmartSound QuickTracks

Access this module using the Audio toolbox—click the music note icon shown in Figure 13.21. The trial version of Studio 8 has only three songs, versus the 25 in the retail version.

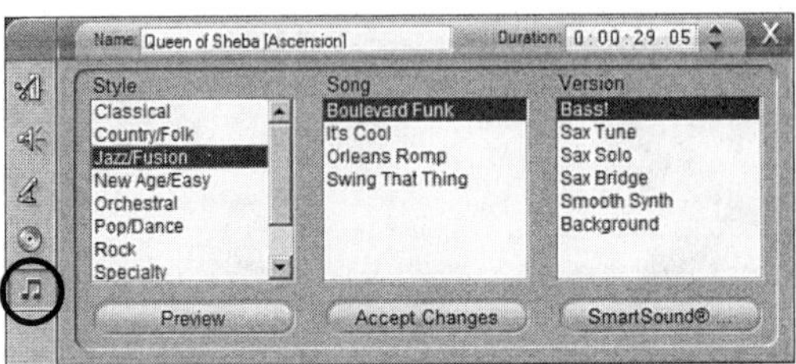

FIGURE 13.21
SmartSound QuickTracks is a fun module that lets you add music that fits your video project's mood and exact length.

Select a tune, preview it, and (if you like it) click Add to Movie. It appears in the timeline's music track. Drag the start/end to the length and position you want, and SmartSound creates a complete tune that begins and ends right on cue.

Adding Sound Effects

Switch to the Sound Effects page, shown in Figure 13.22, by closing the Audio toolbox and clicking the Speaker icon in the album. Studio 8 includes dozens of sound effects. The default group is Animal sounds, but you can access 12 other collections by clicking the file folder icon shown in Figure 13.22.

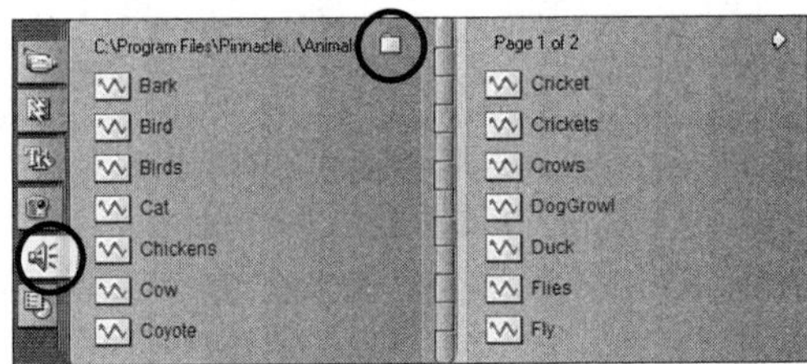

FIGURE 13.22
Studio 8 offers about 200 sound effects.

Select a sound effect, and it plays automatically. To use one, click and then drag and drop it onto an open space in the timeline. You can use either the sound-effect/voice-over track or the background music track. However, you cannot use the main audio track (the one associated with the original video).

Having to choose from only two audio tracks, both of which might have sound files already on them (narration and music) shows the limitations of this and all other entry- to mid-level NLEs. Professional-grade NLEs, such as Adobe Premiere, offer dozens of audio and video tracks. I discuss some of Premiere's features in Hour 14, "Applying Professional Video Editing Techniques with Adobe Premiere."

Using Studio 8's Audio Mixer

Your video might have audio from the original raw video, including a narration, music, and sound effects. You can bring order to this chaos with Studio 8's Audio Mixer.

Task: Use the Audio Mixer to Blend Audio Tracks

▼ TASK

The Audio Mixer lets you adjust each track's sound level in real-time, creating a mix that's pleasing to the ear. Here's how it works:

1. Start a new project by selecting File, New Project. Add several clips to the timeline, as well as music and an informal narration.
2. Open the Audio Mixer by clicking the Audio toolbox icon in the upper-left corner of the timeline and then clicking the Speaker icon shown in Figure 13.23.

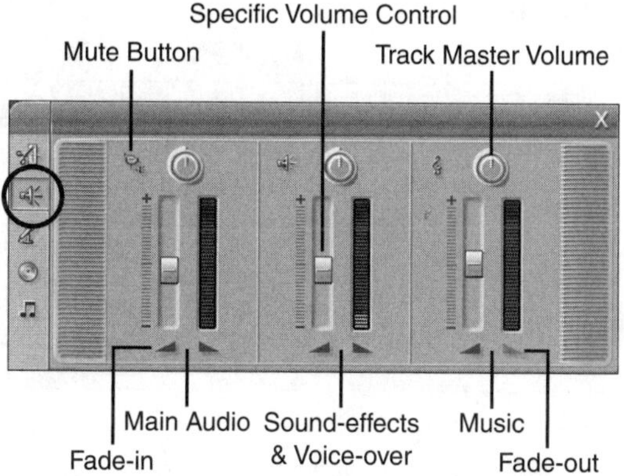

FIGURE 13.23
Use the Audio Mixer to adjust the three audio track volume levels. Each of the three audio tracks has five controls: master volume, specific volume, fade-in, fade-out, and mute.

3. Drag the Timescale slider to the beginning of the video (or press the Home key). Click the Fade-in button (the left triangle) for the main audio track. Do this for the music and narration tracks as well. Next, play your video by clicking the Go to Beginning (Home) button and then clicking Play. Note that the volume increases gradually as the video begins.

When using the Fade-in or Fade-out tool, where that volume change takes place depends on the location of the slider when you click the fade buttons. To fade out at the end, move the slider to the end of the video and then click Fade-out.

▼

4. Experiment with the real-time volume adjustments by playing the video and moving the sliders. Isolate a track by clicking the mute buttons for the two other tracks.

Studio 8's Audio Mixer is *almost* real-time. As you make volume changes or click the mute buttons, the sound levels do not change immediately. It typically takes a second or so for the changes to take effect.

Using the Blue Audio Rubberband

As shown in Figure 13.24, as you move the knobs and sliders or click the Fade-in or Fade–out button, the blue audio rubberbands in each audio track change to match your actions. You can fine-tune your mixer adjustments by clicking and dragging the blue rubberbands to new positions.

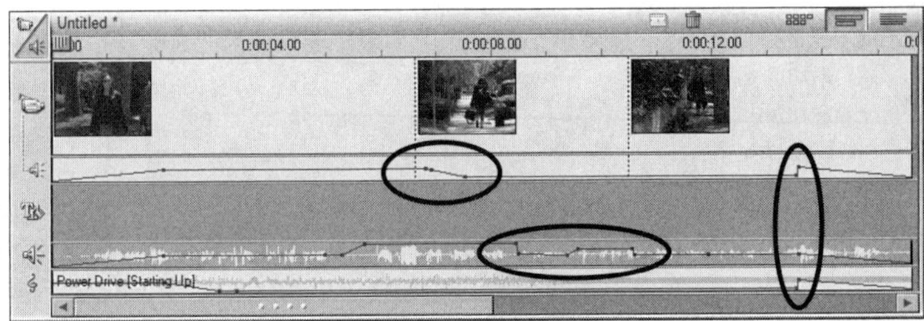

FIGURE 13.24
Blue audio rubberbands show volume level changes for each track.

Select a video clip or audio track. As shown in Figure 13.25, when you roll your cursor over a rubberband in that clip or track, it turns into a dark blue speaker icon. If you roll it over a blue dot—a point where a specific volume change takes place—it turns light blue. Click the rubberband or a volume change dot and drag the volume to a new level.

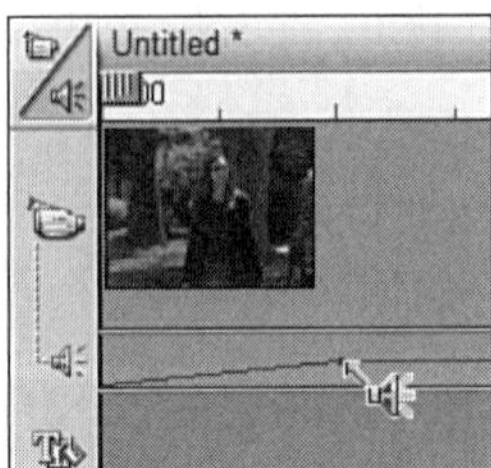

FIGURE 13.25
Use the blue volume level rubberbands to fine-tune your audio adjustments.

Summary

As you gain experience editing videos, you probably will want to move beyond simply creating a sequence of video clips. Adding text or titles to your videos reinforces a message you are trying to get across or reminds viewers of who is on camera or where the video took place. Studio 8 has the best title editor of all the NLEs in its price range.

Still images grabbed from your video or from other sources make great title backgrounds or DVD menus. Use Studio 8 to capture and add them directly to your piece. Or, use Photoshop Elements to touch them up or add special effects before using them in your video or DVD.

As you create your video, you'll add a voice-over, lay in some music, and use sound effects. It's all too easy for those audio tracks to step on each other, so use a mixer to blend them harmoniously and ease in or ease out that audio using fade-up or fade-down tools.

Workshop

Review the questions and answers in this section to reinforce your video, audio, and still image editing and acquisition skills. Also, take a few moments to take the short quiz and perform the exercises.

Q&A

Q I used Studio 8 to create a title over a graphic background, but now I want to use that title over the original video with no background. Is there a way to do this?

A Yes. Simply double-click the selected title in the timeline title track to open the Title Editor. Then, click the Transparency button in the Backgrounds section of the interface. Whether you've saved a title to disk or simply dropped it in the timeline, you can always change its characteristics.

Q I want to add three audio tracks—narration, music, and sound effects—to my video. How do I do that with Studio 8?

A Basically, you can't. Studio 8 has three audio tracks: one for the audio track on the original video and two others for whatever type of audio you want to put on them. You can, however, stop narrating at places where you want to add sound effects and insert that sound on the narration track.

Quiz

1. You've used Studio 8's Title Editor to create an ellipse over which you've added some text. But you want to let the video beneath the ellipse show through. How do you do that?
2. You've added a custom SmartSound QuickTracks tune to your project, but you've decided to remove a couple of video clips and now your video is 30 seconds shorter. How do you shorten the music without cutting off the ending?
3. You used the audio mixer to blend the three audio tracks, but there are still some places where the audio mix just is not optimal. What's another way to fix this?

Quiz Answers

1. When you use the Color Selector window to choose a color for the ellipse, move the opacity slider to something less than 100%—about 50% works well.
2. It's easy: Simply drag the end of the music clip to the left until it lines up with the new end of your video. SmartSound re-creates the music so it has a clean finish at the new out-point. The only caveat is that it might not be the same ending as before. The SmartSound software adjusts the finish to feel right depending on the overall length and the music phrase that precedes the ending.
3. Use the blue audio rubberbands associated with each audio track. Sometimes you can see spikes that are hard to catch with the mixer. Other times, if you hear a noise that just doesn't work, you can drop the rubberband down only at that spot and then return it to the regular level after the noise passes.

Exercises

1. Create your own library of sound effects using your video camcorder. Simply save video clips with interesting audio in a separate file folder. When you want to use a sound, drag the video clip (with its associated audio) to one of Studio 8's two audio tracks. Only the sound portion will show up in the track.
2. Create a standard title look for your videos. Either make adjustments to a Studio 8 template or start from scratch with the text and object tools plus backgrounds. Then, save whatever you come up with by clicking the Favorites tab in the text Looks interface.
3. Record a narration using your camcorder's built-in mic instead of your PC's mic. See which sounds better and use that method in future productions.

HOUR 14

Applying Professional Video Editing Techniques with Adobe Premiere

Studio 8 is the best entry- to mid-level nonlinear video editor (NLE), but it does have some limitations. Depending on how far you want to go with the video editing side of your DVD production, these might be a nonissue or a source of frustration.

If it's the latter, Adobe Premiere is the solution. Premiere is a professional-level NLE with a lot of features and unlimited flexibility. I've included a trial version of Premiere on this book's companion DVD, and you can use it to follow along as I explain some of things it can do.

Premiere lets you layer, or *composite*, videos. Its high-end audio tools and dozens of audio tracks let you sweeten and mix audio to the limits of your creativity. Finally, you can change the characteristics of its special effects gradually over time.

Premiere dominates this market niche, but three contenders are flexing their muscles. I'll present an overview of that group.

The highlights of this hour include the following:

- Touring Adobe Premiere—how it's different from Studio 8
- Looking at layering, or compositing
- Trying three editing tools and two special features
- Evaluating the competition
- Stepping up to a professional video production PC

Touring Adobe Premiere—How It's Different

I want to demonstrate Premiere's power and give you some hands-on time with it, but in one hour I can scratch only the surface of Premiere's features and functionality.

To give you as much value for your time as I can muster, I'll offer only a few detailed, hands-on, step-by-step tasks. Instead, I'll rely primarily on barebones demonstrations.

I suggest you do all the tasks. As for the demos, do what you can to follow along but because I leave out some details and gloss over some steps, something might come up as you try to follow along that will stump you. Not to worry—Premiere is a powerful product that takes a lot of effort to master.

No matter how many problems arise, experimenting with Premiere will open your eyes to its potential.

I am an Adobe Certified Expert in Premiere and the author of *Sams Teach Yourself Adobe Premiere 6.5 in 24 Hours.*

Installing Premiere

Locate the Premiere file folder on this book's companion DVD, open it, and double-click the setup file. Follow the prompts and start Premiere.

The trial version provided with this book is the previous iteration of this product—Premiere 6.0. There is no, and will be no, trial version of the current 6.5 version. Versions 6.5 and 6.0 look alike, but I used 6.5 for this hour because it has some extra features, such as the Title Designer and audio sweetening tools, I want to demonstrate.

The first screen, shown in Figure 14.1, asks you whether you want A/B or single-track editing. Select Single-Track because A/B is a throwback from film editing days.

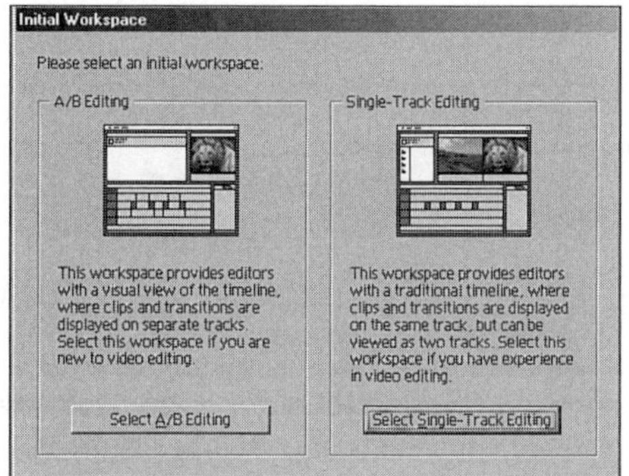

FIGURE 14.1
Your first option is A/B or single-track editing; select Single-Track.

The next screen, shown in Figure 14.2, foreshadows the sometimes mind-boggling array of options available in Premiere. In this case, you tell Premiere the nature of your raw and (ultimately) edited video. Two dozen preset choices, with uncountable options, are available by clicking the Custom button.

For now, keep it very simple. As shown in Figure 14.2, select DV—NTSC (or PAL if that's your country's standard) Standard 48KHz and then click OK.

You might note that Figure 14.2 has a listing for Real-time Preview. This is a new feature in Premiere 6.5, and it is a big deal. It means you can watch special effects and transitions, which are typically processor intensive, in real-time. In the trial version of Premiere 6.0, the only way to see these effects is either to render the effect (you essentially tell Premiere to do the calculations and save the segment as a separate clip) or use the so-called ALT-scrub method (you hold down the Alt key while manually dragging the edit line through the effect).

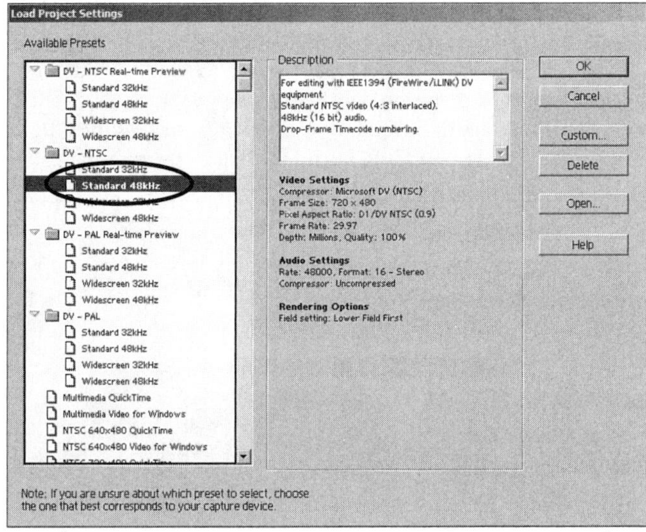

FIGURE 14.2
The first of many collections of options. In this case, keep it simple and select DV.

The 48KHz refers to the audio quality. You should select a high setting at the outset. You always can reduce the audio—and video—quality when you later record your edited project to a DVD, videotape, or other media.

Exploring Premiere's User Interface

As shown in Figure 14.3, Premiere's interface has some familiar elements, including the standard NLE timeline, a Project window for media storage, and two monitor windows (Source and Program). Two monitors let you view your unedited source videos on the left and your edited program videos on the right.

If your Premiere workspace is not as neat and tidy as the one in Figure 14.3, Premiere can fix that. From the main menu select Window, Workspace, Single-Track Editing.

As with Studio 8, I captured images for this application using 800×600 resolution. However, you can run Premiere in 1024×768 or higher if you want more real estate and flexibility. Some specialized PC video cards let you run Premiere (and other programs) in two monitors. I cover more on that later in this hour, in the section "Stepping Up to a Professional Video Production PC."

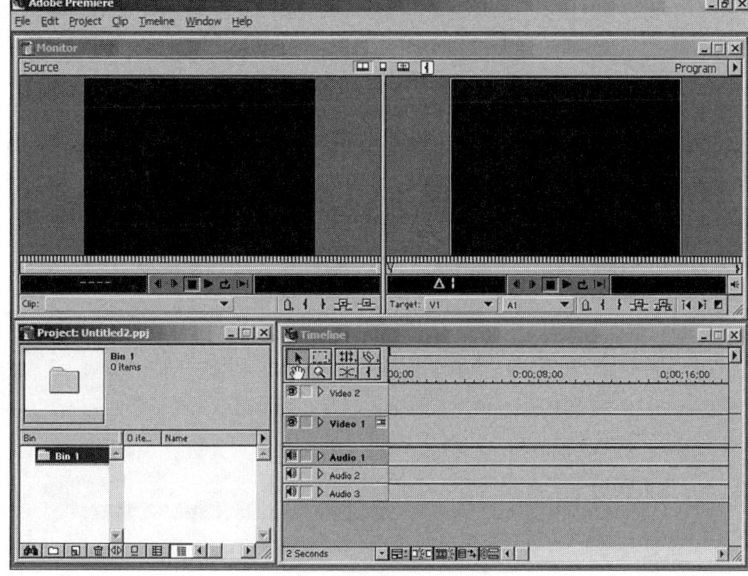

FIGURE 14.3
Premiere's interface has some standard NLE elements.

Task: Start a New Project

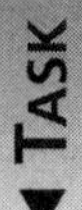

You'll need to load some media into Premiere to try some of its features. You begin by putting those media in the Project folder. Here's how:

1. Right-click in the white area in the Project window's lower-right corner. As shown in Figure 14.4, a menu opens. Select Import, File.

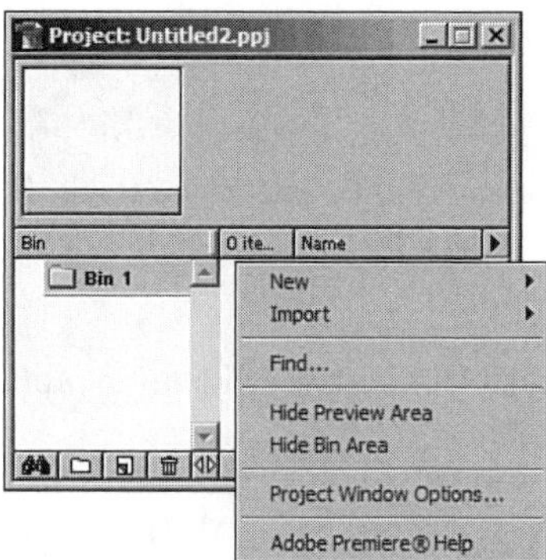

FIGURE 14.4
Use the right-click menu in the Project folder to import files to your video project.

2. Navigate to wherever you've stored your captured videos or to Premiere's sample file folder. Its default location is `C:\Program Files\Adobe\Premiere 6.5\Sample Folder`. Select all the files there, and click Open. The Project window fills with filenames.

3. Drag the right edge of the Project window to the right to see all the information Premiere tracks here (see Figure 14.5).

> One big difference between Premiere and Studio 8 is that in Premiere all media types—videos, stills, audio, and graphics—are accessible in one window instead of as pages in an album.

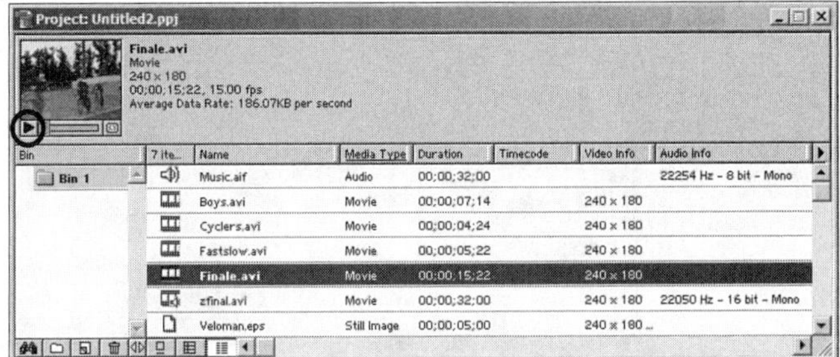

FIGURE 14.5
The Project window displays all the media types in one easily accessible location.

4. Select a video or audio clip and click the Play button below the little monitor screen shown in Figure 14.5. This lets you easily preview your clip.
5. Drag a video clip to the Video 1 track on the timeline. If you're using the Premiere sample videos, select `zfinal.avi` because it has an audio track.
6. In the timeline, click the clip you dragged to the timeline to select it, press the Home key, and click the Play button in the Program Monitor screen.

Looking at Layering

You might notice that Premiere has more than one video track. In fact, you can work with up to 99 video and 99 audio tracks. Figure 14.6 shows how the timeline looks after adding some additional tracks.

On the video side, those extra tracks let you layer video. The following sections present a few brief demos of what that entails.

Adjusting the Opacity

Figure 14.7 shows the result of blending together two clips.

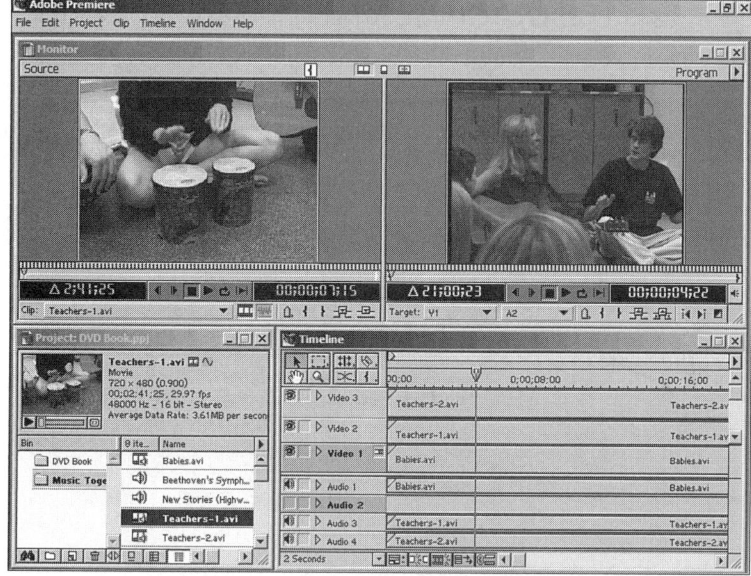

FIGURE 14.6
Multiple tracks means you can combine several video clips into one.

FIGURE 14.7
Two clips blended into one.

Task: Use Opacity Settings to Blend Clips

▼ TASK

This task eases you into several upcoming barebones demos. I've reduced the detail level a bit, and future demos will be even less detailed. Despite this, I think you still can complete this and other demos without problems. Here's how to use opacity to blend two clips into one:

1. As shown in Figure 14.8, drag one clip on the Video 1 track and another directly above the first on Video 2.

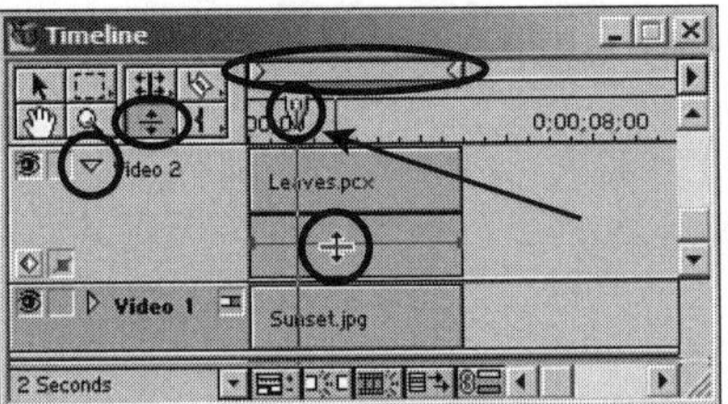

FIGURE 14.8
Place two clips on separate tracks to begin the layering process.

2. Expand the Video 2 track by clicking the little triangle next to Video 2, shown in Figure 14.8.
3. Locate the Fade Adjustment tool (the line with two arrows, shown in Figure 14.8) by clicking the timeline toolbox and selecting it from the pop-up menu.
4. Click the red button on the Video 2 track to display the red opacity rubberband (this is similar to the blue audio rubberband in Studio 8). Then roll your cursor over the rubberband, and note it turns into a horizontal line; then click and drag the rubberband down about halfway—50% opacity.
5. To view your handiwork, you need to use the manual preview method: Hold down the Alt key and click and drag the Edit Line Marker (shown with an arrow in Figure 14.8) through your clip. If you simply click the Play button in the Monitor Program window or press the spacebar, you will see only the clip on Video 1.

▲

Alt-scrubbing through your clip is less than ideal. If, instead you want to see your clip in real-time, you need to render it. To do that, drag the yellow work area bar over the segment you want to *render* and then press Enter (refer to Figure 14.8). Premiere might ask you to name and save the project and then build (render) a file of that segment and play it in the Program Monitor.

Combining Two Clips in a Split-Screen

Another way to combine two video clips is in a *split-screen*. Figure 14.9 shows one example, a screen split down the middle from top to bottom.

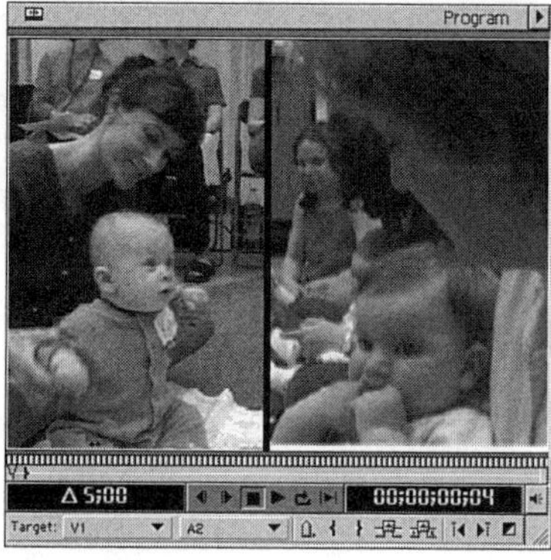

FIGURE 14.9
Split-screens are an easy way to combine two clips.

Here's how you do that:

1. Return the opacity to 100% on the Video 2 clip by dragging the red opacity rubberband back to the top.
2. Right-click the clip and select Video Options, Transparency.
3. In the Sample window in the upper-right corner of the Transparency Settings interface, use the four image adjustment points to drag either the left or right side halfway into the clip (see Figure 14.10). You'll see the clip on Video 1 show through. Click OK and view your work by pressing Enter.

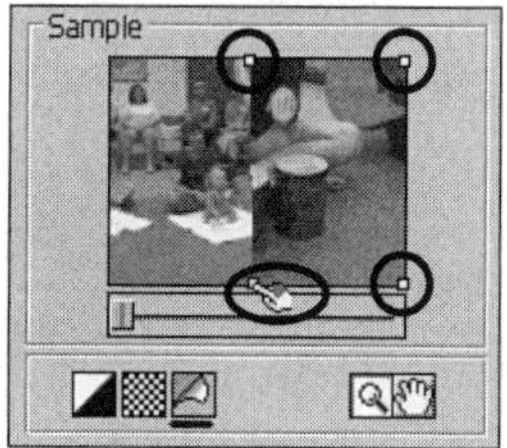

FIGURE 14.10
Create your split-screen by dragging in any of the four corners of the Sample screen window.

Keying Out Parts of a Clip

You might know that TV weather folks do not stand in front of weather maps. As shown in Figure 14.11, they stand in front of solid-color—usually lime green—walls. Technicians electronically *key out* that solid color, turning it transparent, and replace it with weather graphics.

FIGURE 14.11
Matt Zafino, the chief meteorologist at KGW-TV in Portland, Oregon, demonstrates the green screen key.

Premiere offers a green-screen key feature, as well as several other keying methods. Because you probably don't have a green-screen video clip, I'll simply show you how it works.

In the same Transparency Settings window we just used to make a split-screen, I selected 1 of the 14 available keys—in this case, I chose Chroma because it works for any solid color. I used the Eyedropper tool, shown in Figure 14.12, to tell Premiere exactly what color to key out. Then I fine-tuned that using the sliders and noted my results in the Sample window. I set the Sample window to a checkerboard background to get a clearer view of my progress (keying is an inexact science).

> You don't need a true chromakey green or blue screen to use the chroma-key effect. Any solid color backdrop will work. First, you must make sure the color does not match anything in your subject that you don't want to key out; otherwise, your subject will have transparent holes. In any event, it's not easy creating a clean chromakey because uneven lighting causes real problems. You will need to experiment to find the best solution.

FIGURE 14.12
Access keys in the Transparency window.

Putting a Picture in a Picture

Premiere offers a powerful feature called Motion Settings that has many uses. In this case, I've used it to place five video clips over another clip (see Figure 14.13).

FIGURE 14.13
Placing multiple pictures-in-a-picture is an involved process.

Combining six clips is a very involved process and is much more detailed than keying.

Here's a brief demo of how you place one clip over another (Feel free to experiment as you go through this process):

1. Using two clips, one above the other on the timeline, right-click the top clip and select Video Options, Motion. The Motion Settings window shown in Figure 14.14 opens. This figure shows several settings you'll need to change.

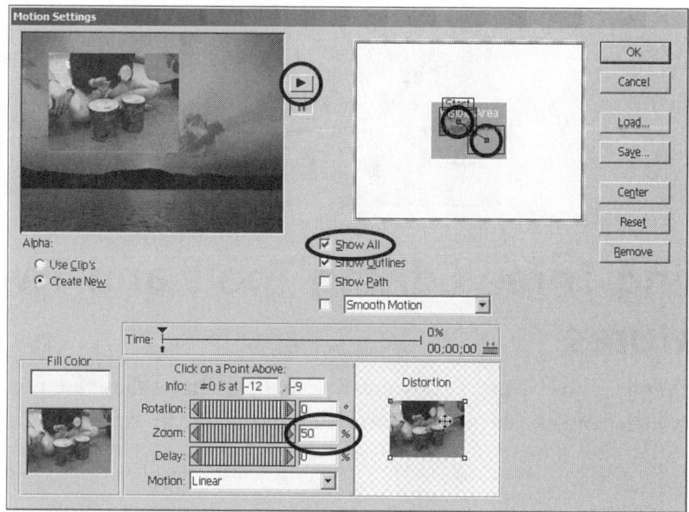

FIGURE 14.14
Use Premiere's Motion Settings interface to create a picture-in-a-picture.

2. Click Show All to see the clip on Video 1.
3. Change the Zoom setting to 50% (press Tab—rather than Enter—to input the value).
4. Move the start and end points to make your clip show up inside the clip on Video 1.
5. Change the End Point Zoom to 50%.
6. Preview your changes by clicking the Play button near the top center of the interface.

You've fixed the start and end characteristics of your clip, but you can add many other points and make your clip rotate, stretch, and change size along the way (see Figure 14.15 for an example). Making all those changes and adding the points is an even more involved process that I'll leave to you to try, if you're willing.

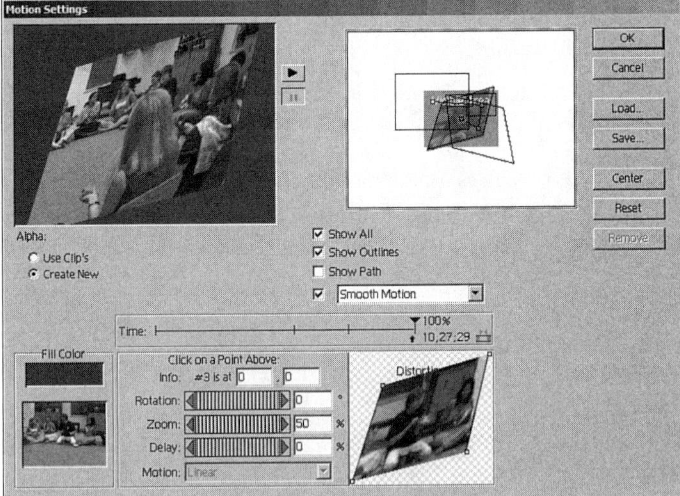

FIGURE 14.15
Use the motion settings to distort, rotate, expand, or contract your clip over time.

Trying Three Editing Tools and Two Special Features

Premiere gives editors extra creative tools. Some of these are remarkably easy to use, whereas others take some effort. All, however, reap exciting rewards.

Playing Clips Backward

Studio 8 lets you slow down or speed up clips. Premiere does all that, and it lets you do it in reverse as well.

Right-click any clip in the timeline and select Speed. As shown in Figure 14.16, your two choices are New Rate and New Duration:

- **New Duration**—Lets you take a clip and change its length without changing its content. That is, you can slow it down or speed it up to make it fit a certain timeframe.
- **New Rate**—Lets you change the clip's playback speed using a percentage. Selecting a negative number, as I did in Figure 14.16, makes the clip run backward. Select a negative number, click OK, and Alt-scrub through your clip (or render it) to see that in action.

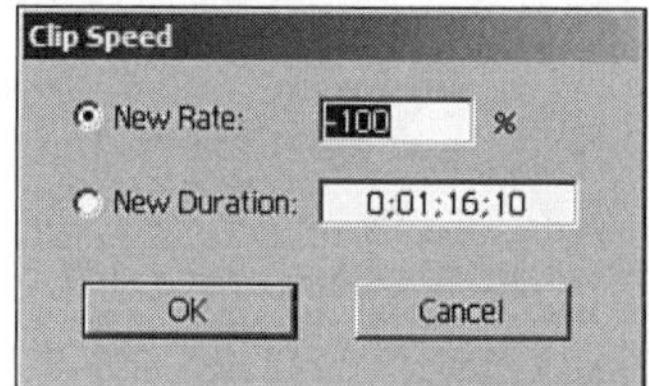

FIGURE 14.16
Clip Speed lets you play a clip in reverse.

News-Style Editing

TV news video editors frequently use techniques called J- and L-cuts. *J-cuts*, sometimes called *audio lead cuts*, quietly play the audio of the next clip under the video and audio of the previous one for a short time, as a means to let their audience know that someone is about to say something or that a transition is coming. *L-cuts*, on the other hand, let the audio trail off under the next video clip.

Here's a brief explanation of how a J-cut works:

1. Place two clips one *after* the other in the Video 1 track.
2. Select the Link/Unlink tool (see Figure 14.17). Click both clips' audio and video tracks to unlink them. This lets you drag their video or audio track end points independently.
3. Drag the audio for one of the clips down to a different audio track.
4. Drag the ends of both audio clips to overlap them.
5. As shown in Figure 14.18, select the Cross-fade tool from the same toolbox location as the Link/Unlink tool. Then, click it on both audio clips. I expanded the audio tracks, by clicking the triangle next to the word Audio, to show you how the red rubberbands display an audio cross fade.

14

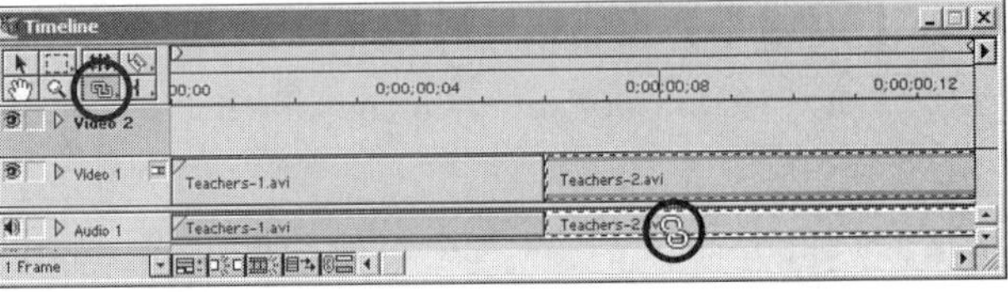

FIGURE 14.17
Use the Link/Unlink tool to disconnect each clip's audio and video tracks.

FIGURE 14.18
Expanding the audio tracks lets you see the audio cross fade in action.

When you use a J- or L-cut, you should add a video cross dissolve to further ease the transition between the two clips (refer to Figure 14.18).

Changing Video Effects Over Time

Studio 8 offers only four video effects—Blur, Emboss, Mosaic, or Posterize—whereas Premiere has more than 90.

In Studio 8, when you apply a video effect, it changes the clip equally for its entire duration. Premiere, however, lets you change the effect over time—a very useful and creative option.

Figure 14.19 shows this process in action. The first frame has no effect applied, the next has a slight change, and the third has a more obvious change.

For in-depth coverage of Premiere, refer to *Sams Teach Yourself Adobe Premiere 6.5 in 24 Hours.*

This is a fairly involved and tricky process, so I'll give you just a barebones demo:

1. From the main menu, select Window, Show Video Effects. A small interface pops up with three tabs: Video, Audio, and History. Select the Video tab.

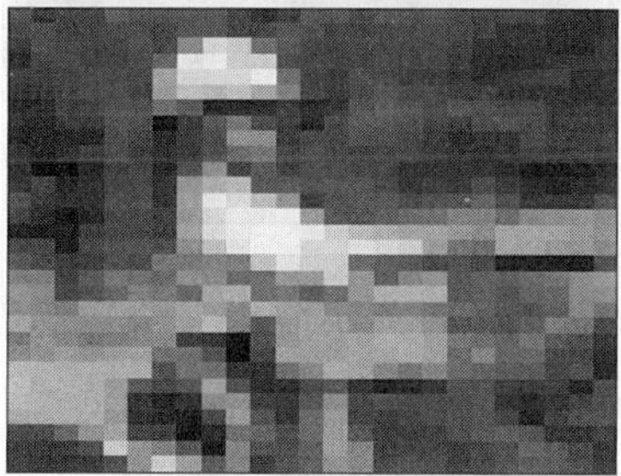

FIGURE 14.19
Changing a video effect over time is an effective creative technique.

2. As shown in Figure 14.20, clicking the small triangle in the upper-right corner, selecting Expand All Folders, and dragging a corner of this window to expand it gives you an eyeful of Premiere's video effects possibilities.

You might note that 79 effects are available—not 90+ as I stated earlier. However, double-clicking the QuickTime Effects icon reveals 15 additional effects.

3. You can use virtually any video effect (see the following caution). For this demo, select Mosaic from the Stylize folder (refer to Figure 14.20). Drag Mosaic to a clip, which opens the Effect Controls palette shown in Figure 14.21.

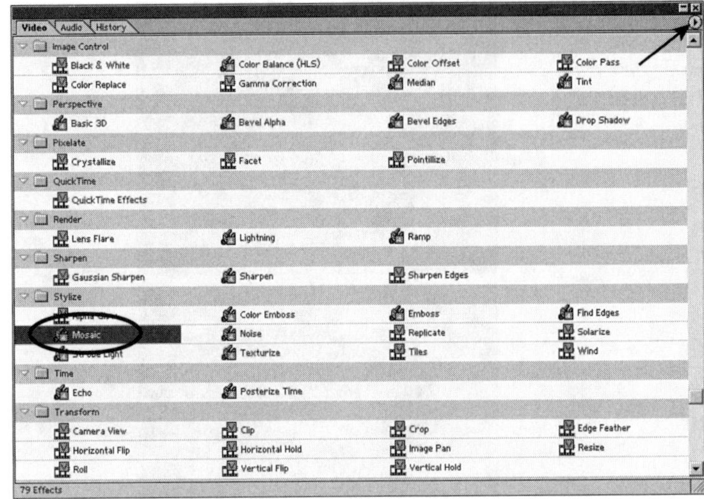

FIGURE 14.20
Premiere's many video effects.

Some video effects, including all QuickTime effects, don't have the capability to change over time. For instance, some are either on or off, and still others have numerous or arcane controls that can stop this process.

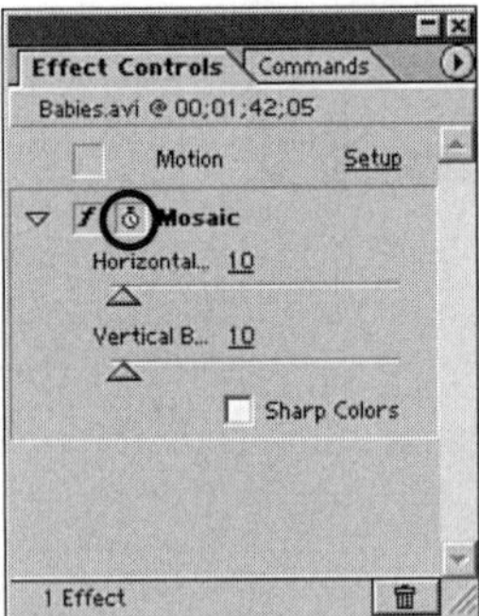

FIGURE 14.21
The Effect Controls palette lets you fine-tune most video effects.

4. I'll gloss over the next steps because they get a little tricky. To change effects over time, you need to turn on keyframes. As shown in Figure 14.21, you do that by clicking the empty box in the Effect Controls palette next to Mosaic, which causes a stopwatch to pop up.
5. You can move your edit control line to various locations in the selected video clip and change the Mosaic settings at each spot. This creates keyframes at each spot. Higher values create smaller mosaic pieces, whereas lower values create larger, chunkier blocks.

6. Alt-scrub through your clip (or render it), and watch as the mosaic pattern shifts throughout the duration of the clip.

Audio Editing

With Premiere's 99 audio tracks, your audio mixing opportunities are nearly endless. As with Studio 8, Premiere uses an audio mixer. Access it through the main menu by selecting Window, Audio Mixer.

As shown in Figure 14.22, you can display controls for every audio track.

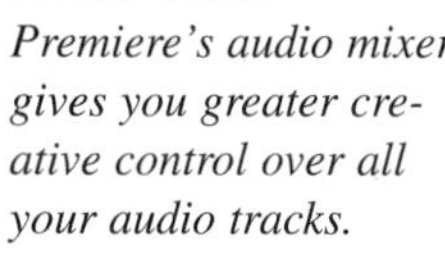

FIGURE 14.22
Premiere's audio mixer gives you greater creative control over all your audio tracks.

What Studio 8 and most others in that market niche do not offer is audio effects—tools that sweeten or dramatically alter sounds. Premiere 6.5, the update to the trial version you have, includes audio enhancing tools from a respected German firm: TC|Works. Figure 14.23 shows one of the three, full-featured products—the Equalizer.

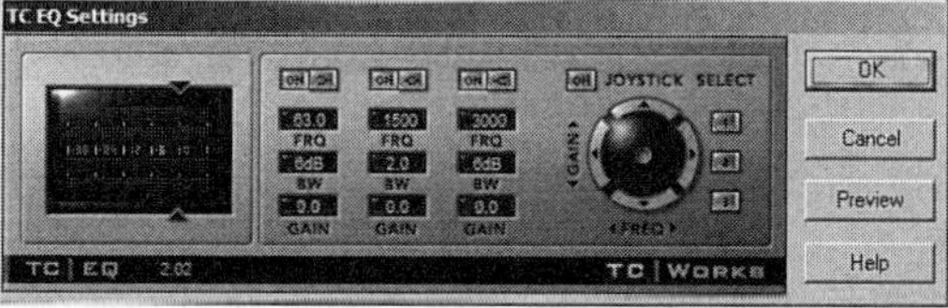

FIGURE 14.23
TC|Works' Audio Equalizer can add some extra presence to your project's sound.

If you have a high-end stereo, you've seen equalizers. They change the volume level in narrow frequency bands, adding more bass oomph or cutting the treble range. Other TC|Works tools let you remove specific frequency ranges to reduce hum, bring down ambient noise while increasing the volume of quiet passages, and add reverb to create the feeling of being in a giant cathedral.

Adding Text with the Adobe Title Designer

This is another new feature in Premiere 6.5. The Adobe Title Designer, shown in Figure 14.24, replaces Pinnacle's TitleDeko that Adobe bundled with previous versions of Premiere, including the trial version on this book's companion DVD.

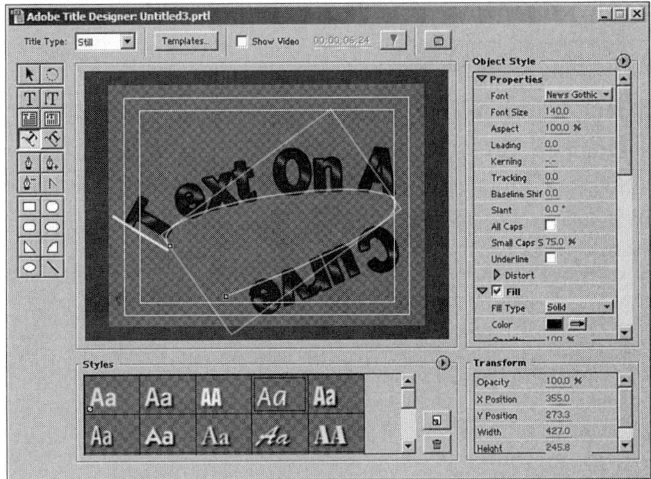

FIGURE 14.24
The Adobe Title Designer even lets you twist text along contorted curves.

With this new module, you can do just about anything you can think of doing with text. Several preset templates are available, as well as customizing tools that let you fine-tune your text to the *n*th degree.

Evaluating the Competition

Premiere is the 400 lb. gorilla in the professional NLE market for PCs. But three contenders give it a run for its money on several fronts: Pinnacle Edition DV, Sonic Foundry Vegas Video, and Ulead MediaStudio Pro 6.5. The following sections give you brief overviews of each one.

Pinnacle Edition DV

Edition DV started its life as FASTStudio DV, a high-end, engineering-intensive NLE from Fast Multimedia, a German electronics company. Pinnacle bought them out in 2001 and has now repackaged an essentially unchanged FASTStudio DV as Edition DV.

At $700, Edition DV takes the high road in price and user skills. Although its nonintuitive interface takes some getting used to, after you figure out its right-click, context-sensitive menus, you begin to see the power under the hood (see Figure 14.25).

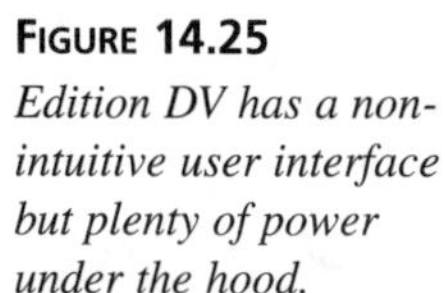

FIGURE 14.25
Edition DV has a non-intuitive user interface but plenty of power under the hood.

But I think it's far too difficult to learn and has too many minor inconveniences to recommend. Each time you add video to your project, Edition creates a separate audio stream—a time-consuming process. It also renders video effects in the background, noticeably slowing performance. If I tried to do too much too fast, it crashed.

Edition DV does comes with an excellent graphics program—Pinnacle's industry-leading TitleDeko—and prosumer-level DVD authoring software, but overall Edition DV is too much in need of a serious overhaul to recommend.

Sonic Foundry Vegas Video

This is the only one of these three professional NLEs that gives Premiere a real run for its money.

Its easy-to-navigate interface, shown in Figure 14.26, uses color codes and icons to help you make sense of its myriad functions. It offers contextual menus that are much more user friendly than Edition DV's. In addition, its audio effects and toolset, including an excellent noise reduction plug-in, are unsurpassed in this field.

It boasts several technical features that boost performance, including a proprietary DV codec that's higher quality and gives smoother playback than Microsoft's and a velocity envelope that lets you change playback speeds in several locations on the same clip. Plus, it takes full advantage of dual-processor PCs and works smoothly on a wide range of PC platforms.

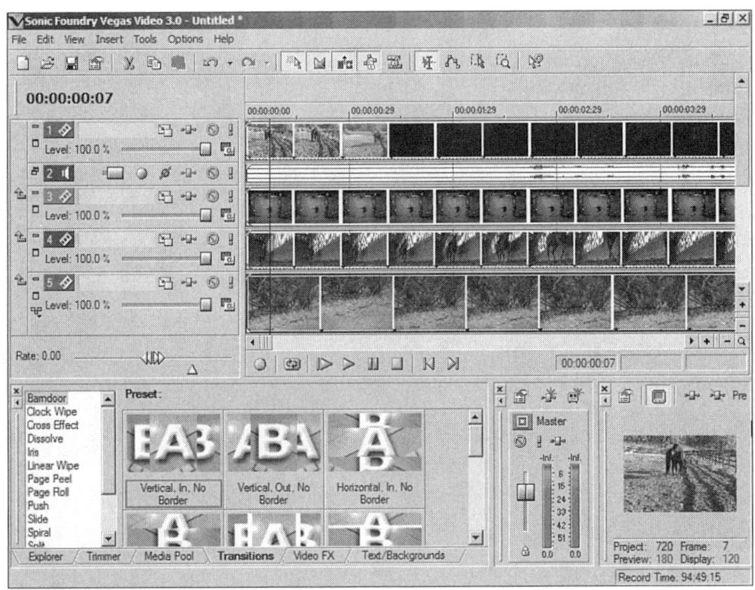

FIGURE 14.26
Vegas Video is Premiere's strongest competitor.

At $480 ($420 for the downloaded version), this is worth checking out. Sonic Foundry offers the full version on a 1-month trial basis at `http://www.sonicfoundry.com`.

Ulead MediaStudio Pro 6.5

This NLE is a hodgepodge of loosely connected modules, all of which are at least reasonably high quality. However, there are too many issues with these various elements for me to recommend MediaStudio Pro.

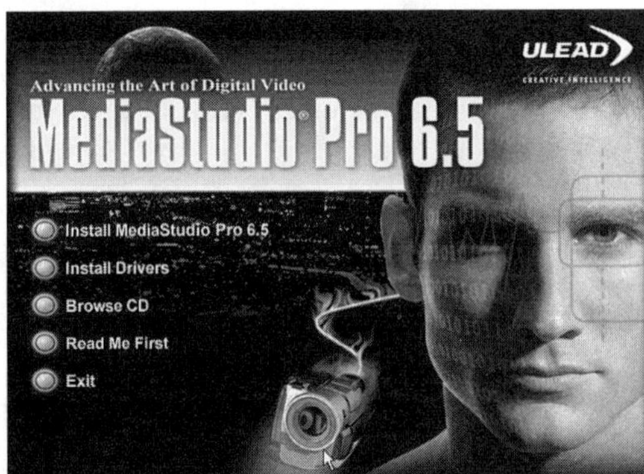

Those modules include Video Paint, which is sort of a Photoshop-Lite, and CG Infinity, which is a graphics animation tool similar to Commotion that's bundled with Pinnacle Edition DV. But they're not integrated well. Creating animated text requires jumping from one module to another and another (see Figure 14.27).

As with Pinnacle Edition DV, MediaStudio Pro renders as you work, slowing things down too much. Finally, the timeline, shown in Figure 14.27, is an odd mix of old A/B style editing and more current single-track. It lists for $500.

FIGURE 14.27
Too much jumping around amongst modules undermines MediaStudio Pro.

Stepping Up to a Professional Video Production PC

If you plan to go beyond basic video editing, you should consider buying a high-end video production PC.

Start your quest by visiting `www.Alienware.com`. This high-end PC maker has an excellent reputation for building super-fast PCs and offering informed, responsive technical support.

Although its Web site likely will look different from the way it did when I wrote this chapter, the home page should have a link to its selection of DV systems, as shown in Figure 14.28.

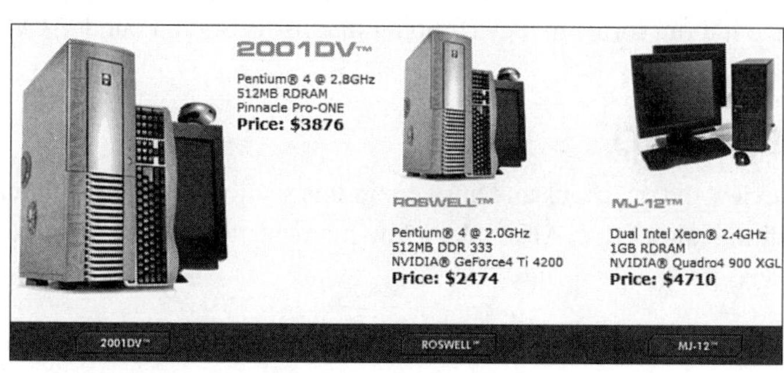

FIGURE 14.28
Alienware has a solid reputation for quality and service.

14

Take a look at the DV models. Alienware will likely have all the features you want included in your dream DV PC, such as

1. One or two fast processors
2. Two 7200rpm (minimum) hard drives totaling at least 100GB
3. 1GB RAM
4. High-performance DVD-RW drive, such as the Pioneer DVR-A05
5. High-end NVIDIA or ATI dual-monitor video card
6. Pinnacle, Matrox, or Canopus high-end video production card
7. Windows XP Professional

The high-end video cards from Pinnacle, Matrox, and Canopus might be new to you. These companies build cards specifically for video editing, and the cards come with video capture ports and high-speed graphics chips that enable real-time display of multiple special effects.

They frequently come bundled with a full version of the latest iteration of Premiere.

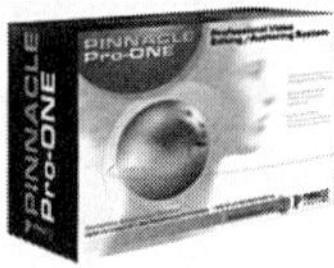

While writing this book, I tried Pinnacle's Pro-ONE card. It's a powerful tool that has its own set of incredible special effects and transitions, and I heartily recommend it. For more information visit `http://www.pinnaclesys.com/`.

Summary

You can take Studio 8 only so far. If you want to raise your video editing bar, you need to step up to a professional-quality editor. Premiere is my top pick.

Its many options mean your video projects are limited only by your creativity, time, and energy. It lets you composite videos, change video and audio effects over time, and create text that rivals the best TV broadcast-quality titlers.

Premiere sits at the pinnacle of the professional NLE market niche, but three contenders give it a run for its money. Topping that list is Sonic Foundry's Vegas Video.

Workshop

Review the questions and answers in this section to reinforce your professional video editing knowledge. Also, take a few moments to take the short quiz and perform the exercises.

Q&A

Q I apply an effect to a clip, but when I press the spacebar to view it, nothing happens. The clip remains in its original state. What's going on?

A Your trial version of Premiere cannot do real-time playback of video effects or transitions. You can either use the manual Alt-scrub method or render the segment by dragging the yellow work area bar over that segment and pressing Enter.

Q When I put video clips on more than one track, I see only the clip in the highest numbered track. Why can't I see the other clips?

A Putting a clip on a higher track is like placing one painting on top of another—whichever painting is on the top of the pile is what you see. The only way to change that is to create some type of transparency in the top clip—using a key or opacity—or change the size of a clip on the higher-numbered track using motion settings or a split-screen.

One minor anomaly exists: If you change the opacity of a clip on Video 2, for instance, even if it's 99% opaque, when you press the spacebar you will see the clip below it as if the clip on Video 2 were completely transparent.

Quiz

1. You want to put your logo over a video clip. How would you use Premiere to do that?
2. How would you slide a small video clip over a different full-screen clip?
3. When would you use keyframes, and how do you start that process?

Quiz Answers

1. As is the case with almost anything you do in Premiere, there are several ways. One way is to place the logo graphic on a track above your clip, open the transparency settings, and use the Chroma Key to take out the logo's background color. If you created your logo in Photoshop, it might already have a transparent background. In that case, you'd open the transparency settings and select Alpha Channel.
2. Use motion settings. Place the clip you intend to shrink and slide in a track above the background clip; then, right-click and select Video Options, Motion to open the motion settings. Click Show All, and then change the start and end point Zoom settings to 30%. Finally, move the start and end points to the desired locations over the clip on Video 1.

3. Keyframes let you change the nature of an effect over time. You can play a clip for a while and then use keyframes and a selected video effect to start changing it. You switch on keyframes in the Effect Controls palette by dragging an effect to a clip. Then, in the main menu select Window, Show Effect Controls and click the empty box next to the effect name to enable keyframing.

Exercises

1. Check out Premiere's full suite of 76 transitions (from the main menu, select Window, Show Transitions). As with video effects, you can open all the file folders and expand the view. In addition, you can select Animate in the pop-up menu to see how they work. As with Studio 8, you simply drag the transition of your choice between two clips and double-click the transition to view the Settings window.
2. Use Premiere's opacity feature on several sets of two clips. Note the opacity settings and combinations of clips that work best. Usually, you need one clip with little detail—for example, a sunset, brick wall, or notepad—or with an area of minimal detail. In these cases, the other clip shows through with more clarity.
3. Experiment with Premiere's motion settings (right-click a clip and select Video Options, Motion). This is a very useful feature and a lot of fun. Rotate, distort, and fly images all over the screen.

HOUR 15

Scanning and Formatting Images for DVD Authoring

This is the final production step before actually tackling DVD authoring. Preparing your still images and graphics for DVDs now will save some minor headaches later.

You might not have given more than a passing thought to how you scan images. But using some basic scanning tips will simplify and streamline that process, give your scanned images a consistent look, and avoid last-minute fixes when authoring your DVDs.

Your photos and graphics—scanned or otherwise—might need some cropping and resolution fixing before importing them into a DVD project. I'll go over how to do that using one of my favorite image grabbing and editing software titles: HyperSnap-DX.

The highlights of this hour include the following:

- Explaining scanner settings
- Scanning images for DVD projects—automated and manual
- Formatting images for DVD projects

Explaining Scanner Settings

The goals for this hour are to help you establish some standard scanner practices and show you how to crop, resize, and set proper DVD/video production resolutions for your images using graphics editing software.

You need to do this because the Sonic Solutions DVD authoring software I cover later in this book expects your menu backgrounds and still images to be a certain resolution or size. If your images fail to match those specifications, you might end up with squashed, stretched, or fuzzy images in your DVD project.

DVD authoring applications from other companies can have different graphic size, aspect ration, or resolution requirements. Because Sonic Solutions' products are the focus of this book, I've used Sonic Solutions' requirements as the foundation for the screen sizes and resolutions presented in this hour. If you use a DVD authoring application from another company, check its documentation to see how it handles still images and menu backgrounds.

Exactly what those resolutions are depends on your country's TV standard (NTSC or PAL), whether you're working in full-screen (4:3 aspect ratio) or wide-screen (letterbox 16:9), and whether you want to compensate for overscan (see the following note).

Overscan is an NTSC issue (PAL and the other TV standard, SECAM, do not overscan). Most consumer NTSC TV sets enlarge the TV signal, pushing it beyond the edges of the TV tube. That means you frequently cannot see about 10% of the original video. For example, when watching CNN Headline News, you might not be able to read all the stock ticker data at the bottom of the screen.

If you're creating an image to play on a standard NTSC TV, you can ensure viewers will see each image in its entirety—top to bottom or left to right—if you create your image using the overscan resolution noted in Table 15.1: 650×490 pixels (this is an approximation—overscan varies from TV set to TV set). You will later place that image within a 720×540 frame. In that way, overscan can cover the frame edges but not the original image.

You certainly can use the full resolution (720×540 pixels), but the edges of your image will not display on most NTSC TV sets. Figure 15.1 shows the approximate overscan area and the visible area, or the so-called *safe zone*.

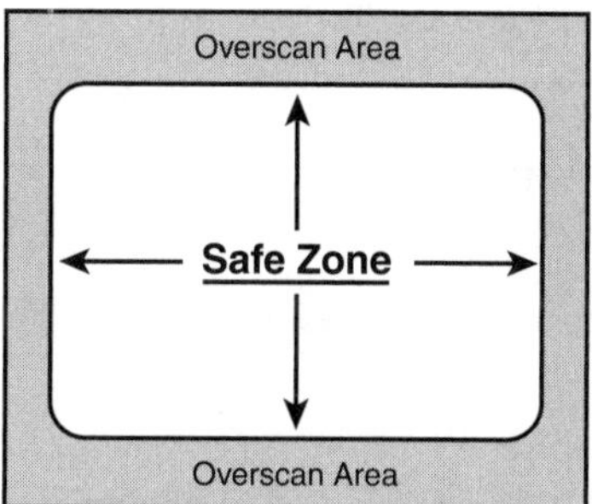

FIGURE 15.1
When working with NTSC DVD projects, you should keep images within the safe zone to compensate for overscan.

To compensate for these differences and ensure that your images retain their original aspect ratios, you should create graphics using the resolutions noted in Table 15.1.

TABLE 15.1 Standard Image Resolutions—Sizes

TV Standard and Aspect Ratio	*Full Resolution*
NTSC 4:3	720×540 pixels or 650×490 pixels*
PAL 4:3	768×576 pixels
NTSC widescreen 16:9**	852×480 pixels
PAL widescreen 16:9**	1024×576 pixels

**To compensate for NTSC overscan resolution.*
***Widescreen resolutions are for images only. Menus cannot display in the 16:9 widescreen format. Create all menus using the 4:3 resolutions for each TV standard.*

In either case, your images should be 72 dots per inch (dpi) or more, the maximum resolution TVs can display (PC monitors typically display at 96dpi). Using a dpi density much larger than 72 is a waste of hard drive space. In addition, the smaller the resolution, the faster the scanning process.

Both NTSC and PAL resolution standards have a 4:3 (4 units wide, 3 units tall) aspect ratio, the standard ratio for PC monitors. 640×480, 800×600, and 1024×768 are all 4:3 aspect ratios. In fact, you can create graphics for NTSC or PAL in any 4:3 aspect ratio. However, if they're smaller than the recommended sizes, DVD authoring software will blow them up and they will lose some definition. Also, if your images are larger than these sizes, DVD authoring software will shrink them and interpolate adjoining pixels in the process, leading to a lower-quality look.

Clearing Up Resolution

Resolution, *dots per inch*, *pixels*, and *image size* are used interchangeably in PC parlance. This leads to some confusion.

First, *dots* equal *pixels* (picture elements).

Second, *resolution* sometimes means relative size and sometimes means dpi.

Most images have a *resolution* stated in dpi, such as 600 horizontal dpi and 600 vertical dpi (usually horizontal dpi and vertical dpi are equal; otherwise, you'd end up with a distorted image).

Sometimes *resolution* refers to image *size* in some number of pixels, as in 800×600 pixels. A 400×300 size resolution image takes up one fourth of your screen if you run Windows at 800×600—that is, if your software doesn't change the viewing area depending on the image resolution.

I said this could be confusing.

An image with 600×600 *dpi* resolution and a 300×300 *size* resolution prints out in postage stamp size, one-half inch on each side (300 pixels—dots—at 600dpi equals one-half inch).

When scanning images, most people think in terms of how the images will look when printed. In those cases, a 300dpi resolution is a bare minimum for decent image quality. Most scanners scan at an even higher dpi, typically 1200×1200dpi, which creates a huge file that is wasted on TV because it cannot display at greater than 72dpi.

As you progress through this hour, you'll see that keeping images slated for DVD and video projects at a low dpi creates manageable file sizes, sharp-looking images, and fast scans.

Scanning Images for DVD Projects—Automated or Manual

Today's scanner software leans toward "idiot proof" status. That's not a good thing because it gives you virtually no direct control over individual scans.

Instead, current software is results oriented, meaning it scans according to your intended use of the image. It typically analyzes an image and notes whether it's a black-and-white document, grayscale picture, or color photo. It asks how you're going to use the image—printout, photo, Internet, or optical character recognition (OCR)—and then cranks out a scan using preset resolutions.

So, what you end up doing, to use a TV production phrase, is *fixing it in post*. After scanning an image, you then set the dpi resolution and size in image editing software, which frequently leads to less-than-optimum visual quality.

You can accept your scanner's defaults and fix everything in post, or you can circumvent the auto-scanning process to save a few steps later and get more predictable results. I'll explain both processes.

Automated Scanning

Most scanners come bundled with software that minimizes user input. I dislike this approach because you have no idea what settings it uses and have virtually no control over the results. Epson's Smart Panel, shown in Figure 15.2, is a leading culprit.

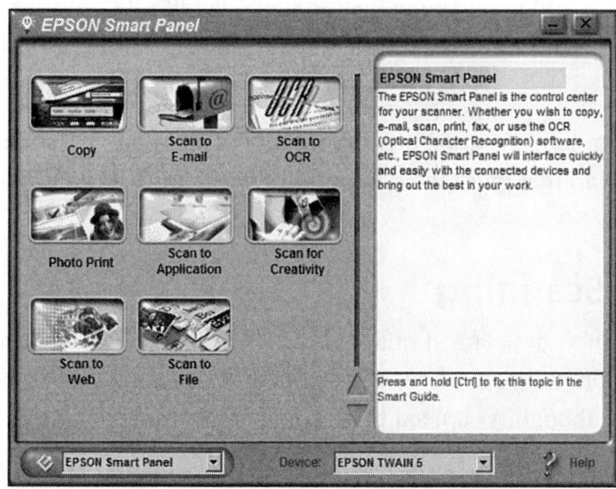

FIGURE 15.2
Epson's Smart Panel automates everything, keeping you in the dark about all scanner settings.

Smart Panel offers eight scanning options. The only one that sort of fits your DVD production needs is Scan to File.

If you're using Epson's Smart Panel, select Scan to File (shown in Figure 15.2) or the equivalent if you have another manufacturer's software. The automated process starts and results in a scanned image that shows up as a thumbnail in the interface shown in Figure 15.3. Double-clicking the thumbnail opens a viewer with no editing controls.

FIGURE 15.3
After the software completes the automated scanning process, it displays a thumbnail of your image.

You either can go back and do more scanning or save this and any other scanned images to your hard disk. When you do save your images, in this case, you'll discover that their image resolution is 300×300dpi. If your original image is an 8"×10" photo, your saved image will be 2,400 pixels wide by 3,000 pixels tall (300dpi times 8" = 2,400 dots—pixels—wide). That's a massive file. If you had selected Photo Print instead, the resolution could have been 1,200×1,200dpi, which is a *monstrous* file.

1,200dpi might look great on a photo-quality printer, but TV screens can't display more than 72dpi so all that extra information is a waste of disk space and takes too much scanning time.

Manual Scanning

Take control of your scanned images. With Windows XP, you have at least one manual scanning option, no matter how your scanner's bundled software works. Before tackling this next task, though, you need to do some calculating.

As I mentioned at the beginning of this hour, your images should have resolutions of 650×490 pixels for NTSC (an approximate compensation for overscan) or 768×576 pixels for PAL. In the case of images for NTSC use, you'll later place those less-than-full-TV-screen-size images in a full-screen frame to ensure NTSC overscan does not cover up their edges.

If you don't mind losing the edges of your images off the sides of monitor then feel free to create images using the full NTSC resolution of 720×540. However, all my upcoming calculations use the overscan compensation resolution of 650×490.

Your original images come in all shapes (also called *aspect ratios*) and sizes. You'll start to overcome those differences by changing your scanner dpi settings depending on the size of the original photo. After the image is scanned, you place it into a 720×540 or 768×576 pixel template before importing into your DVD project.

In Hour 5, "Authoring Your First DVD Project Using MyDVD: Part II," when you used MyDVD to produce your first DVD, you might have created a slideshow using your own images or images from this book's companion DVD. In that case, you did not go to the trouble of placing those images in a template or worry about their dpi resolutions. MyDVD expanded or contracted the images to fit the screen and placed them in its own template, preserving their original aspect ratios.

> So, why bother with all this extra work? Getting the correct dpi ensures that your images will look sharp on TV, and placing them in a template means there will be no overscan issues. In addition, DVD authoring software will not do any automatic adjustments for still images used as menu backgrounds.

Your main task is to find a dpi setting that fits your image. Your goal is to create an image with 72dpi or more (but not too much more) to ensure sharp reproduction on your TV while minimizing file size and speeding up the scanning process.

Calculating Correct Scanner dpi

Photos generally are either vertical format (portrait) or horizontal (landscape). Horizontal matches the general shape of a TV screen but usually does not exactly match the aspect ratio. Photos tend to be a bit wider.

You should make the horizontal format photo fill the TV screen from left to right, meaning the top and bottom will not quite reach the top and bottom of the TV screen.

With vertical format pictures, your goal is to have the top and bottom of the photo touch the top and bottom of the TV screen. That means the left and right sides will be well inside from the edges of the TV screen.

For horizontal format photos, measure the width in inches and divide that into 650 for NTSC or 768 for PAL. For a 5"×7" photo, divide 7 into 650 to get 93dpi. When you scan a 5"×7" horizontal format photo, select a dpi setting of 93 or slightly more.

If you have a 5"×7" vertical format photo, you should divide 7" into the screen height or 490. In this case, 490/7 = 70. That's less than 72dpi, so select at least 72dpi when you scan a 5"×7" vertical format photo.

There are exceptions to this horizontal and vertical format rule. You might have a horizontal format photo that is only slightly wider than it is tall. In that case, the limiting factor might be its height, not its width. That said, this isn't brain surgery. If your dpi calculation is a little off the optimal setting, your image still should end up looking fine.

In general, smaller photos require greater dpi settings and larger photos require smaller dpi settings. But, as listed in Table 15.2, never scan at less than 72dpi because your image will lose fidelity. And it's always a safe bet to round up your dpi numbers—that is, calculate the proper dpi and then round it up to a slightly higher, even number.

TABLE 15.2 Scanner dpi Setting Examples for Photos (NTSC DVDs)

Photo Size	*Horizontal Format*	*Vertical Format*
2"×3"	217dpi	163dpi
4"×6"	109dpi	82dpi
5"×7"	93dpi	72dpi (minimum resolution)
8"×10" or larger	72dpi	72dpi (minimum resolution)

Task: Use the Windows Scanner Wizard to Customize Scan Settings

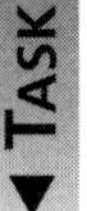

Tucked away in your Control Panel is a wizard that gives you at least two manually adjustable settings that let you create manageable-sized images. Here's how it works:

1. You might find the Scanner or Camera Wizard under Start, Programs, Accessories. If so, start it that way; if not, locate it in your Control Panel. You might have a Control Panel icon on your desktop or find Control Panel under Start, Settings. In either case, open Control Panel, double-click Scanners and Cameras, and double-click your scanner's name in that window. The Scanner and Camera Wizard shown in Figure 15.4 opens.
2. Click Next, and you'll see the Choose Scanning Preferences screen shown in Figure 15.5.
3. Select Custom settings to open the Properties interface. Here you select a dpi setting to match your image size. If it's a 4"×6" horizontal format photo, select 109 dpi, as shown in Figure 15.6. Click OK to return to the Scanning Preferences interface.

FIGURE 15.4
Use the Scanner and Camera Wizard to manually alter your scanner's settings.

FIGURE 15.5
Select Custom settings to adjust the dpi setting.

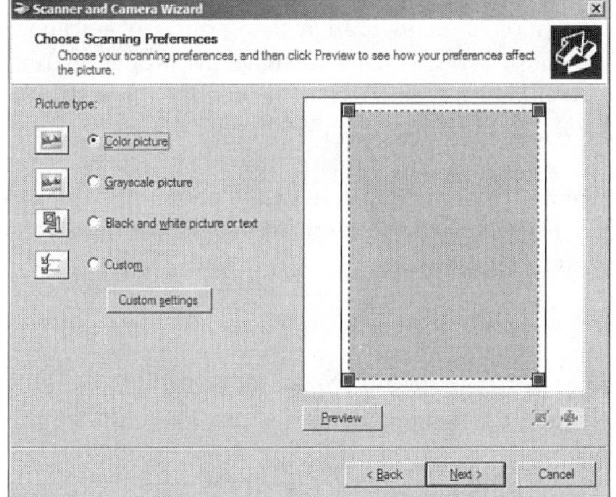

FIGURE 15.6
Use Windows XP's Scanner Wizard Custom setting to select a specific dpi resolution.

4. If you need to do any cropping of your image, now's the time to do it. Click Preview (it might take about a minute for the scanner to make a preview image) to display your image in the cropping window, shown in Figure 15.7. Drag the highlighted box corners to the desired size, and then click Next.

FIGURE 15.7
Crop your image before completing the scan to get the best results later.

You should crop now rather than after you've scanned your image. Doing it now ensures your image will have the proper dpi resolution, whereas cropping it later will lead to a lower-quality image due to reduced dpi (after you blow up the smaller cropped image to fit it on your DVD image template). You should also cut out large borders like those shown in Figure 15.7 and remove unwanted portions of the photo to save you a cropping step later. If you plan to crop, don't include the cropped-out area in the measurement used to calculate dpi.

5. Click Next to open the Picture Name and Destination window shown in Figure 15.8. Name the picture, select a file folder, and choose an image type.

BMP and TIFF are the best of the four image formats available in the Windows scanner wizard because they are uncompressed and retain all image quality. PNG does compress image data, but it only imperceptibly alters its quality. Do not use JPEG because it both compresses and reduces image quality. In any event, Sonic Solutions' DVD authoring products can handle any of these four image types.

6. Click Next to make the scan and store the image. It's now available for further editing.

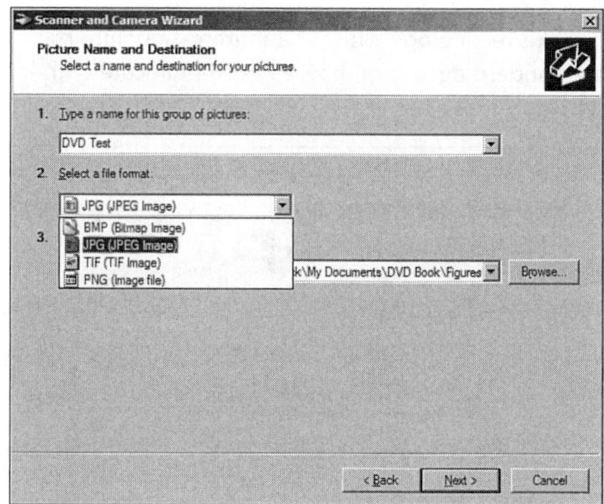

FIGURE 15.8
Select any of the four image formats—BMP and TIFF work best. The DVD authoring software included with this book can handle all of them.

Your scanner software—even Epson Smart Panel—might have a manual option. Finding it can be difficult, but here's how to do it using Smart Panel.

Task: Use Smart Panel's Manual Scanning Option

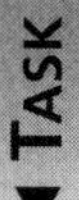

After you use Smart Panel for a manual scan, it defaults to the manual mode. This can end up being the best way to gain control of your scans. Here's how:

1. Start Smart Panel and click Scan to File. That starts the auto-scan process—scanner warm-up and scan.
2. An information screen pops up, shown in Figure 15.9. Click Cancel to stop the auto-scan.

FIGURE 15.9
Switch from auto-scan to manual-scan mode by clicking Cancel.

3. In the next screen, click Manual Mode to open the Epson Twain interface. As shown in Figure 15.10, this screen gives you a plethora of options.

Twain, surprisingly, is not an acronym (some jokingly say it stands for "toolkit/technology without an interesting/important name"). It is an industry standard describing how PCs communicate with image acquisition devices.

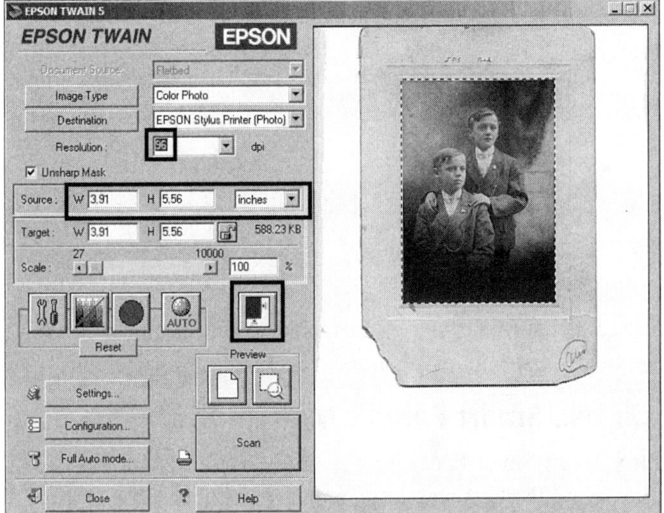

FIGURE 15.10
Epson's Twain interface offers several user-controlled options.

4. Click the Image Type button and select an option, such as Color Photo, Black & White Photo, and so on, from the drop-down box. The Destination button is a little confusing. Select Screen/Web; otherwise, you won't have an option to save the file. If you don't need to crop the photo, select the resolution that matches or is the next number available that exceeds your calculated dpi. Otherwise, wait until after you preview and crop the image to set the dpi number.

Epson has not relinquished full control to the user. It limits resolution settings to only a handful of numbers.

5. Click Preview. Do any cropping by clicking the green crop button shown in Figure 15.10. After the image is cropped, the display lists the new height and width. Use the appropriate number—width for a horizontal format image or height for a vertical format—to calculate the dpi; then select that dpi from the drop-down list. Finally, click Scan.

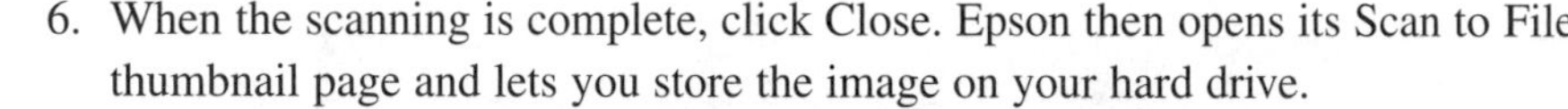

15

6. When the scanning is complete, click Close. Epson then opens its Scan to File thumbnail page and lets you store the image on your hard drive.

Formatting Images for DVD Projects

Few of your scanned images will exactly match the NTSC or PAL screen sizes or 4:3 aspect ratios needed for DVD menus and stills. Therefore, you need to place them on a template of the proper size, effectively adding black (or any other color) edges along the image sides or top and bottom to fill the screen. This way, when the DVD authoring software (or nonlinear video editor) displays them full-screen, the image portion of the file will remain in its original aspect ratio and won't get squashed or stretched to fill the screen.

I've chosen my favorite image manipulation software, HyperSnap-DX, to perform this upcoming task. You can download a trial copy for free or the full version for $35 from `http://www.hyperionics.com/`. Or, you can use any other image manipulation software.

Task: Use Graphics Editing Software to Create a Template for DVD Images and Menus

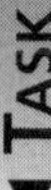

The goal here is to create a template that matches your country's TV standard resolution—NTSC or PAL. Then, in the next task, you'll crop and adjust the resolution size of your images and paste them on the template. Here's how to create a template:

1. Open HyperSnap-DX (or your image editing software of choice).
2. Click the New button, shown in the upper-left corner of Figure 15.11, or select File, New from the main menu. A 400×300 pixel white rectangle displays. Change that into a 720×540 (NTSC) or 768×576 (PAL) black rectangle.

> As a reminder, I've asked you to revert to the full-screen NTSC resolution because you will use this black rectangle as the frame onto which you will place your scanned images. Overscan can push this template's edges off the edges of the TV set, but the image you'll place on the template will display within the TV set viewing area.

3. Click the Resize button shown in the center of the toolbar in Figure 15.11. Doing so displays the Bitmap Dimensions interface shown in Figure 15.12. Change the Width to 720 (or 768 for PAL) and the Height to 540 (or 576). Click OK.

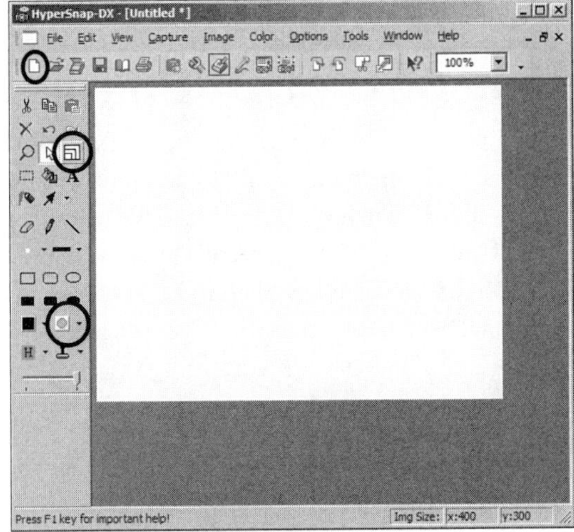

FIGURE 15.11
I recommend using HyperSnap-DX to crop and resize your images.

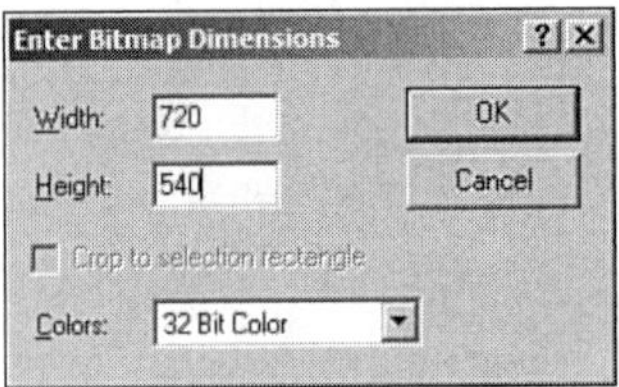

FIGURE 15.12
Change the bitmap dimensions to match your country's video standard.

4. Click the Background/Transparent Color button shown at the bottom of the toolbar in Figure 15.11 to open the collection of 40 color swatches, shown in Figure 15.13.

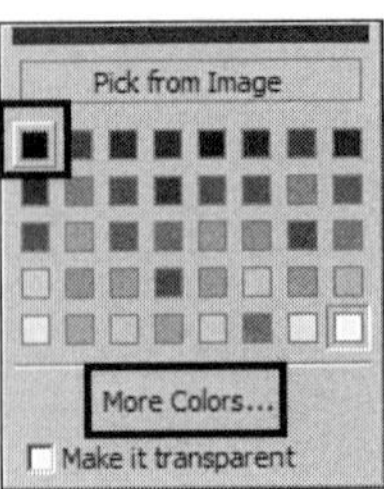

FIGURE 15.13
Use the Background/Transparent Color tool to set the template color to black (or a color of your choosing).

5. If you're creating a DVD for use on PAL TV sets, simply click the black swatch in the upper-left corner and go to step 7. If you are producing a DVD for NTSC (U.S. and Japan), you need to create an NTSC-safe color (see the following caution). Select More Colors(see Figure 15.13) to open the color selection window shown in Figure 15.14.

▼

15

The NTSC TV standard has several drawbacks (versus PAL and SECAM). The one that affects the creation of this task's image frame is color. To avoid the tearing, bleeding, or smearing of colors, NTSC graphics must not be too bright, dark, or saturated. (Those who travel from the United States to Europe, for instance, usually notice how much richer the colors are on PAL or SECAM TV sets.) To ensure that the frame you create for your images is NTSC-safe, keep the RGB (red, green, and blue) values less than 230 and more than 16. If you work with Photoshop or Photoshop Elements, you can use a simple tool to make your colors NTSC safe: Select Filter, Video, NTSC Colors. HyperSnap does not have this tool, so you must use the method described in step 6 to ensure that the black frame is not too black.

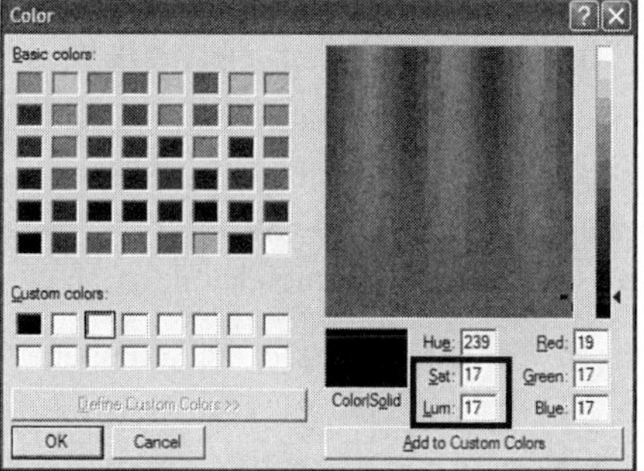

FIGURE 15.14
Select an NTSC-safe template color using this color selection dialog box.

6. To give your template, or frame, an NTSC-safe custom color, select the middle box (Sat—Saturation) shown in Figure 15.14 and type in `17`. Tab to the next box (Lum—Luminance) and type in **17** again. You'll note that the red/green/blue color numbers change to about 17 (the Hue number is inconsequential in this case). The Color|Solid box displays this new color (even though it's NTSC-safe, it still looks very black). Click Add to Custom Colors and then click OK to return to the main HyperSnap interface.

You can create NTSC-safe colors in HyperSnap and other graphics products using many methods. In the case of HyperSnap, you can create a color by inputting separate values that are greater than 16 and less than 230 in the Red, Green, and Blue boxes. Or, you can reduce the Contrast setting and increase the Brightness setting.

7. Click New again, and your rectangle should turn black. Now when you reopen HyperSnap-DX and click New, you'll always get a 720×540 (or 768×576) black rectangular DVD image/menu template.

If a black frame does not work with a particular still, you can change the background color. One approach is to place your image on the black background and use the eyedropper Color Picker tool to select a color from the image itself (perhaps along an edge). You can then use the Paint Can tool to apply the new color to the background, blending it more closely with the image and making it look less like a frame or border.

Task: Use Graphics Editing Software to Crop and Resize Your Images and Paste Them onto Your Template

TASK

Your goal when you scanned your images was to limit the dpi to no less than 72dpi and no more than is necessary to create a full-screen image. In this task, you'll adjust your images so they exactly fit into the 650×490 NTSC or 768×576 PAL viewable area. Then, you'll paste them onto your newly created template. Here's how:

1. Select Open (the icon next to New), and locate and open a scanned (or any other) image file.
2. During scanning, you should have cropped out any extraneous borders or unwanted portions of the original picture. Now you might want to fine-tune that process. Select the Crop tool (see Figure 15.15). Click and drag to define the region you want to keep; then click again within the workspace to tell HyperSnap you've completed setting the cropping boundaries.

If you don't like how you've cropped your image, select Edit, Undo (or use the standard Windows keyboard shortcut Ctrl+Z) to start over.

If you crop too much now, you will end up with a fuzzy, out-of-focus image when you place it in your DVD project. When you made the scan, you limited the dpi to save disk space and speed up scanning. If you crop out a lot after saving the scanned image and then expand the image to fit the template, you will lose resolution. Use this last-minute cropping step solely to make minor fixes.

FIGURE 15.15
Use HyperSnap-DX's cropping tools to trim out unwanted edges from your image.

3. Scale the image to ensure it will fit exactly into the 650×490 NTSC or 768×576 PAL frame by selecting Image, Scale from the main menu. In the Scale dialog box shown in Figure 15.16, either change the width (for horizontal-oriented images) to 650 or 768 or change the height (for vertical images) to 490 or 576. Do not uncheck Keep Aspect (it keeps your image in its original aspect ratio) or Interpolate (it smoothes pixel-to-pixel color changes when you shrink or expand images). Click Done.

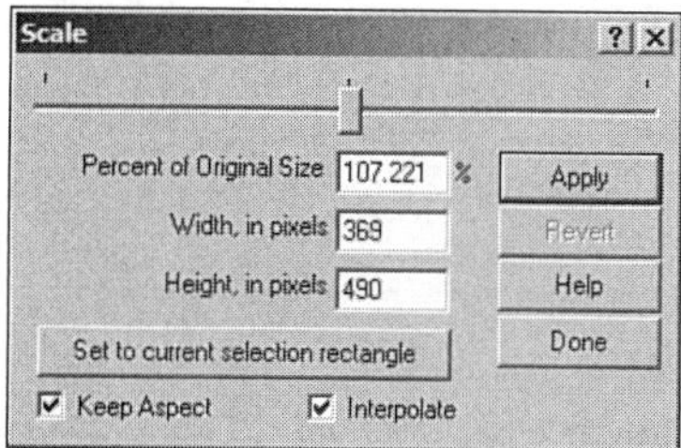

FIGURE 15.16
Use the Scale dialog box to ensure your image fits into the template.

4. Copy your image by selecting Edit, Copy from the main menu (or use the standard Windows keyboard shortcut Ctrl+C).
5. Click New to open your 720×540 (or 768×576) black, rectangular template.

6. Paste the image onto this black background by selecting Edit, Paste (or use the keyboard shortcut Ctrl+V).
7. As shown in Figure 15.17, drag the image so it's centered over the template. When you're satisfied with its location, click outside the image to anchor it.

The NTSC template, shown previously in Figure 15.4, should have room around the image for overscan. For the PAL template, the top and bottom of a vertical image will be flush with the top and bottom of the template and the left and right edges of a horizontal format image will be flush with the left and right edges of the template.

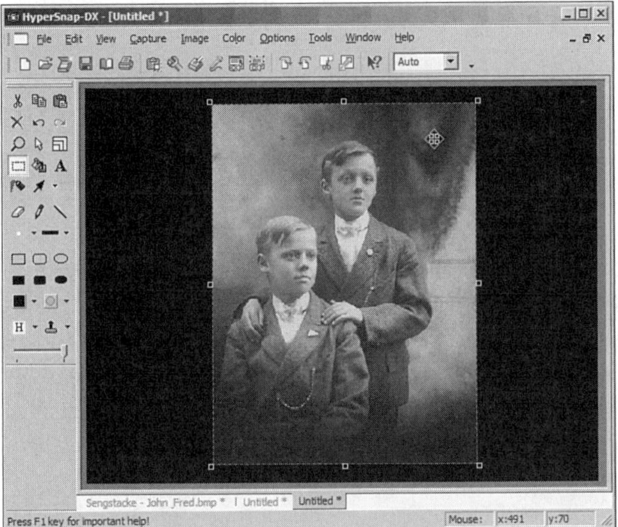

FIGURE 15.17
Paste your image onto the template to complete this process.

8. Save your DVD-ready image by selecting File, Save and choosing a name and file folder location.

You have numerous image types from which to choose. BMP, GIF, JPG, and TIF (the top four in the HyperSnap-DX list) are your best options for compatibility and image quality retention.

The end result of all this effort is a collection of images with dpi resolutions, size resolutions, and aspect ratios that will lead to a high-quality, consistent look for all the stills and menu backgrounds in your DVD projects.

Summary

Creating consistently sharp-looking and distortion-free images for a DVD project takes some extra effort. The process begins by scanning images using the proper dpi settings for the original image size. While scanning, you should crop your images to remove borders or unwanted areas around the edges. When creating projects for NTSC, compensate for overscan by calculating the scanner dpi using the NTSC safe zone 650×490 size resolution.

You need to create an image background template that fits your local TV standard: NTSC or PAL. You then can use image manipulation software—I recommend HyperSnap-DX—to do some final cropping and size resolution adjustments to your images and then drop them onto the template.

This somewhat detailed process guarantees your images will maintain their aspect ratios and quality, which would be lost if the DVD authoring software had to expand or shrink the images to fit its screen.

Workshop

Review the questions and answers in this section to reinforce your image scanning and formatting techniques and skills. Also, take a few moments to take the short quiz and perform the exercises.

Q&A

Q As long as I capture images using dpi settings that match or slightly exceed the 72dpi resolution of TV sets, why do I need to put the scanned images in a template?

A When you use still images on DVDs or as backgrounds for menus, most DVD authoring software stretches or squeezes them to fit into a TV screen's 4:3 aspect ratio. Most images do not have that perfect shape, so if you don't place your images on a 4:3 (720×540 NTSC or 768×576 PAL), they'll look distorted in menus. Even though some authoring software, such as MyDVD, creates slideshows by automatically placing images of all aspect ratios onto a template and not distorting them, overscan probably will push the edges of those images off the screen.

Q Your overscan illustration in Figure 14.1 does not note the exact size of the safe zone. I'd like to create a safe zone template to ensure my images fit exactly on an NTSC TV screen. Which dimensions should I use?

A There is no hard and fast rule. NTSC sets are notorious for their safe zone inconsistencies. In general, TV producers think in terms of a 10% action safe zone and a 20% title safe zone. That is, for any onscreen action, assume 10% (or 36 pixels on each side and 27 pixels on the top and bottom) will fall into the overscan area and not display. To ensure that all text or titles show up, assume 20% (or 72 pixels left and right and 54 pixels top and bottom) will not display. NTSC is a poorly implemented technology. Engineers blithely refer to it as "never the same color."

Quiz

1. In general, when scanning large photos—8"×10" or larger—which dpi setting should you choose?
2. You plan to scan a 4" tall oval photo and want to make sure it fits into the NTSC safe zone. Which dpi setting should you use? If it's 4" wide (a horizontal format oval photo), which dpi setting should you select?
3. The black template is just too jarring for some scenes. Sepia would work better for some older photos, or dark blue might fit some landscapes. How do you change the template color using a color from the photo or one of your own?

Quiz Answers

1. Depending on the format, a photo dimension of 7" or more in height or 9" or more in width means your calculated dpi starts dropping below 72, the standard TV resolution. So you should scan any image with these dimensions or higher at no less than 72dpi.
2. The NTSC safe zone height is 490 pixels, so you divide 490 by 4" for a dpi setting of 123, which you round up to 130. The NTSC safe zone width is 650. So, for the horizontal format image, you divide 650 by 4" for a dpi setting of 163, which you round up to 170. Before making the scan, you should use the rectangular cropping tool to trim as close to the oval as you can. You later can use HyperSnap-DX's paint and erase tools to trim out the rest of the unwanted areas between the oval and the corners of the rectangular cropped area.
3. To lift a color from the photo, use HyperSnap-DX's Foreground Color tool. Then, select Pick from Image, use the Eyedropper tool to click a color on the image, click the Paint Can icon, and click it on the black border to change its color to the selected color. To create a custom color, click the Foreground Color tool icon and select a standard color. Or, click More Colors to open the color palette and create one there.

Exercises

1. Set aside a day to organize photos you might want to scan for use in DVD projects. Use a batch scanning process—that is, scan by photo size, doing all 5×7 prints in one sitting, for instance. That way, you won't need to change the scanner settings each time you scan. Come up with a standardized naming convention for your photos and store them in descriptively named file folders.
2. Instead of using a nondescript black template for your photos, use greeting card software or a simple paint program to create a set of borders to use in different types of DVD projects.
3. Use more than one photo on a template. This can take some experimenting because you must calculate each dpi setting to make the image fit exactly within the template parameters. However, if you want to create a montage, you can either scale down your already scanned images (losing some image quality in the process) or scan them at smaller dpi settings knowing you're going to combine them into one grouping.

Part IV
Authoring DVDs

Hour

HOUR 16

Evaluating Competing DVD Authoring Products

DVD authoring software is a newly emerging and rapidly changing market niche. A labyrinth of features and usability issues confound the evaluation and buying process. Confusion reigns.

In this hour, I'll attempt to bring clarity to the confusion. I'll present an overview of what features to look for in authoring products. And I'll review all the major, standalone, entry- to prosumer-level DVD authoring products I could get my hands on.

Because this book's companion DVD contains several products from industry leader Sonic Solutions, clearly my tilt is in its direction. But competition ensures that this industry continues to push the envelope while reducing prices. Sonic Solutions is not the only player to watch.

The highlights of this hour include the following:

- Selecting DVD authoring features that work for you
- Evaluating entry-level DVD authoring products
- Evaluating prosumer-level DVD authoring products

Selecting DVD Authoring Features That Work for You

In Hour 17, "Designing DVD Projects and Authoring with DVDit!: Part I," you'll start working with DVDit!, a prosumer-level DVD authoring product. Then, in Hour 23, "Professional DVD Authoring with ReelDVD: Part I," I'll introduce you to Sonic Solutions' higher-end product: ReelDVD. Before taking those steps, though, I want to clarify what makes these products and other prosumer DVD authoring titles several cuts above entry-level products such as DVDit!'s sibling MyDVD.

Standalone DVD authoring products come in all sizes and prices—from mindlessly simple, $25 products to insanely complex systems selling for thousands of dollars. My guess is that you will eventually settle on a prosumer-level title in the several hundred dollar range. But you might choose to get your feet wet using something in the $25–$100 field.

You also could opt to use DVD authoring modules included with virtually all the latest video editing software products. However, as I noted in Hour 11, "Crafting Your Story and Selecting Video Editing Software," none of the entry- to mid-level video editing software titles offers a full range of authoring tools. Here's a quick reminder of my take on each product's DVD authoring module:

- **Pinnacle Studio 8**—Rudimentary and virtually undocumented
- **Ulead VideoStudio6**—Barely functional
- **Roxio VideoWave 5**—The best of these four but far from MyDVD or other Sonic Solutions products
- **Cyberlink PowerDirector 2.5**—Easy to use but has few menu design options

In Hour 3, "Burning Data DVDs," my view of the DVD authoring module in Roxio's Easy CD & DVD Creator 6 was that this first attempt at adding DVD authoring to what had been a straightforward CD/DVD data burning title looked good but had too many kinks to recommend.

The better standalone products tend to give you the full enchilada. However, figuring out what you want rolled up in your authoring title might not be very obvious. As with word processors and their myriad untapped features, authoring products frequently offer tools that might appear at first glance to have no intrinsic value (or might make no sense to you). But later, as you become more proficient, you might wonder how you could have worked without them.

DVD Authoring Tools and Features

To give you an overview of the possibilities, I'll divide DVD authoring functionality into these subsets:

- Input file types
- Menu features, including navigation options
- Button and text creation options
- Extras, including video editing, still frame grabs, slideshows, video capture, DVD-ROM data, and more
- Playback options, including extra audio tracks, subtitles, and widescreen
- Output file types and MPEG/Dolby video/audio compression/encoding

Input File Types

You should select a DVD authoring product that can handle several video formats. At the very least, it should handle MPEG, Microsoft AVI, and Apple QuickTime MOV files. Some MPEG files you create or obtain will be encoded in *elementary* streams—that is, separate audio and video files. Better DVD products handle these as well.

Common audio files include WAV, MP3, WMS, AIFF, MPEG-1 Layer II (also known as MPA), and Dolby digital AC-3. Standard image, graphic, and still formats include BMP, GIF, JPG, PCT, PNG, WMF, PCD, TGA, and TIF.

Extra credit goes to authoring software that can handle layered Photoshop files, which is a great way to create buttons that change color or appearance depending on viewer actions).

Menu Features

Entry-level DVD authoring products typically provide rigid menu templates with specific looks and button placements. Prosumer products, on the other hand, let you fully customize your menus.

The best DVD authoring products let you create hundreds of menus per DVD with the option to use audio and video backgrounds. Other features some entry-level applications and many prosumer products offer include

- Adding chapter points in videos is a powerful tool. These allow viewers to jump to predefined segments in a movie.
- The ability to use chapter point thumbnail images selected from the chapter scenes.
- Full menu navigation controls, such as what happens when a video ends or the viewer presses Next on the remote.

- Timed events such as still images or menus that automatically move to a default video or another menu if the viewer does nothing with the remote

Button and Text Creation Options

Here's where prosumer authoring tools really shine, and options can be nearly endless. The best products let you change the shape, color, size, placement, and drop shadow characteristics for all buttons and text. Other features include

- Photoshop layers for buttons
- Text creation tools that allow extra features, such as borders and gradient colors
- Transparency option for buttons and text
- The capability to link text to video and still images (most DVD authoring products let you link only buttons to video clips or stills)
- Animated or video buttons
- User-selected video frames for buttons, typically when linking buttons to video chapters
- Multiple button states (as viewers use the remote to navigate among the buttons, the buttons change color or animate differently)

Extras

Some of the following options might seem like afterthoughts, but most can make or break a DVD project:

- **Slideshows are a fun feature**—The best DVD authoring products let you add transitions between images, such as dissolves (a surprisingly complex technical feat on a DVD), plus add music or narration.
- **DVD-ROM files**—This is a powerful option that lets users with PC DVD-ROM drives access data files, word processing documents, spreadsheets, PowerPoint presentations, or even the original video and image files used on the DVD. Only some prosumer-level products offer this option.
- **A flowchart or timeline project layout**—Instead of using only a collection of menus, this makes producing a DVD more intuitive. No entry-level product uses either of these toolsets.
- **Video capture**—This lets you bypass an NLE and go directly from your camcorder to DVD authoring.
- **Video editing tools**—No DVD authoring product offers anything approaching a full-featured video NLE. But some give you the option to trim, combine, or split clips as well as add transitions such as wipes and dissolves. If your project needs are simple, these features let you circumvent the NLE editing process.

- **Still frame capture that enables you to use any video frame as a menu background.**
- **First-play option**—Lets you choose what happens when viewers insert the DVD into a player (the default First-play is the main menu). This could be a brief video, previews of coming attractions, or an animated logo.
- **DVD/CD and jewel box and disc label creation tool.**

Playback Options

Most DVD feature films offer several extras ranging from subtitles to Dolby digital surround sound. At first blush, you might think you won't tap these features, but they offer several creative options:

- **Extra audio tracks**—Let you add a narration/commentary, foreign language dubs, different quality audio streams, or music.
- **Subtitle tracks**—Let you offer onscreen text language translations and extra textual references that you can use instead of an NLE's title editor. Some authoring products expect you to import subtitle files, but a couple go the extra mile and offer built-in subtitle editors.
- **Links to URLs**—Give PC users viewing your DVD direct access to associated Web pages. This is a great way for businesses to tie their DVDs to their Web site promotions.
- **Multiple camera angles**—This rarely used option can enhance a DVD movie. For instance, the feature film *Moulin Rouge* has some dazzling dance numbers that use rapid-fire edits and multiple camera angles. To give viewers a different feel for those scenes, the DVD's producers let viewers choose one or more camera angles.
- **Widescreen or 16:9 aspect ratio**—Puts your video into a letterbox on a standard TV set or expands it to fit a widescreen display. No entry-level authoring applications have this functionality.

Output and Encoding

Most DVD authoring products will convert your media into MPEG video files.

That's the only video format that can play in consumer set-top DVD players. The level of user control over that conversion process is what sets the top authoring products apart from the rest of the field. Here are some options to consider:

- **User adjustable encoding output quality settings, such as bit rate.**
- **Dolby digital stereo encoding**—Top-end DVDs offer Dolby AC-3 5.1 discrete six-channel surround sound (two front speakers, two rear, a center, and a subwoofer).

Or, viewers can choose from other standard, uncompressed or minimally compressed audio streams, such as PCM or LPCM (linear pulse code modulation).

- **Audio sampling rate conversion**—Changes various quality audio files to standard DVD 48KHz quality.
- **Output formats**—These include video CD (VCD), super video CD (SVCD), or VR (rewritable CDs/DVDs that allow later reauthoring).

Sonic Solutions has begun shipping DVD products that incorporate the company's new Open DVD format. DVDs created with Open DVD-compliant products can be opened in other Open DVD authoring software and completely reworked.

- **Media formats**—These include all recordable DVDs (DVD+R/RW, DVD-R/RW, and DVD-RAM), all recordable CDs, and digital linear tape (DLT) as well as DVD images created on the hard drive.

Evaluating Entry-level DVD Authoring Products

In the competitive PC software universe, you usually get what you pay for. In the DVD authoring sphere, at the entry-level at least, you get more. Even though entry-level products don't offer much flexibility or customizability, they still produce impressive-looking DVDs.

DVDs are a novelty (for how much longer is anybody's guess). That means there's still a significant "gee whiz" factor to simply using DVDs as your presentation media. If you use an entry-level authoring product, your audience won't know that your DVD creation software wouldn't let you change the size or location of the menu buttons or let you use animated menus.

True, you soon might grow frustrated with the lack of features, but entry-level products are a great introduction to the technology.

Five Entry-level DVD Authoring Products

I evaluated five entry-level products: Sonic Solutions MyDVD 5, Dazzle DVD Complete, MedioStream neoDVD Plus 4, Pinnacle Expression 2, and Ulead DVD MovieFactory.

Sonic Solutions MyDVD 5 is the best of this field. If you completed the exercises in Hours 4–6 and created a DVD using MyDVD, you have a reasonably solid grounding in what it and other entry-level products offer. I'll briefly review MyDVD's features here and cover the four other products in more detail.

Sonic Solutions MyDVD 5

Several things set MyDVD 5 apart from and above the competition. You've already seen that its logically laid-out and icon-driven interface eases users into DVD creation. In addition, MyDVD gives users much more flexibility and customization opportunities than other DVD authoring applications in this entry-level niche.

It offers video capture and editing, animated menus and buttons, and slideshows with transitions. You easily can create multiple, nested menus; change menu and button characteristics; and go so far as to build a personalized menu and button template from scratch.

What you might not have checked out yet are its tutorial and help files. You can access them from the opening dialog box (shown in Figure 16.1) or by selecting Help in the main interface. Sonic Solutions has gone the extra mile to offer detailed yet accessible documentation and well-organized lessons.

FIGURE 16.1
Access MyDVD 5's excellent tutorial and help files from this opening window.

Sometimes it's what you don't see that makes a product line robust. In the case of Sonic Solutions, that hidden strength is its DVD creation engine. Over the years, Sonic Solutions software engineers have fine-tuned the software for the entire product line—professional to prosumer to consumer—to ensure it can handle a wide variety of media file types and create DVDs that are compatible with a full-range of set-top players.

That under-the-hood expertise is evident in MyDVD 5. Despite its reasonable price tag and consumer-friendly design, it provides the tools to create customized and complex DVDs with multiple menus and media types. And you can proceed with a high degree of confidence that the burning process and playback will go smoothly.

Dazzle DVD Complete Deluxe

This $100 product shows promise but has too many miscues for me to recommend it.

On the plus side, it has some elements you'd expect to see only in higher-end DVD authoring products, such as extra button and menu navigation options, button highlight color and opacity selection (shown in Figure 16.2), and extra text controls. It also offers chapters, animated video buttons, jewel case and disc label creation tools (shown in Figure 16.3), and smooth simulation.

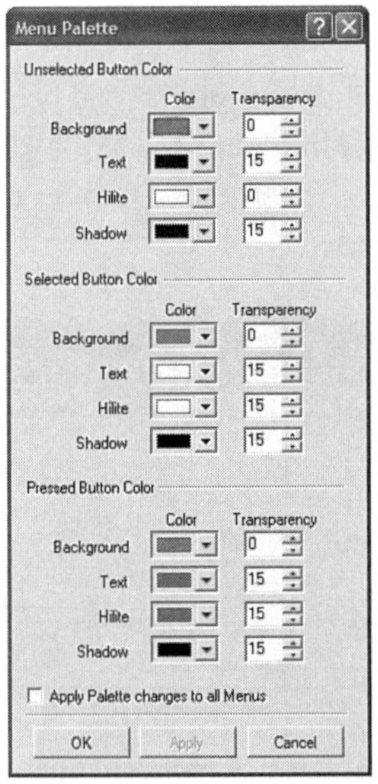

FIGURE 16.2
You typically find this detailed level of button highlight color and opacity control only in professional DVD authoring applications.

On the other hand, inconsistent implementation and a poorly crafted interface detract from Dazzle's overall performance. It runs only at 1024×768 or more but does not take advantage of all that real estate (see Figure 16.4). Its menu editing window does not fill the screen, and it uses non-resizable windows that cover up filenames, creating cluttered template displays.

It is not user friendly, especially for an entry-level product. It sports arcane acronyms and nonstandard terminology; its manual is skimpy; and its tutorial barely begins to explain Dazzle's feature set.

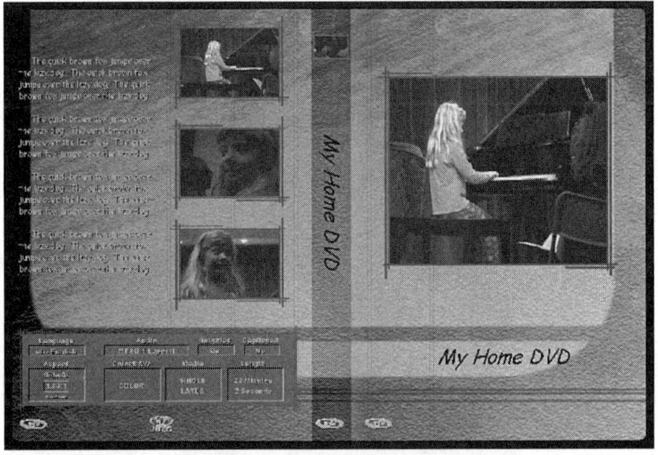

FIGURE 16.3
Dazzle's DVD jewel case and disc label editor is a feature few authoring applications offer.

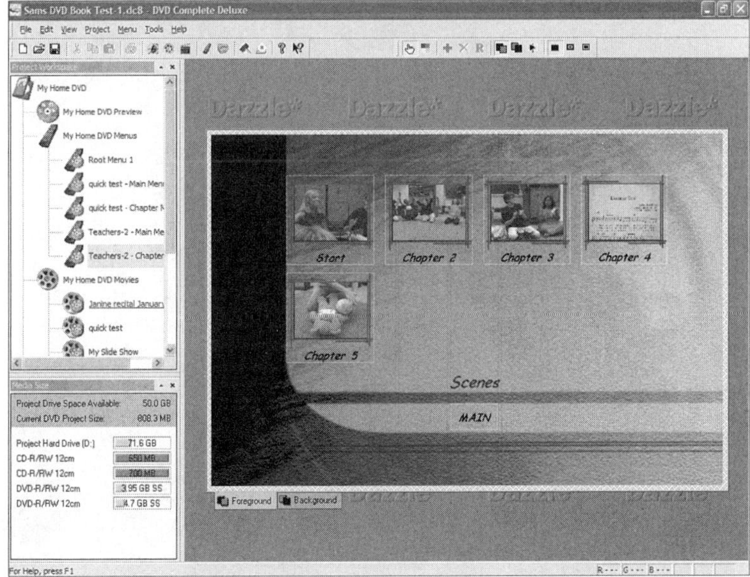

FIGURE 16.4
You work in a 1024×768 resolution or higher but Dazzle's interface does not use that space well.

It encodes image and video files to MPEG as you add them to your project rather than when burning the disc. Therefore, after you've added a movie or slideshow you cannot trim the video or rearrange the photos. Dazzle works only with AVI (video), WAV (audio), and MPEG (video and audio) files—no WMV or QuickTime video or AIFF or WMA audio files.

It sports the high-end feature of graphically representing button navigation (the horizontal arrows shown in Figure 16.5), but you can't use that window to drag and drop those links to rearrange them. Finally, its menu editor, shown in Figure 16.5, lets you move and resize button frames, but the thumbnails inexplicably stay in place.

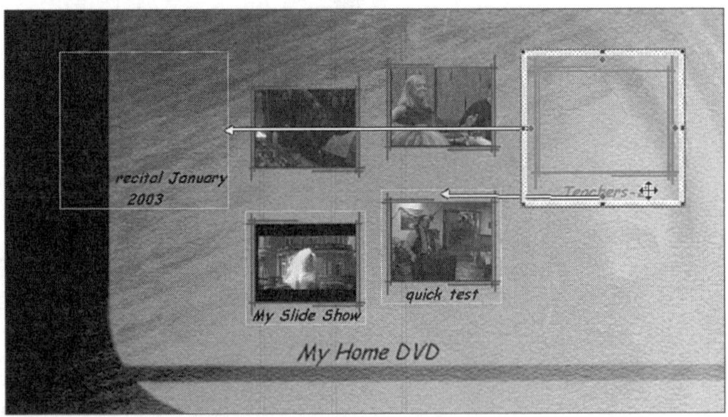

FIGURE 16.5
You can move or resize a button frame, but the associated thumbnail does not follow suit.

MedioStream neoDVD Plus 4

You might recall in Hour 10, "Capturing Video—Transferring Videos to Your PC," that neoDVD Standard comes bundled with the AVerDVD EZMaker analog video capture card. As a standalone product, newDVD 4 lists for $30. For this section, I tested the $50 Plus version shown in Figure 16.6. For the extra $20, you get a slideshow feature, Dolby digital audio, a narration recording mode, and moving menus (these are actually animated/video buttons).

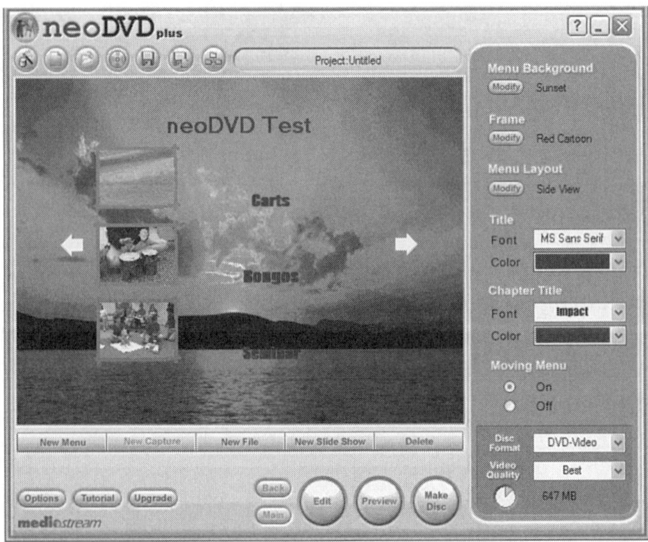

FIGURE 16.6
MedioStream's neoDVD Plus 4 is the runner-up in this five product field (MyDVD 5 tops them all).

For a product that suffers from poor layout and clumsy user interface design, neoDVD still beats Dazzle DVD Complete, Pinnacle Expression and Ulead DVD MovieFactory in terms of features and customizability. If MyDVD's slightly higher price tag is a deterrent, neoDVD 4—Standard or Plus—is your best entry-level choice.

It's a fairly full-featured DVD-oriented product offering video capture, simple video editing, and data disc copying.

On the authoring front, it uses wizards to step you through the process, although you can dive right in. It comes with 29 generic menu backgrounds, but you can use any of your own BMP or JPG images.

As shown in Figure 16.7, neoDVD limits you to only four menu/button layouts. As you add video files or image collections, it adds buttons in these predetermined locations.

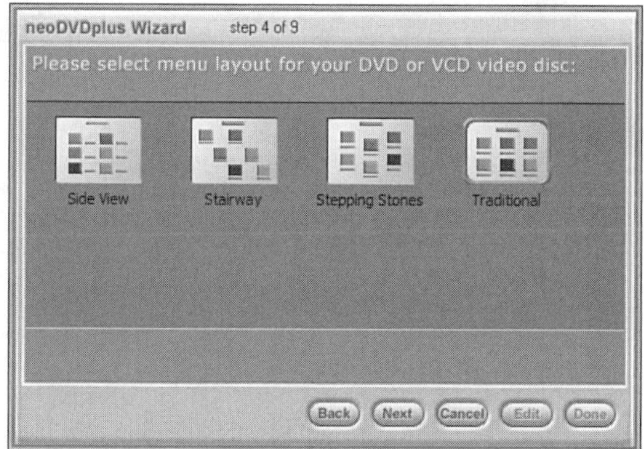

FIGURE 16.7
Choose from only four narrowly defined menu layouts.

Menu text options are very limited. NeoDVD ships with only a handful of fonts and offers only text color as an option. No font size, underline, bold, or italic features are available.

Instead of graphic buttons, neoDVD uses 16 frames (see Figure 16.8). Each frame holds a user-selected image from its associated video. As with the three other also-rans in this authoring category, no options are available to change button features such as location, size, or color or to add drop shadows.

The video edit mode shown in Figure 16.9 lets you trim video clips, insert transitions at the start and end of clips, add audio, select chapter points, and grab still frames to use as menu backgrounds. Two nifty features are its built-in narration recording function and two simplified menu navigation options.

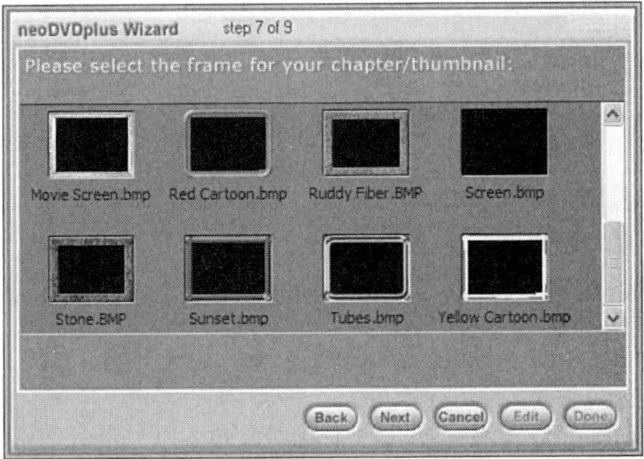

FIGURE 16.8
Sixteen buttons—chapter/thumbnail frames—highlight video still images.

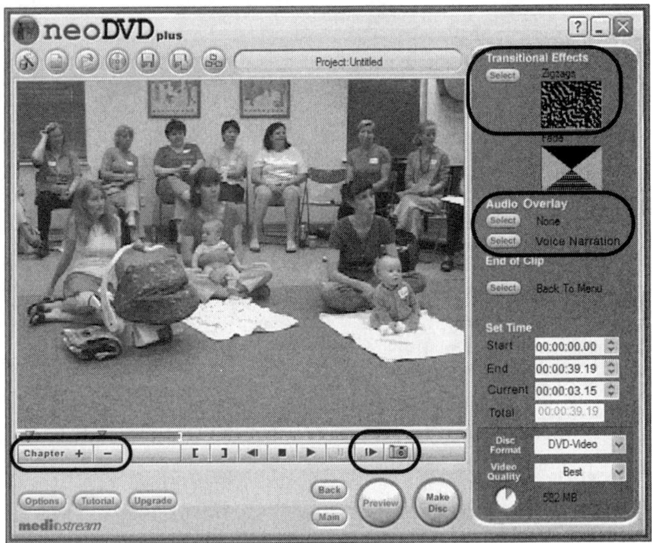

FIGURE 16.9
The video edit module has a solid feature set.

When you select chapters, neoDVD automatically creates a submenu, but accessing it is awkward and unpredictable.

The moving menu option is a kludge. When you play the completed DVD, each button in turn plays its associated video for 15 seconds; it's distracting and useless. It would work much better if a button played a video only when you rolled the cursor over it or if they all played at the same time.

Pinnacle Expression

This, too, is a $50 product, but Expression doesn't quite match neoDVD's functionality and options. Its interface has a polished look, but some quirks keep me from recommending this product.

The oddest quirk is its menu/button layout. As shown in Figure 16.10, if you have fewer videos and slideshows than buttons in a particular template, Expression leaves a blank or empty button frame on the menu.

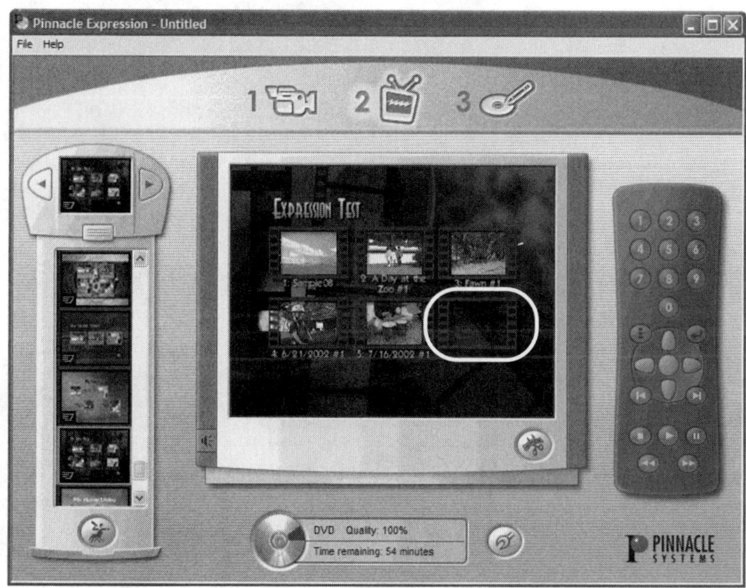

FIGURE 16.10
If you have fewer videos and slideshows than buttons, Expression leaves blank frames on the menu.

The video capture module offers scene detection, which is one way to add chapter points to a video, but it's an inexact process. The slideshow feature offers four transitions and user-added audio.

No nested menus are available, but the collection of menu backgrounds does offer a few with animation. You can add music, and the small collection provided is adequate.

Although you can't customize fonts, Expression offers a nice collection of font styles. As shown in Figure 16.11, however, problems arise when larger fonts crowd menu and button layouts.

FIGURE 16.11
Text style templates give you more options, but unchangeable font sizes sometimes lead to crowded menus.

The simplified video edit mode shown in Figure 16.12 lets you split videos and insert transitions between consecutive videos.

FIGURE 16.12
Minimal video editing options let you split or combine clips, plus add transitions.

Expression also has an image editing tool, shown in Figure 16.13, that lets you crop, rotate, and change color characteristics. Its encoding options are reasonably customizable, and it comes with a CD/DVD label and jewel case layout template. Finally, its documentation—an 80-page printed manual—is the best of this bunch.

FIGURE 16.13
Expression's image editing tool is a nice plus.

Ulead DVD MovieFactory

Take a pass on DVD MovieFactory; it is by far the weakest of the five entry-level authoring tools. Poor workflow, limited options, and a counterintuitive interface keep this $25 product out of contention.

The main user interface, shown in Figure 16.14, is a number-crunching nightmare. You don't need to see all that data when all you want to do is add videos to a menu.

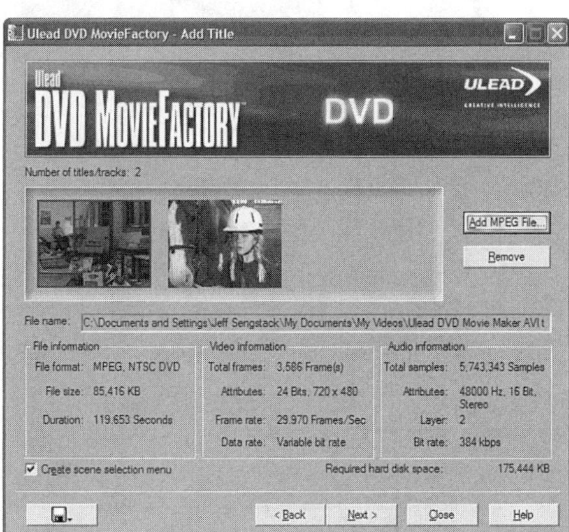

FIGURE 16.14
DVD MovieFactory's user interface is a data-cluttered mess.

DVD MovieFactory works only with MPEG files. So, you need to use its capture module, select Edit, select your video files, and select Export to DVD to see the prompt in Figure 16.15, which tells you to convert your video files to MPEG (otherwise, you can't do any DVD authoring).

This is a horribly slow process, and it interrupts the workflow. As with other DVD authoring products, it should be part of the final DVD burning step.

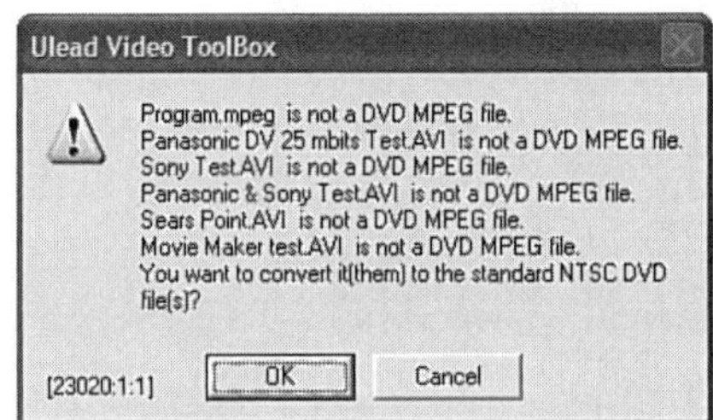

FIGURE 16.15
Stop the presses! You must convert all media to MPEG files before you can author a DVD with DVD MovieFactory.

Unlike the competition, DVD MovieFactory offers no button attributes—simply video stills with no borders or drop shadows. Your only option with text is color selection.

Its preview function, shown in Figure 16.16, works smoothly, but the remote control's limited button options demonstrates what little customizability this product offers.

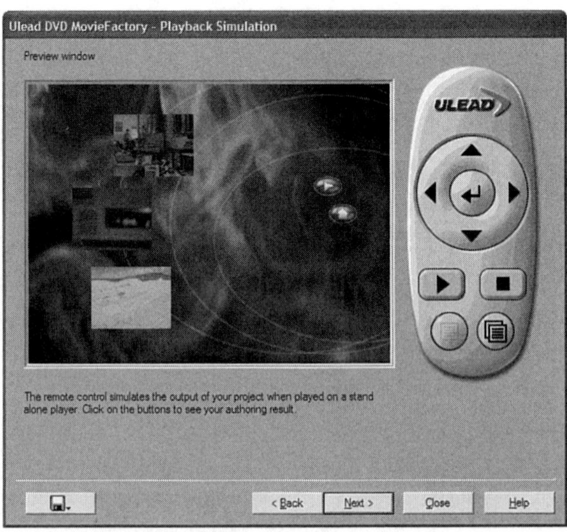

FIGURE 16.16
The preview mode works smoothly but offers few options.

In its favor, you can create chapters (doing so automatically adds a nested menu with those chapters buttons). Additionally, it has a First Play feature, and you can input your own images for menu backgrounds and add music to menus. However, the songs included with DVD MovieFactory are mainly tinny, cheesy MIDI files you'll want to avoid. Its edit function lets you split and trim clips.

Testing Prosumer-level DVD Authoring Products

As you develop your DVD production skills, you likely will end up using a prosumer-level authoring product. When I was putting this book together, four such products were available for PCs: Sonic Solutions DVDit!, Pinnacle Impression DVD-PRO, and Ulead DVD Workshop. The fourth, Adobe Encore DVD, fits in a unique market niche of top-end prosumers and professionals who are adept at using Adobe Premiere and Photoshop.

I strongly favor DVDit! On this book's companion DVD, I've included a copy of DVDit! 2.5. You'll use it in Hours 17–20.

By the time this book arrives on shelves, the prosumer-level DVD authoring landscape might have shifted yet again. This is a rapidly growing business niche; analysts says that by 2006 there will be 50 million PC DVD recorders. Therefore, Sonic Solutions, Adobe, Pinnacle, and Ulead might improve their products; prices probably will drop; and other companies might join the fray.

Regarding the rapidly changing DVD authoring marketplace, Pinnacle Impression is a case in point. A company named Minerva first released Impression in 1998 with a retail price of $10,000. Pinnacle bought it out, dropped the price to $1,000, dropped it again to $600, and as of mid-2003 sold Impression for $400.

For this moment then, here's a snapshot of the prosumer side of the DVD authoring market.

Sonic Solutions DVDit! 2.5

DVDit! 2.5 is the prosumer DVD authoring tool of choice. You will get a lot of hands-on experience with it in the upcoming hours. For now, I'll give you a brief overview.

Its interface, shown in Figure 16.17, facilitates a typical DVD authoring workflow. Menu backgrounds, buttons, text, and media are all instantly available at the press of a button. It ships with an extensive palette of graphics to give your DVD a truly customized look.

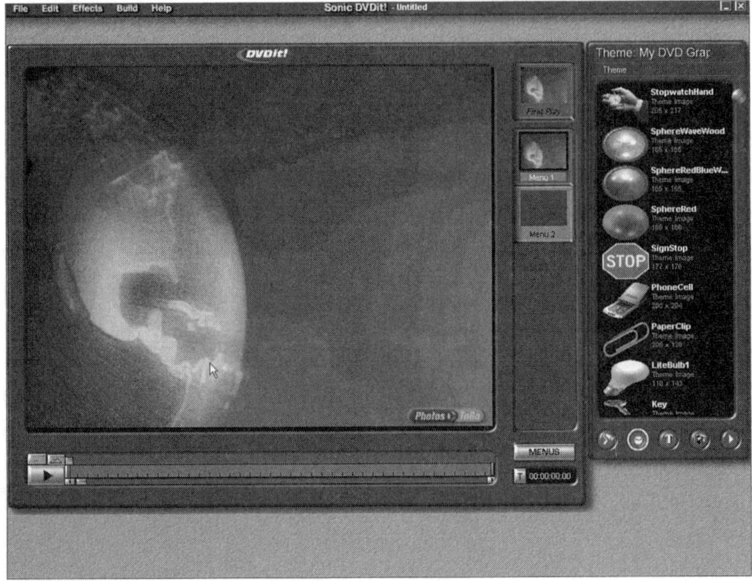

FIGURE 16.17
Sonic Solutions DVDit! 2.5 features a logical interface layout that follows a typical DVD authoring workflow.

DVDit! 2.5 streamlines the button and menu linking and attribute processes using right-click accessible Menu and Media Properties windows and drop-down lists. As shown in Figure 16.18, it displays all button hotspot boundaries, letting you drag and drop media or menus to the right locations to automatically create links.

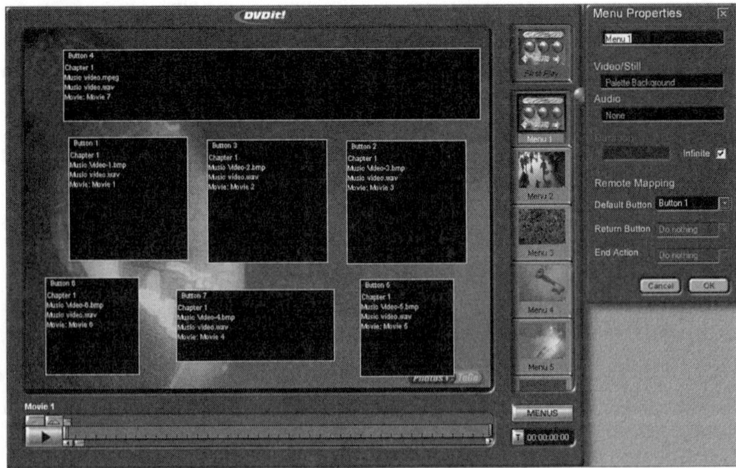

FIGURE 16.18
DVDit! 2.5 gives you direct access to all button elements, providing direct control over virtually every aspect of DVD creation.

Setting chapter points in videos is a simple double-click process. After points are set, you can easily drag those chapter points to a scene selection menu.

Finally, switching to preview mode takes only a moment, and the built-in transcoder—the software that converts media into MPEG files—is top-notch and works in real time. All in all, DVDit! is a powerful yet easy-to-use DVD authoring tool.

16

Pinnacle Impression DVD-PRO

Impression fails to fit comfortably into any market niche. It's too complicated and counterintuitive for entry-level users, and it lacks the flexibility and tools that would satisfy prosumer users.

Its timeline, shown in Figure 16.19, creates unnecessary complexity by spreading media assets out in a long line. Also, its apparent sequential order does not reflect the interactive flow of a typical DVD.

FIGURE 16.19
Pinnacle Impression uses a professional-looking but cumbersome timeline.

Most professional-level DVD authoring products assume you'd prefer making menus and buttons using third-party graphics software such as Adobe Photoshop, Photoshop Elements, Corel PhotoPaint, MicroGrafx Picture Publisher Pro, or any other paint program that can create layers and save them in the PSD (Photoshop) file format. This is a bit of a trade-off: You get much more control over the look and feel of your DVD but much less convenience.

But some prosumer or mid-level DVD authoring products accommodate producers who want to use built-in menu and button creation or editing tools. Impression has no such tools, so you must make and later edit all menus and buttons in third-party software. That can be very tedious.

Impression does ship with some templates, shown in Figure 16.20, but they barely begin to make up for the lack of built-in tools.

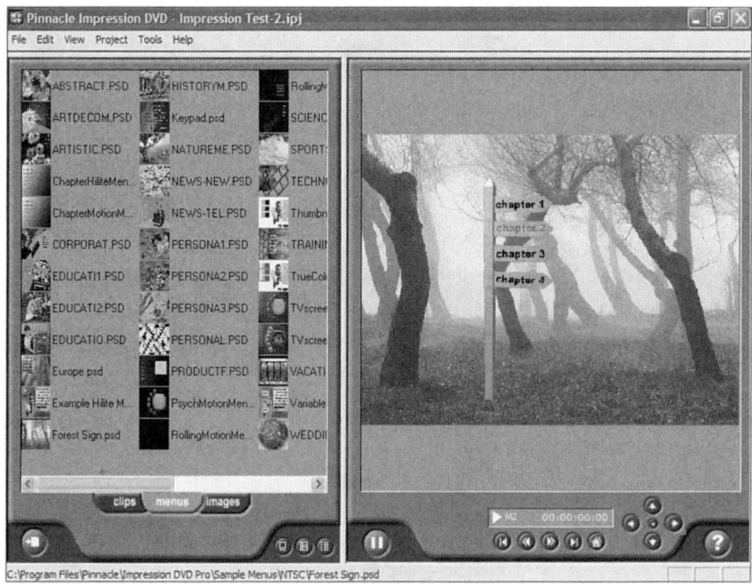

FIGURE 16.20
A few templates do not make up for the lack of menu and button creation or editing tools.

Creating chapters is simple and linking them to buttons is equally easy, but there's no built-in way to use a video scene in a button.

Impression supports motion menus but expects you to go through a convoluted process to accomplish that. Unfortunately, as shown in Figure 16.21, following those steps leads to a bizarre composite of two menus: one in motion, the other static.

Background graphics must be 24-bit color BMP files *only*! Even worse, importing 16-bit color graphics crashes the program.

To its credit, Impression supports subtitles, but adding them is laborious and tedious. You must use a graphic program such as HyperSnap-DX or Microsoft Paint to create each subtitle phrase as a separate 16-bit, 720×480, 1MB BMP file. Then, you must create a text file similar to the one shown in Figure 16.22, listing each BMP file and noting its duration and exact frame location in the video. As a final insult, you can't preview subtitles—you must burn a DVD before you can see how they'll look in your video.

FIGURE 16.21
The motion menu option failed to work and created a static/motion menu sandwich.

```
00:00:05:00·00:00:09:29·C:\Subtitles\SimpleT1.bmp
00:00:10:00·00:00:14:29·C:\Subtitles\SimpleT2.bmp
00:00:15:00·00:00:19:29·C:\Subtitles\SimpleT3.bmp
00:00:20:00·00:00:24:29·C:\Subtitles\SimpleT4.bmp
00:00:25:00·00:00:29:29·C:\Subtitles\SimpleT1.bmp
00:00:30:00·00:00:34:29·C:\Subtitles\SimpleT2.bmp
00:00:35:00·00:00:39:29·C:\Subtitles\SimpleT3.bmp
00:00:40:00·00:00:44:29·C:\Subtitles\SimpleT4.bmp
```

FIGURE 16.22
The completely manual subtitle process, including this text file listing of all subtitles, is terribly tedious.

Impression does offer a few high-end features, such as multiple video angles, eight audio tracks, and Dolby digital. But this DVD authoring software has too may issues to warrant a recommendation.

Ulead DVD Workshop 1.3

At $300, DVD Workshop falls at the low-price point of this mid-range market niche. As shown in Figure 16.23, its look and feel is similar to its entry-level sibling, DVD MovieFactory, and it suffers from some of its lower-priced sibling's inconsistencies. Ulead overcomes some of those issues and journeyman functionality by packing DVD Workshop with buttons, menus, and templates.

Its video capture is basic. In addition, its video edit section, shown in Figure 16.24, is very limited, only letting you trim and split clips, add audio, set chapter locations, and grab thumbnail images for use as menu buttons or backgrounds. It has no between-scene transitions for its slideshow feature.

Where DVD Workshop shines is in menu and button creation. It ships with dozens of menu and text templates, menu backgrounds, buttons, thumbnail frames, and text fonts. When creating or altering text or buttons, you can easily change sizes, colors, and transparency.

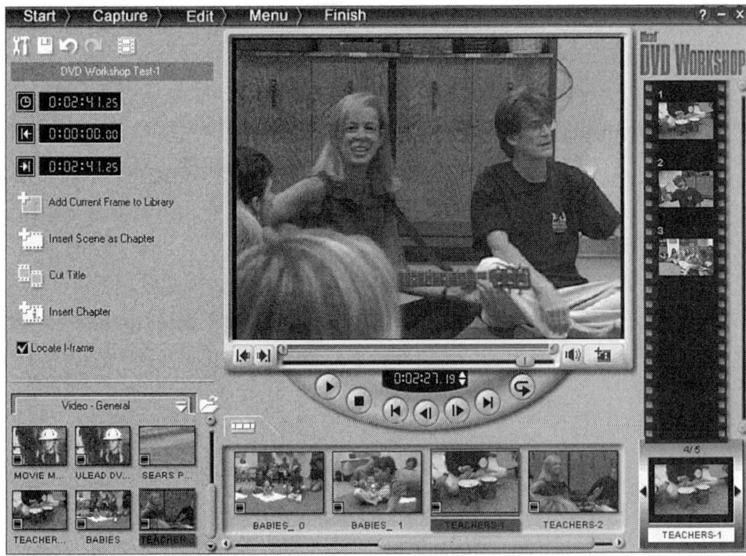

FIGURE 16.23
DVD Workshop has the same look and feel as its lesser sibling, DVD MovieFactory.

FIGURE 16.24
The video edit module offers the bare minimum in functionality.

It also offers some context-sensitive, right-mouse-click menus that let you align buttons and text plus copy and paste their attributes (size and style) to achieve a consistent look. And an easy-to-use menu lets you customize button *states*—how they change appearance when you navigate the remote over them.

You might opt to use the Menu Wizard, shown in Figure 16.25, but it's poorly implemented and confusing. It's much easier to create menus in the regular user interface.

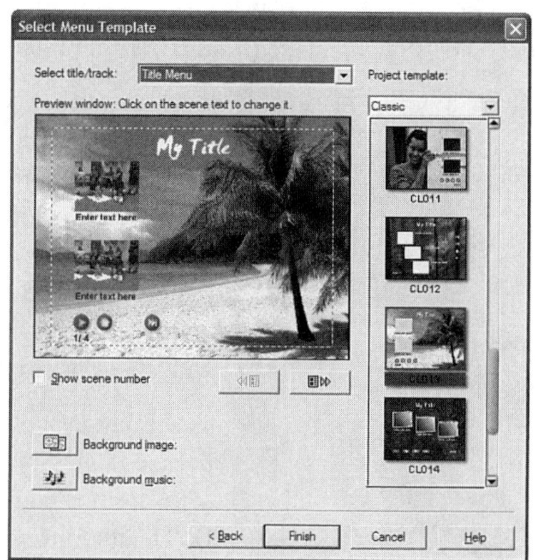

FIGURE 16.25
The Menu Wizard is more trouble than it's worth.

Menu navigation controls are limited. DVD Workshop supposedly lets you use videos for menu backgrounds and to create animated buttons, but I could not get either feature to work. It doesn't provide any widescreen option or capability to add data files, URLs, Photoshop files, Dolby AC-3 audio, subtitles, or additional audio tracks, all of which are elements you'd expect at this price. Finally, DVD Workshop's output choices, shown in Figure 16.26, give you some data rate flexibility but not much more.

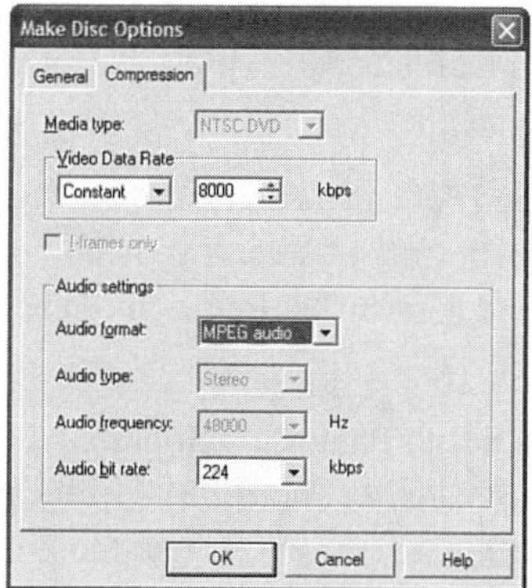

FIGURE 16.26
Output choices are limited to data rate.

Adobe Encore DVD 1.0

As I began wrapping up this book, Adobe invited me to test an early alpha and later a more solid beta version of Adobe Encore DVD. This newest addition to the Adobe product line is an excellent, carefully engineered, and powerful product with a solid pedigree.

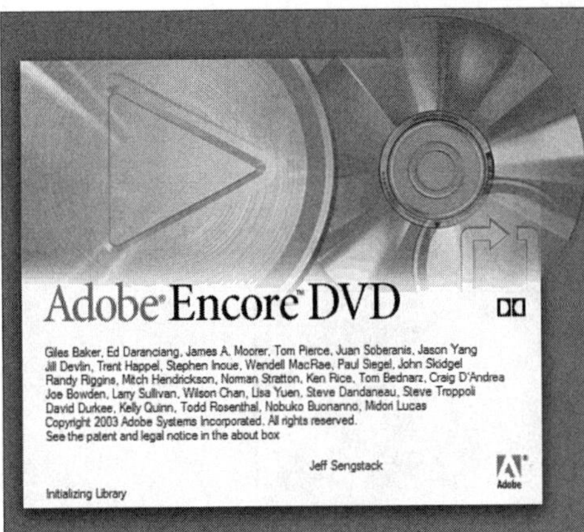

Its roots go back to the development of Premiere 6.5. Adobe foresaw the explosive growth in the DVD authoring market and wanted to add a DVD authoring module to its immensely popular Non-Linear Editor. It worked out a deal with Sonic Solutions to provide Premiere 6.5 users with an LE version of DVDit! 2.

That relationship led to another deal to license Sonic Solutions' AuthorScript software (the engine that drives all its DVD authoring products) as well as provide a team of Sonic Solutions engineers to help Adobe develop the standalone, DVD authoring product called Adobe Encore DVD.

At $549, it fits into a unique market niche: above Sonic Solutions' prosumer DVDit! and below Sonic's professional products: ReelDVD and DVD Producer.

As shown in Figure 16.27, Encore DVD's user interface has the look and feel of other Adobe products, such as Photoshop. Much like Sonic Solutions DVD Producer (see Hour 24, "ReelDVD: Part II, DVD Producer and DVD Trends," for an overview), Encore DVD uses a timeline for each video or slideshow. Plus, it incorporates an Explorer-like window to list menus and buttons, and their attributes. It also has monitor windows with tabs to readily move from one menu to another.

It offers some neat twists on setting DVD menu and button navigation. Figures 16.28 and 16.29 show two such schemes: a collection of arrows you can drag and drop to change button navigation within a menu and a means to drag a button action from a navigation window to a menu.

Another strength is Encore's close ties to Photoshop. It has a built-in means to recognize menus and other graphic elements created in Photoshop specifically for Encore and uses the Photoshop layer approach to facilitate customizing button highlight colors and other characteristics (see Figure 16.30).

One other strength is Encore's menu design capabilities. These include extensive menu layout options; text style editing; and Photoshop design tools that let you make immediate, easily reversible changes.

FIGURE 16.27
Adobe Encore DVD uses a Photoshop-like interface as well as file lists and tabs in windows to easily access and update menus and buttons.

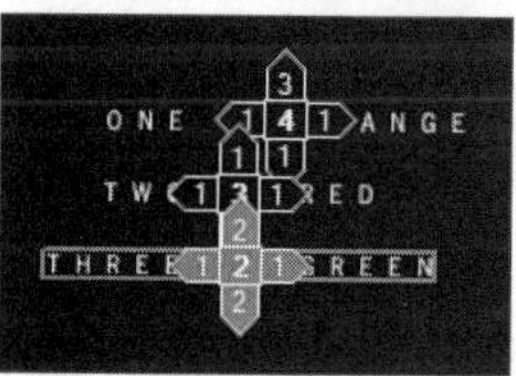

FIGURE 16.28
Use this four-point tool to drag and drop new button navigation within a menu.

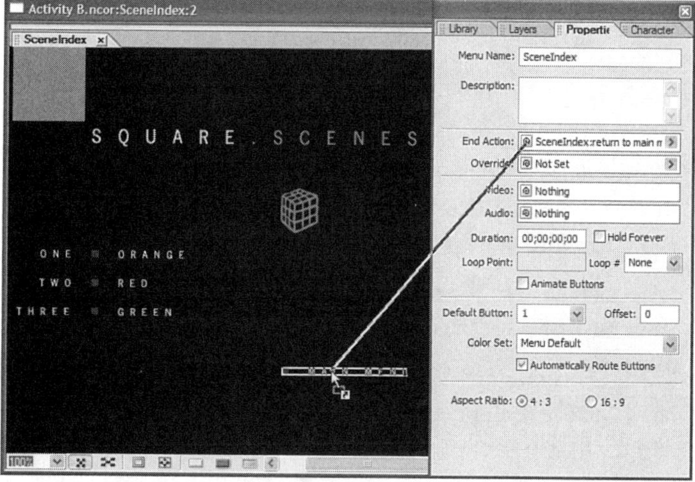

FIGURE 16.29
Set button attributes by dragging and dropping them from a dialog box to a menu.

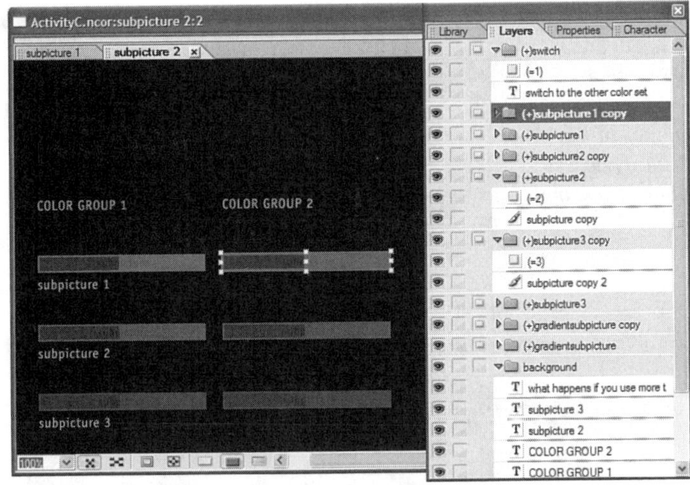

FIGURE 16.30
Encore emulates Photoshop layers to give users direct control over setting button highlight attributes.

It also offers features you'd expect to see only in more expensive professional authoring applications, such as chapters, 8 audio tracks, 32 subtitle tracks, and the option to add DVD-ROM data files to your project. For those who have doubted the value of DVD authoring applications, Adobe's entry into this market should make believers of them.

Summary

DVD authoring software does much more than add menus to movies. Entry-level products ease you into the process by using templates and offering a lot of button, menu, and text creation options. Prosumer products give you deep flexibility in menu layout.

Better products let you import myriad graphics, video, and audio formats and let you output your authored DVD using adjustable bit rates or to a variety of media.

When making your authoring purchase decision, consider the value of special features such as subtitles, extra audio tracks, URL links, chapters, motion menus, video capture, and video editing.

DVD authoring software is a newly emerging and rapidly changing market niche. A labyrinth of features and usability issues confound the evaluation and buying process.

The DVD authoring software market is changing rapidly, and new products arrive regularly. Quality is climbing while prices are dropping. In this hour, I presented a snapshot of the industry as of early 2003. By the time you read this, things will look different, so use the information here to help guide you in your DVD authoring software purchasing process.

This hour's material doesn't lend itself to the typical Q & A section of other hours. But I would suggest a couple of tasks.

Exercises

1. The DVD authoring software business does not stand still. Check out the Web sites of the major players in this field: Adobe (`www.adobe.com`), Cyberlink (`http://www.gocyberlink.com/`), MedioStream (`http://www.mediostream.com/`), Dazzle (`http://www.dazzle.com/home.html`), Pinnacle Systems (`http://www.pinnaclesys.com/`), Roxio (`http://www.roxio.com/`), Sonic Solutions (`http://www.sonic.com/`), and Ulead (`http://www.ulead.com/`).
2. I recommend visiting DV Direct at `http://www.dvdirect.com/TSS/charts/DVDAuthoringComparison.htm`. It's an online retailer that has posted feature-by-feature comparisons of most DVD authoring products. Web sites come and Web sites go. So, there's no telling whether this helpful site will be around by the time you read this. When I last looked, it did not have the absolute latest versions posted, but it still is an excellent place to start your comparison shopping.

HOUR 17

Designing DVD Projects and Authoring with DVDit!: Part I

I've organized this book to build to this point. Up until now, you've created content and dabbled in DVD authoring using My DVD 5. In this hour, you'll step up to a full-featured DVD authoring application and start the process of creating highly customized DVDs.

In this and most of the next four hours, you'll work with DVDit! 2.5, Sonic Solutions' prosumer DVD authoring application and the best DVD authoring application you'll find at its $299 (Standard Edition) to $399 (Professional Edition) price point.

I'll start this hour by offering some general planning concepts. Then we'll take a first look at DVDit! 2.5.

The highlights of this hour include the following:

- Planning your project—what's the message and who's the audience?
- Presenting your media in its best light
- Organizing your DVD's menu structure
- Checking out the DVDit! interface

Planning Your Project

This hour serves as a general introduction to project planning. Later, in Hour 20, "DVD Authoring with DVDit!: Part IV," I'll offer specific anecdotal examples of a wide range of DVD projects. I'll use them to show you how others have organized their assets, planned their projects, and found creative and exciting ways to make their DVDs more interesting and entertaining for their audience and more practical and useful for their clients.

In this and the next four hours, my objective is to let you see firsthand what DVDit! 2.5 can do and help you achieve a comfort level when working with this prosumer-quality authoring tool. Once accomplished, it'll be much easier to create plans to suit your specific project needs.

What's the Message, and Who's the Audience?

In my view, when beginning to plan your project, you need to ask only two questions: "What's the message?" and "Who's the audience?"

By *message*, I mean what are you trying to accomplish with your DVD? Will you archive videos, create video vacation albums, educate employees, or sell a product?

In each case, you'll need to take a different approach. For instance, if your goal is only to archive videos, you'll probably use just one simple opening menu with buttons linking viewers to each separate video. On the other hand, if you're trying to sell something—real estate, wine, travel, and so on—you'll want to increase your production values by adding music and animation to your menus; creating custom highlights for buttons; and giving viewers virtual tours, fast-paced edits, and many choices.

Who's your audience? What are their needs and sophistication level? The beauty of DVDs is that just about anyone can use them. You just stick them in a DVD set-top player and, with remote control in hand, sit back and enjoy.

That said, you still need to create an intuitive flow to your project as a means to direct viewers to areas you want to emphasize. And if you want viewers to use PCs to access data files, you need to ensure they'll have the means and resources to access those files.

On the other hand, if you're creating a review of your child's soccer team's season, consider that parents and kids will view your DVD. They'll want to jump directly to individual games, specific highlights, and player statistics. To make that work well, you'll need to create a logical organization for your DVD.

Presenting Your Media in Its Best Light

By purchasing this book, you've made a commitment to do more than simply throw media on a DVD. You want to create a quality product—something that will have an impact, send a message, or create an impression. Something you can be proud of.

Keep It Simple

In general, at this early stage in your DVD production development, I'd suggest you keep it simple. Avoid busy menu backgrounds and too many buttons. Keep the number of menus down to a manageable size, and ensure the navigation of your project is intuitive. That is, at the completion of a video clip, take viewers to what would logically follow that clip—for instance, the main menu or the menu from which they accessed that clip.

Keep It Short

Look at your video and image assets, and think in terms of keeping things short. If you absolutely insist on including every second of video you took at a wedding, do two things: First, edit a highlights video and let users access it through the main menu. Second, set chapter points in the original, unedited video and let viewers jump directly to those specific wedding moments using a scene selection submenu.

Take It for a Test Drive

Test your final product on a colleague, friend, or spouse. If you plan to mass produce your DVD, burn a single copy and play it on a set-top device. If you will make only one or a small number of DVDs, let your PC be your test bed.

In any event, you want your test subjects to navigate through and view your DVD without prompting from you. Then ask them what they think you were trying to accomplish. You'll want to note whether they had any trouble moving through the menus or making media selections.

This peer review can be an uncomfortable process. Nevertheless, it gives you a chance to step back from your work and take a viewer's perspective. I find that when I do this, I almost always see glaring issues that need repair. Something that made perfect sense when working on it fails to resonate with my test subject—and usually for good reason.

Organizing Your DVD's Menu Structure

DVDs are interactive. That's one of their real fortes. You should therefore organize your DVDs to exploit that strength. To do that, you use nested menus, intuitive navigation, and clearly labeled buttons.

Nested Menus

Nested menus, as I explained in the Hour 1, "Discovering What DVDs and DVD Authoring Software Can Do for You," are menus within menus. If you worked through Hours 4–6, you've already encountered nested menus. You built a DVD with a main menu that had a button link to a *chapter* menu, which is a menu within a menu.

Consider a family history DVD. Its main menu might have button links to a video overview, specific family lines, immigration stories, photos, documents, and living history interviews. The link to the video overview could take viewers to a scene selection—chapter—menu. Each of the other subcategories also could have its own menu.

To ensure a logical flow to your DVD, organize it using a flowchart similar to the one in Figure 17.1. This follows the basic family tree DVD structure I touched on in the first hour. The main menu gives you access to submenu topics, which in turn might give you further access to additional submenus or nested menus. In this case, your viewer would be able to drill down quickly to his specific area of interest or take a more linear and leisurely stroll through the DVD's contents.

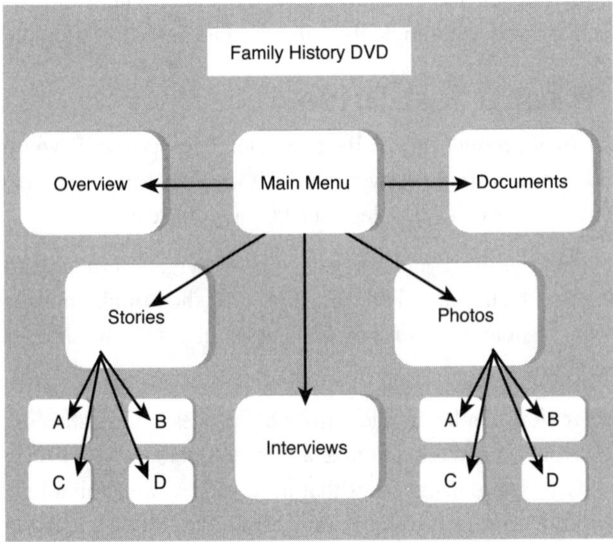

FIGURE 17.1
One possible DVD menu flowchart for a family history DVD.

Buttons Should Do What They Say

This might seem like I'm stating the obvious, but I've seen too many DVDs that don't make sense. It's unclear what will happen when you press a button or where a click will take you. Many times DVD authors try to use only icons for buttons. Although they might make sense to the author, they might not make sense to the viewer.

One advantage of DVDit! is that not only can you place text on or near a button, but you also can use that text as a link to the same media as its associated button. The bottom line is that clear navigation trumps clever artwork.

Checking Out the DVDit! Interface

17

My view of DVDit! 2.5 is that it's a prosumer-quality DVD authoring application with some consumer-friendly features. It offers some of the power and flexibility found in professional products that expect a higher level of expertise on menu creation from the user, but it eases users into authoring by offering plenty of templates, an organized workflow, and easy access to customization tools.

In the remainder of this hour, you'll do some file management housekeeping and then take a brief tour of DVDit!'s interface and preference settings.

Task: Combine DVDit!'s Graphics into One File Folder

TASK

Sonic provides two sets of backgrounds and buttons—Default and Corporate—with this SE version of DVDit! By default, the DVDit! install places them in separate folders, forcing you to switch between them if you want to access objects from both groups. Follow these steps to combine them all into one personalized location:

1. Install DVDit! 2.5 SE by navigating to the DVDit! 2.5 SE folder on this book's companion DVD and double-clicking its `Setup.exe` file. To make following the coming tasks easier, accept the default storage location. You'll need to reboot to complete the installation process.

> This is a trial version of DVDit! 2.5 that will time-out (stop functioning) in a month. You can purchase it online at `www.sonic.com`. Its minimum specs are Windows 98, 2000, Me, or XP; 300MHz CPU; 128MB of RAM; 18GB hard disk; 24-bit color graphics display; 1024×768 resolution; and Direct Show 8.0.

2. After rebooting, open My Computer or the Windows Explorer to navigate to the Sonic Solutions Themes directory—`C:\Program Files\Sonic Solutions\DVDit! SE\Themes`.
3. Create a new folder in the `Themes` folder by selecting File, New, Folder. As I've illustrated in Figure 17.2, name it something such as **`My DVD Graphics`**.

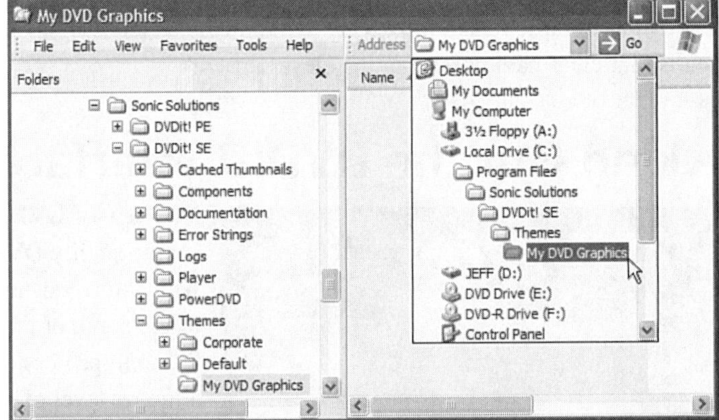

FIGURE 17.2
Create a new file folder to store all your graphics and media for use in DVDit!

4. Open the `Themes/Default` file folder and copy/paste its three folders into your newly created folder. (You can use the click/drag method to select all three at once.)

After you've copied these three subfolders into your newly created `My DVD Graphics` folder, you need to add files to them, one folder at a time. Thus you must perform the extra, manual labor in steps five and six.

5. Go to the `Themes/Corporate/Backgrounds` folder and copy/paste its one file—`Corporate_Backgrounds.SonicTheme`—to your `My DVD Graphics/Backgrounds` file folder.
6. Do the same with the `Themes/Corporate/Buttons` folder and copy/paste its one file—`Corporate_Buttons.SonicTheme`—to your `My DVD Graphics/Buttons` file folder.

Be sure you do not change the DVDit! `Themes` file folders naming convention: `Backgrounds`, `Buttons`, and `Media`. These names cue DVDit! to handle files from those folders in specific ways.

Listing DVDit!'s Attributes

Your copy of DVDit! 2.5 SE is loaded with features and functionality. It works with both Microsoft AVI and Apple QuickTime MOV audio and video files as well as MPEG-1 and 2. Its MPEG transcoder (the software that converts media files into MPEG) works in real time and produces high-quality output.

Your projects can have unlimited menus with up to 36 buttons per menu, chapters for scene selection options, video and chapter point thumbnail images for buttons, and timed menus with audio; you can even add ROM data to the disc.

DVDit! lets you directly control the menu and movie navigation properties, button and text placement, and the appearance of text and buttons including setting color and drop shadow characteristics.

On the output side, DVDit! SE lets you burn projects to virtually all DVD and CD recordable media and save disc images on your hard drive.

DVDit! PE (Professional Edition) costs $100 more ($399) than the SE version. It offers widescreen (16:9 aspect ratio) support, Dolby audio import and export (transcoding), and Digital Linear Tape (DLT) exporting. I discuss DLT in Hour 21, "Burning DVDs and Dealing with Mass Replicators." Basically, if you want to use a mass replicator to stamp hundreds of copies of your DVD, it's not absolutely necessary to export your project to a DLT. However, DLT is more reliable than using the other industry-standard replication master medium: DVD-R.

Touring DVDit!'s Interface

You'll venture into actual authoring in the next hour. Before taking that step, I want you to examine the DVDit! interface as a means to get comfortable with the workflow to come.

DVDit! divides your authoring tasks into three basic steps:

- Menu layout
- Adjusting links and properties
- DVD burning

Routine drag-and-drop methods simplify most of these steps. You'll drag menu backgrounds to a window, drag buttons onto that background, add text (adjusting its location and other characteristics), and drag movies and chapters to those menus creating buttons and links in the process.

Task: Examine DVDit! 2.5's Interface

That work will be a lot easier if you first get acquainted with the workspace. To do that, follow these steps:

1. Start DVDit! 2.5 by double-clicking its icon on the desktop or selecting Start, Programs, DVDit! SE, and then clicking its icon. After the small splash screen disappears, click Start a New Project from the opening menu, shown in Figure 17.3.

In the future, instead of clicking Start a New Project, you can click Open an Existing Project to access any work you've saved during the upcoming tasks.

FIGURE 17.3
The DVDit! opening menu.

2. The opening menu changes appearance and offers two drop-down lists. Select NTSC or PAL—depending on your country's TV standard—and MPEG-2 (DVD Compliant). Click Finish. That opens the DVDit! interface shown in Figure 17.4. In a few moments, you'll use its TV screen to create menus.
3. Test to ensure that your file management housekeeping worked smoothly by clicking the small-print Theme in the Theme window (see Figure 17.5), selecting Open Theme, selecting your newly created file folder in the dialog box shown in Figure 17.6, and then clicking OK.

As a result of your file management housekeeping, you will have 30 backgrounds and 40 buttons in your theme—all readily accessible. You won't need to switch from the default theme to the corporate theme to search for a graphic. Simply stick with your `My DVD Graphics Theme` folder.

FIGURE 17.4
The DVDit! interface streamlines the production process.

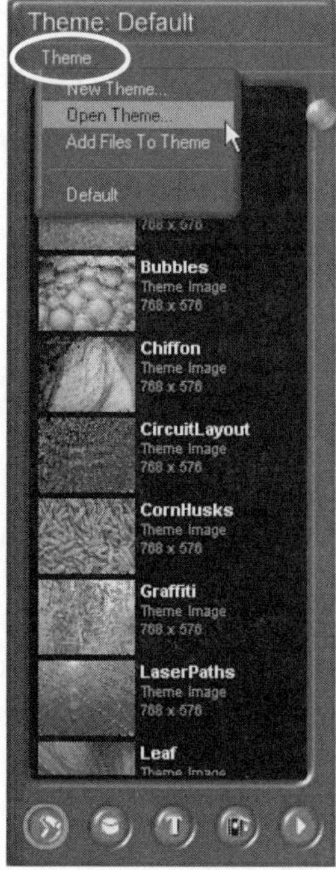

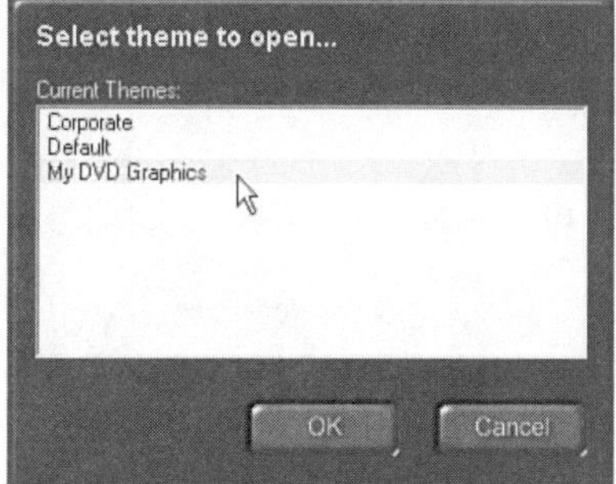

FIGURE 17.5
Click the small-print Theme to open the dialog box shown on the right.

FIGURE 17.6
Select your newly created theme as the one you'll use to access menu backgrounds and buttons.

4. Check out the Theme palette, shown in Figure 17.7. Here you will find all the elements—menu backgrounds, buttons, text, and media—to create your DVD project. Note the five buttons highlighted in Figure 17.7. Click each in turn to open its respective palette:
    - **Background**—You can use any of the images in this palette as menu backdrops.
    - **Buttons**—You use the buttons in this palette to create links from menus to media or other menus.
    - **Text**—You use this palette to access any font installed on your system and to add text to any menu or button.
    - **Media**—Any video clips, audio cuts, and images you've stored in the Sonic Solutions media folder show up in this palette.
    - **Play (or Preview)**—Clicking this button opens a push button remote control–like interface, shown in Figure 17.8, that you'll use to test your DVD project before burning a DVD or CD. Click the × in the upper-right corner to close this interface.
5. Add your tutorial media assets to this project by clicking the Media button; selecting Theme, Add Files to Theme (see Figure 17.9); navigating to the DVDit! 2.5 Tutorial Assets file folder on your hard drive; selecting all but the menu files; and clicking Open. Your Theme window should look similar to Figure 17.10.

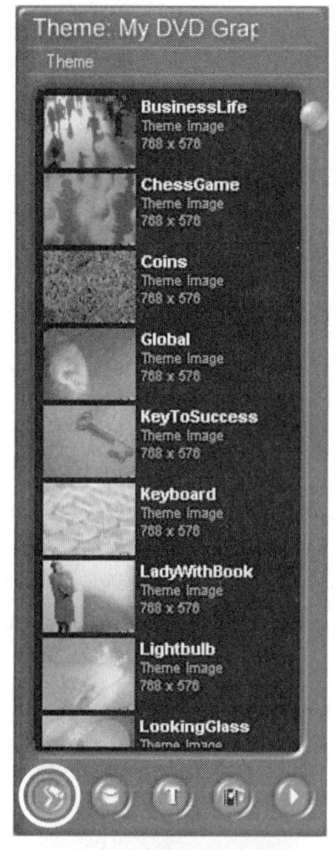

FIGURE 17.7
The DVDit! Theme window displays background images, buttons, text, and media (audio/video/stills).

FIGURE 17.8
You use this remote control–like interface to test the functionality of your DVD project before burning a DVD or CD.

FIGURE 17.9
Add media files to your project by selecting Theme, Add Files to Theme.

FIGURE 17.10
After adding media to your project, the Theme window should look like the image on the right.

Summary

In Hours 4–6, you worked with a purposely consumer-friendly DVD authoring application. MyDVD makes taking your first steps into DVD creation easy.

On the other hand, DVDit! 2.5 gives you much more control over your DVD production. You can create menus, buttons, and navigation that exactly suit your DVD design.

In this hour, I covered the first steps in that process, suggesting ways to think through your DVD's navigation and introducing you to DVD flowcharting and layout. I took you through the DVDit! authoring workflow and briefly explained how DVDit! 2.5 gives you direct access to button and menu attributes.

Workshop

As a means to review the DVD design and DVDit! features covered in this hour, review the following questions and answers, take the short quiz, and perform the exercises.

Q&A

Q I like the flowchart approach to DVD design, but wouldn't a simple outline suffice?

A First, an outline doesn't give you the visual feedback of a flowchart. And second, the flowchart model matches the workflow of ReelDVD, a professional-level product I introduce in Hour 23, "Professional DVD Authoring with ReelDVD: Part I," and Hour 24, "ReelDVD Part II: DVD Producer and DVD Trends." That product uses *objects*, icons that you place in a workspace and link using a simple drag-and-drop method. It's an elegant paradigm (I had to use that word at least once in this book).

Q I have two hours of video I want to put on a DVD. Should I accept the default 8.0Mbps setting?

A No. A two-hour video, transcoded to MPEG-2 at 8.0Mbps would require 7.2GB of DVD space (120 minutes × 60 seconds per minute × 8Mbps ÷ 8 bits per megabyte = 7200MB, or 7.2GB). DVDs hold about 4.7GB, so you'd need to reduce the transcode rate approximately 65% (4.7 ÷ 7.2) or to about 5Mbps—the Normal setting in DVDit!'s Video Compression Settings.

Q Why all the organizing and planning? All I want to do is create family vacation and kids' sports DVDs.

A Even in that case, a little planning will go a long way to creating an effective, engaging DVD. Plus, I know that, once you make a few DVDs, you'll start seeing the limitations of the consumer-level MyDVD and begin to recognize the creative potential of the prosumer-level DVDit! 2.5. A little planning now will facilitate bigger plans later.

Quiz

1. Explain the term *nested menus*.
2. How do you add media to your project theme?
3. What are the four principal concepts to consider as you plan your DVD?

Quiz Answers

1. These are menus within menus. A main menu, for instance, could have buttons with links to several other menus: scene selections, slideshows, or interviews. Each of those submenus is a *nested menu*. This is not to be confused with *chained* menus, which is a MyDVD menu approach used when a menu has more than the maximum number of buttons allowed by that menu's template (usually six). When that happens, MyDVD adds a link to the next chained menu, which basically serves as a continuation of the previous menu.
2. Click the Media (film icon) button at the bottom of the Theme window. At the top of the window, select Theme, Add Files to Theme; then navigate to the files you want to add (video, audio, images) on your hard drive, select them, and click Open.
3. What's the message, who's the audience, keep it simple, and keep it short.

Exercises

1. Create some flowcharts of DVD projects, real and imagined. As you consider the myriad possibilities—music, data files, scene selection via video chapters, and customized buttons and backgrounds—all sorts of possibilities will present themselves.
2. As you consider those possibilities, you might want to tap a couple of ideas presented on the Sonic Solutions Web site: `http://www.dvdit.com/dvdit_inaction.html`. There you'll see an interactive overview of training and marketing DVDs created using DVDit!

HOUR 18

DVD Authoring with DVDit!: Part II

DVDit! 2.5 is a prosumer authoring application that gives you much more control than a consumer authoring application over how your DVD looks and behaves. It affords you the opportunity to create DVDs that match your creative vision.

You can place buttons and text anywhere on the menu that suits you. You also can enhance those buttons and text with special attributes and customized linking capabilities.

In this hour you will start producing a DVD using media provided on this book's companion DVD. You will first adjust a few project settings and then start working with menu backgrounds, buttons, and text.

The highlights of this hour include the following:

- Adjusting project settings
- Beginning to work on menus
- Adding buttons
- Working with text

Adjusting Project Settings

Before starting your first project, you should take a look at the handful of user-selectable preferences. In general, you'll likely accept the defaults, but I think you'll want to see what's under the hood.

Task: Examine DVDit! Preferences

Settings are limited to output medium, project name, whether you'll add ROM data to the final DVD disc, and the transcode quality. Here's how you adjust those preferences:

1. Access Project Settings by selecting File, Project Settings. Doing so opens the dialog window shown in Figure 18.1.

An alternative means to access Project Settings is by selecting Build, Video Settings. In that case, the same Project Settings window opens with the Video tab selected instead of the Project tab.

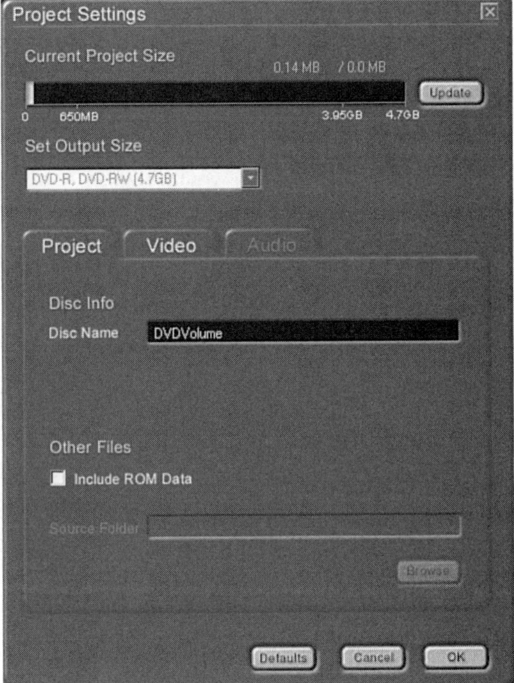

FIGURE 18.1
Project Settings offer only a few options: primarily transcode quality and whether to add ROM data to your DVD.

2. The Current Project Size setting is for your information. If you plan to put a video on a CD, it lets you know whether you've exceeded a CD's size limits.

DVDit! lets you create what Sonic Solutions calls a *cDVD*: an MPEG-1 or MPEG-2 video on a CD that displays on your PC using a software player automatically added to that CD by DVDit! when you burn your project.

3. Set the output size by opening that drop-down list and selecting the medium you plan to use. Usually, you'll select DVD-R, DVD-RW (4.7GB). DVDit! offers more output options than other similarly priced DVD authoring applications.
4. Change the Disc Name setting by selecting it and typing in a new name. **`My DVD Test`** suits the upcoming tasks.

This Project tab has an Include ROM Data check box. This lets you add data files to your DVD for use by persons on their PCs. I discuss this in Hour 22, "Creating Custom MyDVD Templates with Style Creator."

5. Click the Video tab. As shown in Figure 18.2, your only option here is either to accept the best quality transcode (conversion of AVI and MOV videos and all images into MPEG video) or manually set the transcode quality. For now, accept the default value (in Figure 18.2, I clicked the manual setting to display the five setting points). Click OK to close the Project Settings dialog box.

At the default 8.0Mbps (million bits per second), you can store about 1 hour and 20 minutes of video on a DVD. Hollywood films typically use lower data rates or double-layer DVDs to fit a full-length feature film on one side of a DVD. Because films run at 24 frames per second (versus 29.97 for NTSC), they have about 20% fewer frames to compress. That gives producers a bit of wiggle room when deciding how much compression to use. The more compression, the lower the video quality.

The Audio tab is grayed out (disabled) in your SE version of DVDit! In the PE version, however, you can access that tab to import or export Dolby digital audio.

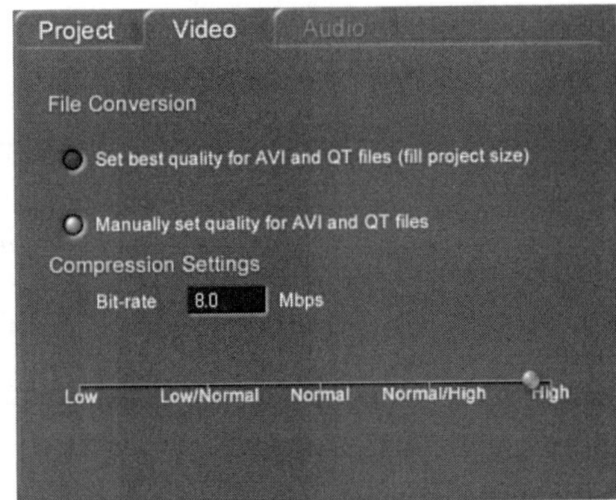

FIGURE 18.2
The File Conversion section lets you adjust the size and quality of your MPEG file.

Beginning to Work on Menus

Now you'll start to see that DVDit! is a major departure from MyDVD. Instead of relying on narrowly defined styles and templates, creating menus in DVDit! 2.5 is a free-form process.

Your first choice is whether to start your DVD with a video clip or a menu. This is called your *first play*. For this series of tasks, your DVD will start with a menu. I explain how to add video clips to your project in the next hour.

Task: Create a Menu

TASK

The opening menu is the foundation of your DVD project and sets its tone. To create that first impression, follow these steps:

1. Open DVDit! The Theme palette opens by default to the Backgrounds palette.
2. Add four menu backgrounds to your theme collection by selecting Theme, Add Files to Theme. Then navigate to the DVDit! 2.5 tutorial assets folder, select the four Music Video Menu files, and click Open.
3. Scroll down through the Backgrounds thumbnails, locate the four newly added Music Video Menu thumbnails, select the first one, and drag it to the First Play placeholder screen (see Figure 18.3). This does three things:
   - It places the chosen background into both the First Play and Menu 1 placeholder screens.
   - It displays the background in the main screen.
   - It adds a new, blank Menu 2 placeholder screen next to the main screen.

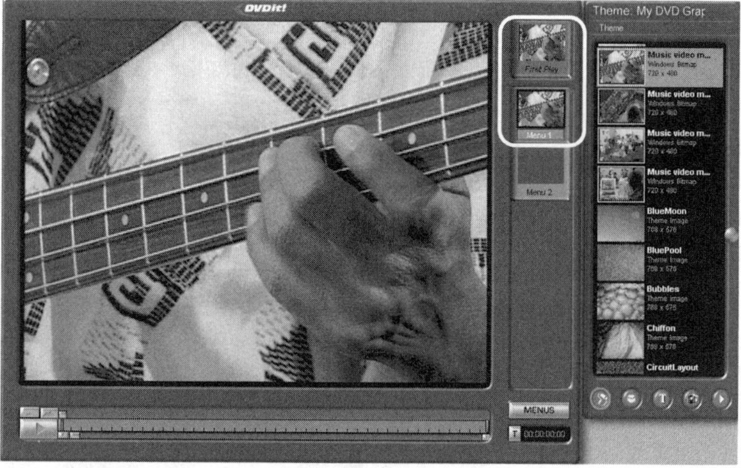

FIGURE 18.3
Dragging a background to the First Play placeholder adds it to the Menu 1 placeholder and the main screen.

4. Give your menu a descriptive name. To do that, simply click Menu 1 below the small thumbnail placeholder screen to the right of the main screen. This highlights the Menu 1 name. Type in something such as **`Main Menu`** and press Enter.

If, instead, you use one of the menu backgrounds that come with DVDit!, you might notice an icon in the lower-right corner identifying the company that provided those menu graphics. Not to worry. The logo falls outside the NTSC safe area, so most people watching DVDs created with these backgrounds won't see the logos.

You are not limited to using only the backgrounds that ship with DVDit! SE. You can use your own graphics, too. DVDit! supports the following graphic file types: BMP, RLE, JPG, PIC, PCT, PSD, PNG, TIF, TGA, VDA, ICB, and VST.

To import your background graphics to DVDit!, open the Background palette and select Theme, Add Files to Theme. Navigate to your graphics file(s), select one or more, and click Open.

For best results, your image should have a 4:3 aspect ratio such as 640×480. If it's less than 640×480, DVDit! will expand it and some sharpness will be lost. If it's significantly greater, DVDit! will shrink it and that will also result in some quality loss. Also, if the source graphic does not have the proper aspect ratio, DVDit! will distort it to fit its 4:3 aspect ratio.

5. Create a submenu by dragging the second Music Video Menu to the newly created Menu 2 placeholder screen. Doing so automatically creates a new Menu 3 placeholder for your next menu. Name this menu **`Photos`**.
6. Create two more submenus by dragging the third Music Video Menu to the Menu 3 placeholder and the fourth Music Video Menu to the automatically created Menu 4 placeholder. Name these menus **`Clips`** and **`Scenes`**. Your project window should look similar to Figure 18.4.

To replace any menu background, simply drag a new background to the menu thumbnail placeholder you want to replace.

FIGURE 18.4
How your project should look after adding three more menu backgrounds.

7. To save your project, select File, Save from the DVDit! main toolbar. Then navigate to a file folder of your choosing (as shown in Figure 18.5, I used the `Tutorial Assets` folder). Finally, give your project a name and click Save.
8. Later, if you want a different menu to be the first play item, simply drag that menu's thumbnail placeholder to the first play placeholder at the top of the column. However, that does lead to a minor inconvenience (see the following caution).

DVDit! SE has a small idiosyncrasy. If you change the name of Menu 1, that also changes first play to the name you give Menu 1. That is as it should be. However, if later you drag a different menu or movie to the First Play window, it does not display the name of this new menu; instead, it retains whatever you typed in originally for Menu 1.

This can become confusing. Just keep in mind that whatever menu or move is in the top first play screen (no matter what name this screen has on it) is the menu or movie that plays immediately after the DVD is inserted into the drive.

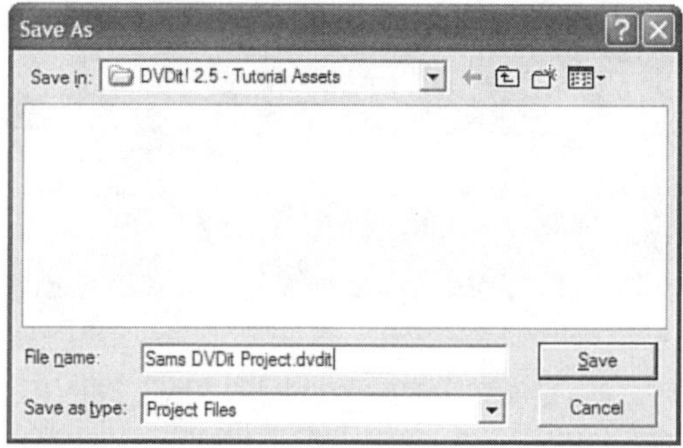

FIGURE 18.5
Save your project in the DVDit! `Tutorial Assets` *folder.*

18

Adding Buttons

TASK

DVDit! offers a full palette of button types: arrows, coins, business icons, a stop sign, and even a brick wall. Plus, you can make your own buttons in programs such as Adobe Photoshop Elements and then import them.

Task: Place Buttons in Your Menus

This task is fairly easy; it focuses only on selecting, placing, and resizing buttons. Hour 19, "DVD Authoring with DVDit!: Part III," covers changing their other attributes: colors and drop shadows. I'll also limit work in this task to the main menu. The next hour explains how to place video and photo thumbnail buttons on the Photos and Clips menus. Follow these steps to add and resize buttons:

1. Open your main menu by clicking its placeholder thumbnail.
2. Open the Buttons palette by clicking the Button icon. Scroll down the button collection to find a button you think fits on the menu background. As shown in Figure 18.6, I opted for the blue GlassWave button.

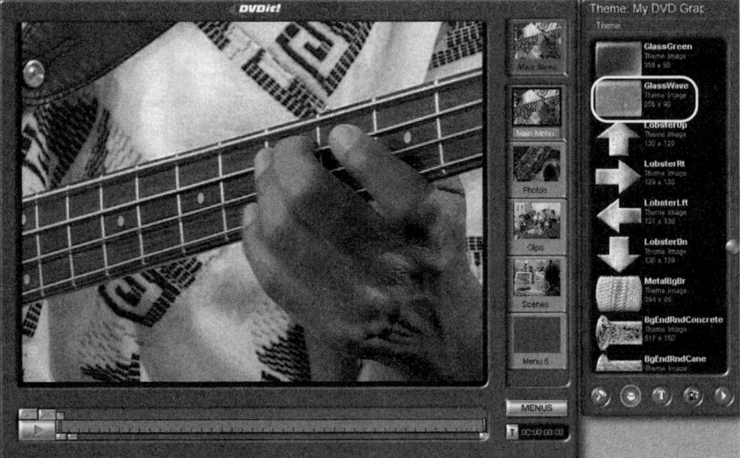

FIGURE 18.6
The Buttons palette has 40 button types, all of which can be placed anywhere on the menu and be resized.

3. Drag and drop your selected button onto the main menu screen. As shown in Figure 18.7, the proportions of the button do not match the icon in the palette.
4. Click the button to put a frame with red handles around it. Now you can resize and change its proportions by dragging the corners or edges.

In a slightly different twist to graphic resizing, Sonic Solutions has added a center handle. Click and drag it around to see how this changes the shape of the button while holding its position over the menu using that center handle as an anchor point.

Holding down Shift while using a corner or center point to resize maintains the button's current aspect ratio.

5. Delete this button by selecting it and pressing the Delete key.
6. Drag four other buttons to the main screen. As shown in Figure 18.8, I chose the GlassGreen button.
7. To change the shape or move all four buttons uniformly, select all of them by Ctrl+clicking them one at a time. Then drag them or resize them by dragging the corner or center handle of the one with red handles. All will act in unison, as illustrated in Figure 18.8.

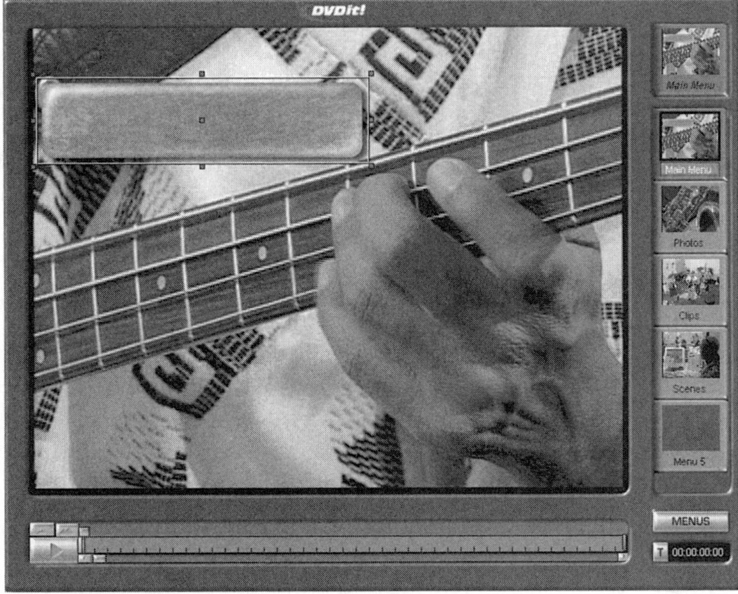

FIGURE 18.7
Use the button bounding box handles to change the size and proportion of the button.

As you add buttons to your menus, they show up in the menu thumbnail placeholders to the right of the main menu display.

FIGURE 18.8
Uniformly change the shape of selected buttons by Ctrl+clicking them one at a time and then dragging an edge or handle of one button.

As with most graphics programs, you can copy and paste these buttons. Here's how:

1. Create one button to your satisfaction and select it by clicking it.
2. Select Edit, Copy from the main menu.
3. Select Edit, Paste. Doing so places a duplicate button on top of and slightly to one side of the original. Drag this duplicate to a new location.

Although you can uniformly resize several buttons simultaneously, there are no tools such as a grid to help you line them up perfectly.

Here's a workaround: Create a dummy button first. Place it on the screen and expand its size to fit where you want to line up your buttons. Then add your buttons on top of the dummy button using it to line up their edges. After this is completed, select the dummy button and delete it.

Dragging your buttons with the mouse does not lead to pixel-specific placement. However, you can nudge them to line them up reasonably well by clicking them to select them and then using the keyboard arrow keys.

The order in which you add buttons to a menu is the order in which they will play if your DVD viewers press Next or Previous on their remotes. In general, this order can be a non-issue. If you do have a sequence in mind, create your buttons accordingly. If you want to keep the buttons but change the sequence, you can drag and drop the buttons on the interface to change the appearance of the sequence or change whatever movie, still image, or menu you link to particular buttons. I cover how you link buttons to media and menus in the next hour.

Working with Text

DVDit! lets you create text using any font on your PC. You can position that text anywhere on the screen; make it bold, italic, or underlined; give it any color or luminance; and add a drop shadow.

In addition, the ability to use text as a button sets DVDit! apart from most other authoring applications in its price category. You can link it to media or menus, which is a handy feature.

Task: Enhance Your Menu with Text

TASK

In this task, I've limited the work to only adding text to a menu or button and changing its size and shape. I cover changing its other attributes in the next hour. Here's how to work with text in DVDit!:

1. In the main menu, adjust the four buttons so they leave room at the top for a title.
2. Add a fifth button to the menu to serve as a background for the menu title. Use Figure 18.9 as a guide.

FIGURE 18.9
Adjust your menu buttons to make room for a title.

3. Add a title to your menu by clicking the Text icon in the Theme palette, selecting a font, clicking that font, and dragging it to the main screen above the buttons. As shown in Figure 18.10, doing so adds the word `Text` to the menu.

> By default, your text has a drop shadow. I show you how to adjust or remove this in the next hour.

4. Click Text to display its red-handled bounding box. Click again to highlight the word `Text`. If you click a third time, DVDit! places a cursor within the word `Text`. All three types of text displays are shown in Figure 18.11. Now, type in **`Music Video`**.

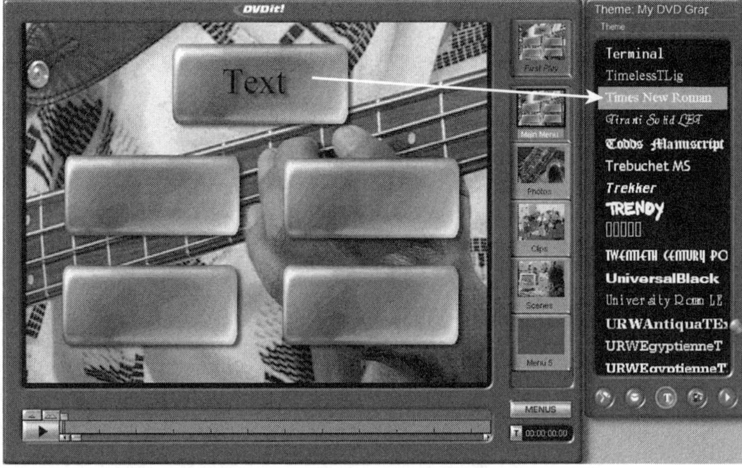

FIGURE 18.10
Selecting and dragging a text font to a menu adds the word `Text` *to the menu.*

As you type, the bounding box doesn't expand to accommodate additional text. Instead, the text scrolls by. When you finish typing, click outside the box and all your text will appear.

FIGURE 18.11
The three types of text displays, depending on the number of times you click the text you've added to the menu.

5. When you're done typing, click outside the text area to accept and view what you've typed.

6. Click your text to bring up the red-handled bounding box again. You now can move the box to reposition the text or drag the corners or center to change its overall size and shape.
7. To edit your text, click inside to highlight the text, click again to place the cursor within the text, and then type your new text.
8. Give your main menu's four buttons text descriptors by dragging a font to each and typing **`Main Movie`**, **`Scenes`**, **`Photos`**, and **`Clips`**. Use Figure 18.12 as a guide.

FIGURE 18.12
DVDit! lets you add labels, which also can act as button links, to your buttons.

9. Add the same title background button to the other three menus by clicking it to select it. Then select Edit, Copy (the keyboard shortcut is Ctrl+C); click the Photos menu placeholder thumbnail; and select Edit, Paste (Ctrl+V). Add that button to the other three menus.
10. Follow the same procedure you used in step 9 for the main menu text. Copy it to each of the other three menus; then select that text in each menu and replace `Main Menu` with the titles of those menus. Your project should look similar to Figure 18.13.
11. Save your project by selecting File, Save. Select your project name and click Save. DVDit! might ask whether it's okay to overwrite the current save. Click Overwrite.

18

FIGURE 18.13
Give your project a consistent look and feel by copying the title text and button background to the other three menus.

Summary

It might take more time to plan your project than to create its menus. The menu-building process is straightforward and consists primarily of dragging menu backgrounds to menu placeholder screens and then adding buttons and text.

DVDit! 2.5 gives you more options than MyDVD. Where you place buttons and text is wholly up to you. The sizes, shapes, and colors of those menu objects are under your direct control. In addition, you can use all the typeface fonts on your PC.

In this hour you created menu backgrounds and added buttons and text. What you did not do was add movies or photos to your project, nor link buttons in those menus to media and other menus. Those tasks are the focus of the next two hours.

Workshop

Review the questions and answers in this section to reinforce your DVDit! menu layout knowledge. Also, take a few moments to take the short quiz and perform the exercises.

Q&A

Q I'm having trouble working with text in DVDit! I drag a font to a menu, but when I type nothing happens. I click the word `Text` and start typing, and again nothing happens. What should I do?

A Click one more time. Changing the DVDit! text is a three-click process: one to select and place the text on the menu, another to place a frame around the text to change its shape and move it, and another to highlight the text for editing. When you're done typing, click outside the box to complete your work. Later, if you want to edit the text, it's another three-click process: one to create the frame, a second to highlight all the text, and a third to put a cursor within the text.

Q I create buttons and select all of them by Ctrl+clicking to change their shapes all at once, but when I click a center handle only that button changes shape. What's going on?

A You probably clicked a white center handle as opposed to a red handle. Even though all the buttons are highlighted with frames, only one will have a red center handle. This is a conscious design decision by Sonic that I think is counterintuitive. I think you should be able to click any handle on any button to change all their shapes simultaneously. Nevertheless, if you want to use a center handle to change the buttons' shapes, select the button with the red center handle. One other possibility is that you did not click directly on the red center handle. Even if you missed it by a pixel, DVDit! thinks you're selecting the button, not grabbing its handle.

Quiz

1. How do you change the menu background for the first play menu?
2. How do you set a different menu as the first play menu?
3. How do you make menus using your own background graphics?

Quiz Answers

1. Drag a new background to the menu placeholder screen for the menu that is currently the first play menu. Don't drag that new background to the first play screen (even if it's no longer named `First Play`). Doing so will turn that background into a new menu and put it in the first play screen, kicking out the former first play menu.
2. Drag any menu from the column of menu placeholder thumbnails to the first play placeholder screen (always the top menu placeholder). The new menu will show up in the first play screen, and the former first play menu will remain in its place amongst the menu placeholder thumbnails.

3. First, import your background(s). Open the Background palette and select Theme, Add Files to Theme. Navigate to your graphics file(s), select one or more, and click Open. Scroll down through the Background palette, and you should see your backgrounds in that collection. Select and drag one or more to separate menu placeholder screens.

Exercises

1. Create four blank menus by dragging different backgrounds to the menu placeholders. Select each menu in turn (by clicking its placeholder thumbnail), and drag the three other menus to it in turn. Doing so automatically creates buttons, linking each menu to the other three. To see how that works, open the Remote Control interface. The first play menu should be in its display. Start clicking the menu icons to see how easily DVDit! lets you jump from one menu to another.
2. Use your own graphics to create menu backgrounds. You can use any image as a menu (a 4:3 aspect ratio, more or less, is best). DVDit! will expand or contract it to fit a standard 720×480 (NTSC) screen size, so an image of that approximate size is best to ensure no quality loss. Use Photoshop Elements to create some buttons. A typical button size is 72×60 pixels or 300×80 for larger buttons.

HOUR 19

DVD Authoring with DVDit!: Part III

In this hour you will complete the menu creation process using DVDit! You'll add videos and photos to your project, automatically creating buttons as you go. You'll mark chapter points in a video and use them to create a scene selection menu. And you'll adjust text, button, and menu characteristics, including color, drop shadows, and other properties.

The highlights of this hour include the following:

- Adding videos and stills to your menus
- Working with chapters and a scene selection menu
- Editing button, text, and menu attributes

Adding Videos and Stills to Your Menus

This is the exciting part of DVD production. You have various types of media in hand and are ready to pull them all together—combining your videos, photos, and music into a cohesive whole.

Task: Add Images to a Menu

▼ TASK

In this task, you add still images to the Photos menu, automatically creating buttons in the process. In the following task, you add video clips to their menus. Follow these steps:

1. Start DVDit! and open your project.
2. Select the Photos menu by clicking its thumbnail placeholder.
3. Select the title text button background and use its handles to adjust its size so it more suitably fits the word `Photos`. That graphic should look similar to Figure 19.1.

FIGURE 19.1
Change the size of the menu title button background to fit the text.

4. Click the Media button in the Theme palette to display your videos, audio files, and photos.
5. Click and drag `Music Video-1.bmp` to the Photos menu. As shown in Figure 19.2, this adds a beveled button of that photo to the menu.

▼

The beveled thumbnail look might not suit your tastes. You can resize it, change its shape somewhat, alter the drop shadow, and adjust its overall hue. But you're stuck with that beveled button look. If you want something else, the best option is to use a video editor, grab stills from your videos, open them in Photoshop, and create buttons there using any 3D or other type of look at your disposal. Then, import them into DVDit!

FIGURE 19.2
Dragging a photo to a menu automatically creates a button using that photo as a thumbnail.

6. Drag the remaining seven `Music Video.bmp` images to the Photos menu, arranging them as you see fit. You can use Figure 19.3 as a guide.

As you arrange the title text and buttons on this and any other menu, leave a little room around the edges to ensure that all the elements are inside the NTSC Safe Area, meaning they will not fall outside the edges of most consumer TV sets.

You might wonder why you should go through all this work to give viewers access to a collection of still photos. Basically, it's because DVDit! 2.5 does not have a slideshow capability. But, as you can see, that doesn't preclude you from including still images. It's just not as elegant a solution. One workaround is to create a video of a collection of images—complete with music or a narration—and include it on your DVD.

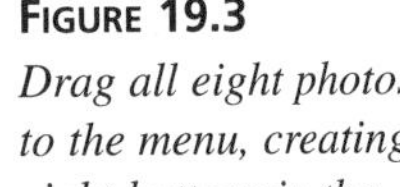

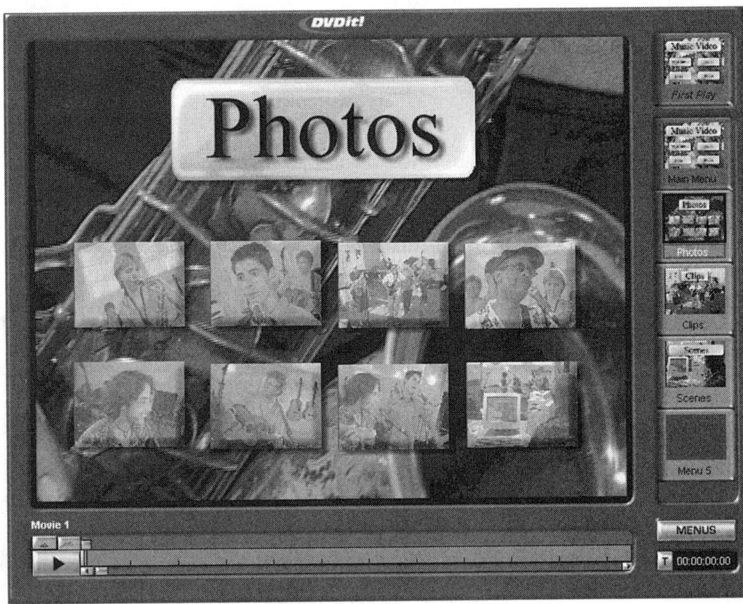

FIGURE 19.3
Drag all eight photos to the menu, creating eight buttons in the process.

Task: Add Videos to a Menu

This works similarly to adding images. The principal difference is that you need to add audio as well. Here's how:

1. Open the Clips menu and adjust the size of its title text button background to fit the word `Clips`. You might also move both elements off to one side so as to not cover up any of the musicians.
2. From the Media collection, click and drag the `Behind the Scenes.mpeg` file to the menu. As with the still images, doing so adds a button to the menu.

> This file is an *elementary* MPEG file, meaning it's video-only with no audio. You'll add its associated audio to it, and to the other two clips, in step 4.

3. Drag and uniformly position the other two MPEG clips—`Sax solo` and `Scat vocal`—to the Clips menu. Your menu should look similar to Figure 19.4.

Because this is a busy menu background, these thumbnails tend to blend in with the background. I show you how to make the buttons stand out later this hour in the "Editing Button, Text, and Menu Attributes" section.

FIGURE 19.4
As with still images, adding video clips to a menu automatically adds thumbnail buttons.

4. None of these three clips has an audio track. Add audio tracks to them one at a time by dragging and dropping their associated WAV files to them. As you add each WAV audio file, DVDit! plays a little chime.

You might have noticed that the phrase `Movie 1` has appeared below the lower-left corner of the video monitor window. As you add still images and videos (DVDit! calls all media *movies*), DVDit! adds them to a drop-down list that is accessed by clicking `Movie 1`. By this point, you have added 11 movies; click `Movie 1` to see that list.

Working with Chapters and a Scene Selection Menu

Most Hollywood movie DVDs have scene selection menus. They let viewers access points within the movie. The DVD specification refers to them as *chapters*. They're not actually separate video clips—they're only locations in the main video designated with markers.

You might recall that, when you created chapters in MyDVD, the authoring application automatically built a chapter—or scene selection—menu, with buttons linked to the chapters. DVDit!, being a prosumer/professional product, assumes you want to take charge of the menu creation process and does not take that automatic step.

DVDit! lets you set those chapter markers in a timeline, name them, and then drag them one at a time to a menu—in this, case the Scenes menu.

Task: Create Chapters

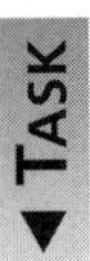

In this task, you add a video to your project and then set chapter points in it. Follow these steps:

1. Click the silver Menus button, shown in Figure 19.5, to open a small drop-down list. Select Movies to change the thumbnail placeholders from Menus to Movies and keep the first play main menu thumbnail at the top. Your project should look similar to Figure 19.6.

FIGURE 19.5
Use this small drop-down list to switch to the Movies palette.

2. If the Media palette isn't open, click the Media button at the bottom of the Theme column to display thumbnails of all your media files.
3. Scroll down to the `Music video.mpeg` file; then click and drag it to the empty movie placeholder (probably Movie 12) at the bottom of the column. As shown in Figure 19.7, the text `Movie 12` then displays in the lower-left corner.

FIGURE 19.6
Selecting Movies displays the 11 movies and stills (up to 5 at a time) added to your project.

FIGURE 19.7
Adding a video to a placeholder adds its name to the drop-down list below the video playback screen.

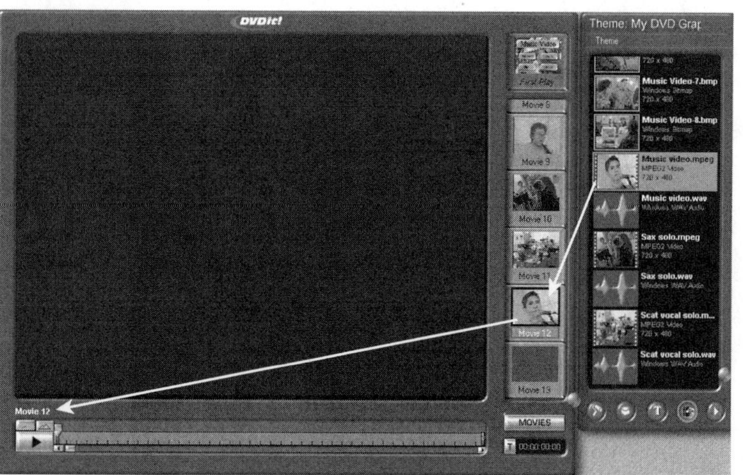

4. Add audio to this video-only, elementary MPEG file by dragging `Music video.wav` to that video's thumbnail placeholder (if you've followed the tasks exactly, it will be Movie 12). In this case, as you add audio, DVDit! does *not* play a little chime.
5. Rename Movie 1 the same way you renamed menus. Click `Movie 1` below the thumbnail and type **`Main Movie`**. Doing so changes its reference in the drop-down list.
6. Play this video by clicking the Play/Stop button below the video monitor. Stop it by clicking that same button.

No fast forward or rewind buttons are available, and you can't scrub through a video, watching the images flash by as you slide the cursor. Instead, if you're looking for a particular scene, you must move the small square cursor to an approximate location in the clip and release the mouse button. Then, look at what appears on the screen and adjust your location accordingly.

7. To set chapter points in the timeline, simply double-click the timeline at the point you want to add a chapter marker. As shown in Figure 19.8, this adds a small triangle with a yellow `Chapter 2` label (by default, all videos on DVDs have Chapter 1 set to the first frame).

Chapter points cannot be too close to each other. The reason is that MPEG videos are divided into groups of pictures (GOPs), which are typically 4–18 frames long (1 second of NTSC video equals 29.97 frames). The DVD specification requires that chapter points be only at GOP headers. So, if you attempt to add a chapter point too close to an existing chapter point (that is, between GOPs), DVDit! jumps to the next GOP and adds a new chapter there. If the movie being edited has no GOPs (for example, it's an AVI or MOV file), a new chapter is added at 1 second beyond the previous chapter point.

8. Add five more chapters (more or less, if you choose). Do this by playing the video to an obvious scene change, clicking Stop, and then double-clicking at that point on the timeline. Or, you can simply drag the square cursor into the video to find places to add chapter markers.

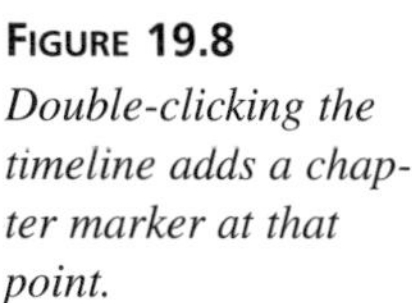

FIGURE 19.8
Double-clicking the timeline adds a chapter marker at that point.

You can adjust the timeline interval display. The default is to display the timeline for the entire length of the selected video. The longer the video is, the more time interval hash marks are displayed and the closer they are together. You can expand or contract the hash mark spacing by clicking the small or large mountains above the Play button shown in Figure 19.9. Spreading out the hash marks makes finding frames easier.

To delete a chapter marker, right-click it and select Delete. You can rename a chapter using the same right-click method. To move a chapter marker, Ctrl+click it and drag it to a new location.

FIGURE 19.9
Clicking the small or large mountain icon changes the time interval hash marks on the timeline.

Task: Add Chapters to the Scene Menu

Just as you dragged media to other menus, in this task you drag chapters to a menu. Once again, DVDit! automatically adds thumbnail buttons to that menu. Here's how it works:

1. Switch back to the Menus view by clicking the silver Movies button and selecting Menus.
2. Select the Scenes menu and adjust its title text and background button to fit better together.
3. Click the drop-down list below the lower-left corner of the monitor and select Main Movie. As shown in Figure 19.10, this displays all the chapter points you created in the previous task.

FIGURE 19.10
When in Menus view, use the Movies drop-down list to display a movie's timeline and any chapter points.

4. Click and drag chapter 2 to the Scenes menu. As happened when you dragged a video to the Clips menu, dragging a chapter point adds a thumbnail image button.
5. Drag the remaining buttons to the Scenes menu and arrange them. Your menu should look similar to Figure 19.11.
6. Save your project by selecting File, Save.

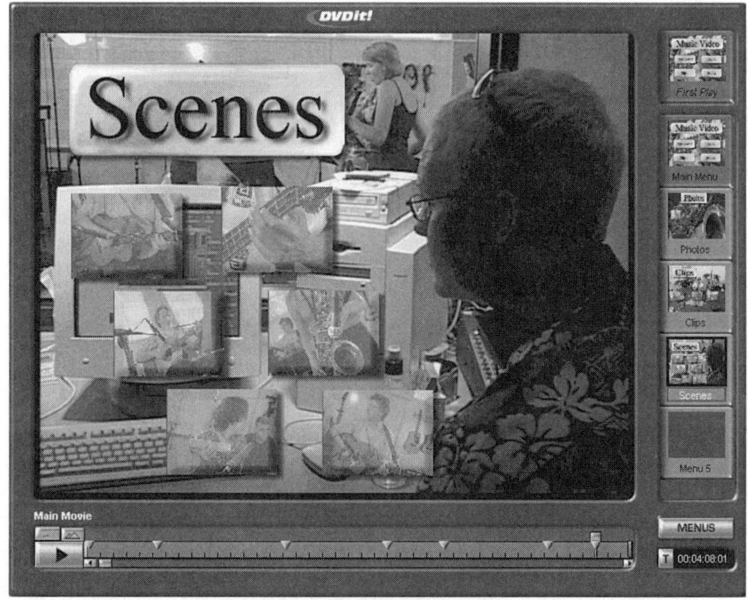

FIGURE 19.11
Drag and drop all the chapter points (except chapter 1) to the preview window to create a scene selection menu.

Editing Button, Text, and Menu Attributes

DVDit! lets you do much more to text and button elements than moving them around the menu and changing their sizes and shapes. You can adjust the color, saturation, and brightness of all menu elements (including the menu background) as well as add drop shadows using an intuitive and customizable tool.

Task: Change Text Properties

Text has its own properties dialog box, so we'll start there. In it you can make some standard text appearance adjustments. Here's how:

1. Select the Clips menu and then select the `Clips` text title by clicking it.
2. From the main DVDit! toolbar, select Effects, Text Properties. This opens the Text Properties dialog box, shown in Figure 19.12.

By first selecting a text object, you ensure that the Text Properties dialog box references that text string. In Figure 19.12, my text is 70-point Times New Roman.

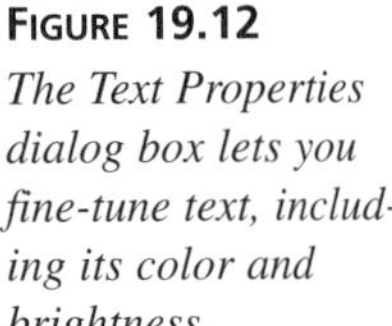

FIGURE 19.12
The Text Properties dialog box lets you fine-tune text, including its color and brightness.

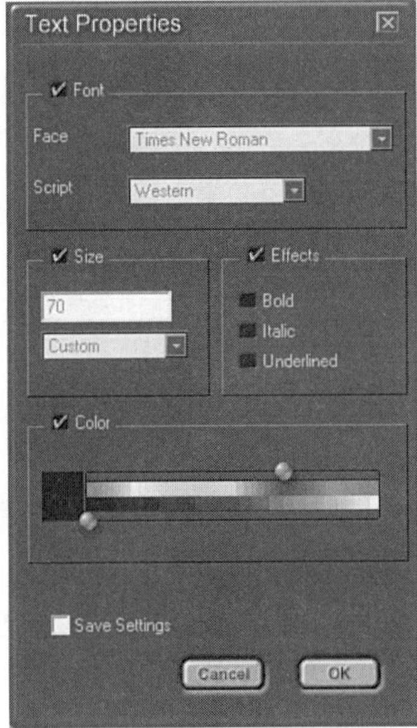

3. The Text Properties dialog box lets you make the usual text changes, including font typeface, size, and bold/italic/underlined. It does have two interesting functions:
   - **Script**—This drop-down menu lets you change the lettering from Western to five other alphabets.
   - **Color**—This area controls both the color and brightness. The top slider changes the text color, and the bottom slider changes the brightness or luminance.
4. Make some changes to the word `Clips`, be it underline, bold, italic, or a different color. When you're done, click the Save Settings check box and click OK. Save Settings is one way to give your project a consistent look.
5. Go to any other menu; select that menu's text object(s); and then select Effects, Text Properties. As shown in Figure 19.13, DVDit! automatically changes the selected text attributes to match those you created in step 4. If this is what you want, click OK.
6. Apply those text changes to the rest of your menus; then save your project by selecting File, Save.

FIGURE 19.13
Use the Save Settings check box to make the text settings uniform throughout your DVD.

Task: Adjust Drop Shadows

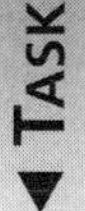

Drop Shadows is a separate dialog box that works with both text and buttons. All buttons have drop shadows by default. You can turn those drop shadows off if you want or change the drop shadow attributes, including distance, blur, opacity, color, and direction. Here's how to make changes:

1. Open the main menu and use the Ctrl+click method to select all five text items.
2. Select Effects, Drop Shadow to open the Drop Shadow dialog box shown in Figure 19.14.

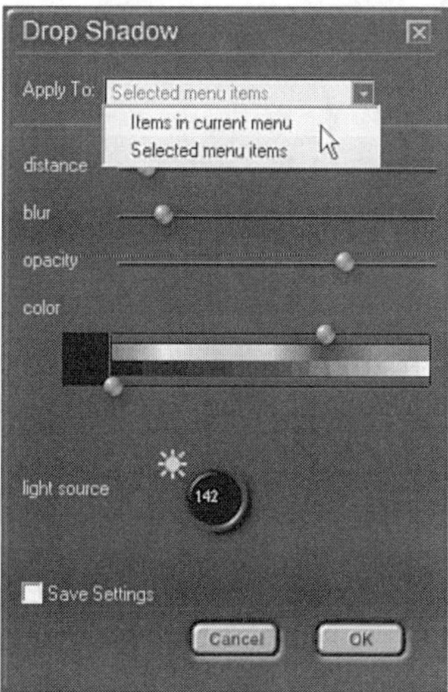

FIGURE 19.14
This interface makes adjusting drop shadow characteristics easy.

3. The Drop Shadow dialog box is an intuitive and fun toy. Using simple sliders you can adjust a shadow's color, blur, opacity, and distance and direction from the object. As shown in Figure 19.15, from the Apply to drop-down list, choose the Selected Menu Items option. This ensures that drop shadows for all the text objects will have the same characteristics, giving your menu text a consistent and more realistic look.
4. Make adjustments to the drop shadow characteristics, including changing the color to more closely match the buttons' color (shadows in real life have color). As shown in Figure 19.15, those changes show up in real time on the selected text objects (I purposely made the shadows a bit excessive to demonstrate the process). Click Save Settings and then click OK.

FIGURE 19.15
As you adjust drop shadow settings, those changes show up immediately in the menu window.

Changing a shadow's color is a two-step process. First, you slide the bottom luminance button to the right a bit to give the shadow some brightness (otherwise, it remains black/gray). Then, you slide the color button to select a color you like.

Text in DVDit! has drop shadows by default. You can turn off those drop shadows by selecting a text object, opening the Drop Shadow dialog box, and sliding the distance and blur buttons all the way to the left.

5. Using the same process described earlier in steps five and six of the task titled "Change Text Properties," apply these drop shadow characteristics to the text objects in the three other menus. As a reminder, click the text object in each menu and select Effects, Drop Shadow.
6. Give your menu buttons—both the graphic and thumbnail buttons—drop shadows in the same fashion. If a menu has a distinctively colored background, adjust the drop shadow color accordingly.
7. Save your project by selecting File, Save.

Task: Adjust the Color

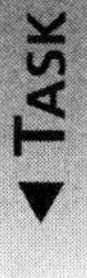

Finally, you can change the color of all the menu items, such as text, buttons, and even the background. In particular, because this project has two busy menus, I show you how to change the background while emphasizing the buttons. Here's how to make those changes:

1. Open the Clips menu and select Effects, Adjust Color to open the Color Adjustment dialog box (see Figure 19.16).

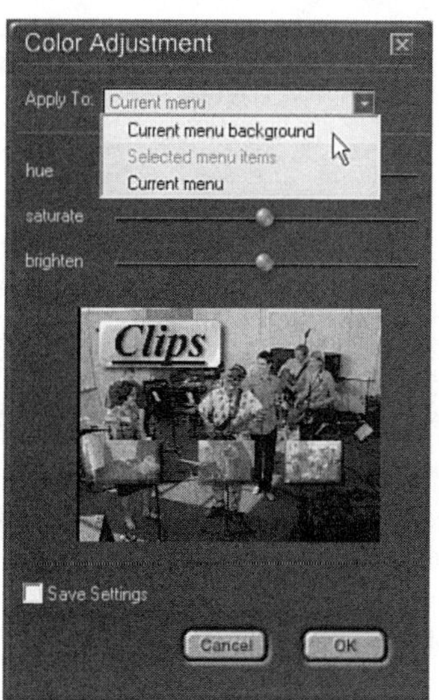

FIGURE 19.16
The Color Adjustment dialog box lets you change the color characteristics of any object, menu, or combination of items you select.

2. Open the Color Adjustment dialog box's drop-down menu. You have three selections: Current Menu Background, Selected Items, and Current Menu. Select Current Menu.

19

3. I suggest making the menu a bit darker and less colorful. Do that by sliding both the Saturate and Brighten buttons to the left a bit (in step 5, you'll add color to the menu buttons to make them stand out from the darkened background).

> Experiment with the three sliders. You can create some wild-looking graphics using them. My menu darkening approach is somewhat conservative. Feel free to come up with another means to make the buttons stand out from the background.

4. Click Save Settings. Next, click OK and then apply this color shift to the other busy background in the Scenes menu.
5. While in the Scenes menu, select all the chapter thumbnail buttons. Then open the Color Adjustment dialog box, increase the saturation, and adjust the brightness to make these buttons stand out.
6. Return to the Clips menu and apply those new settings to its thumbnail buttons. For consistency's sake, apply those changes to the Photos menu even though it does not have as busy a background as Clips and Scenes do.
7. Save your project.

Summary

DVD authoring involves designing the project's menu structure, creating the menus and buttons, adding media, and linking all the elements. In this hour, you added media to your menus (automatically creating buttons in the process), set chapter points, used them in a Scenes menu, and gave buttons and menus some pleasing visual characteristics.

In the next hour, you'll connect the dots by linking media and menus to buttons. You'll also fine-tune some movie and menu properties noting how they behave when the DVD is playing in a set-top player.

Workshop

Review the questions and answers in this section to sharpen your DVDit! authoring skills. Also, take a few moments to tackle the short quiz and perform the exercises.

Q&A

Q I try to give my drop shadows color to match the background, but when I move the color slider, the shadow stays black. What's going on?

A You first need to adjust the luminance or brightness. The default setting is no luminance. Slide the luminance button over to the right a bit before adjusting the color slider.

Q I created a set of layered buttons in Photoshop and saved them directly into the DVDit! `Buttons` folder. But all I see is a white frame with one button in it. When I drag that to a menu, there is no way to remove the white frame or use the other buttons. Why is that?

A You need to let DVDit! import the graphic—it's not enough to put it in the proper file folder. In the Button palette, select Theme, Add Files to Theme; select your Photoshop PSD file; and then click Open. DVDit! recognizes that this is a layered Photoshop file with an alpha channel transparency and splits up the graphic file into its constituent buttons.

Quiz

1. You've marked a chapter on a video but now want to move it elsewhere on the timeline or delete it. How do you do both of these tasks?
2. You drag a video to a movie placeholder and click the Play/Stop button below the video monitor. The video plays, but there's no sound. What went wrong?
3. Your menu background has very distinct and obvious shadows. You want any drop shadows you create to fall in the same direction. How do you do this?

Quiz Answers

1. To move a chapter point, Ctrl+click it and drag it to a new location. To delete it, right-click it and select Delete.
2. The video you added to the project is an elementary MPEG, video-only file. You need to add its associated audio to it by dragging the file to its thumbnail placeholder.
3. In the Drop Shadow dialog box is a light source setting—a circle with a number in the center and a sun symbol outside it. The drop shadow falls away from the sun symbol, so simply move it to make the drop shadows match your menu's background shadows.

Exercises

1. Work on button and text attributes. Try creating a menu with five buttons, using the same shape for each but giving each different color and saturation characteristics. Then apply text to the buttons. Now give the buttons one set of drop shadow characteristics, making them look as if they are floating far above the menu. Then give the text different drop shadow characteristics, making it look close as if it's close to the buttons. Finally, adjust the color of each text object's drop shadow to approximate its button background.
2. If you're versed in Photoshop, create some buttons. Then import them to DVDit! by selecting the Button collection in the Theme palette and selecting Theme, Add Files to Theme. You can create those buttons in layers. You also need to create a transparency layer; otherwise, the buttons will appear as full-screen boxes, with the buttons' graphics within the boxes.
3. Create a family tree DVD using still images as menu backgrounds and buttons. Add sound effects or period music to enhance still images. Also, use text to create banners and titles for menus.

HOUR 20

DVD Authoring with DVDit!: Part IV

In this hour you will wrap up authoring with DVDit! You'll link all your project elements together, add audio to menus, and set some menu and media properties. You'll do the final testing and burning of your project in Hour 21, "Burning DVDs and Dealing with Mass Replicators."

Before you record your DVD project, I will conclude this hour with some reminders of project ideas you might consider. To whet your appetite, I'll relate four anecdotes about how others have used DVD authoring applications: from consumer-level to professional.

The highlights of this hour include the following:

- Linking media and menus
- Adjusting menu and media properties
- Creating personal projects
- Designing business-oriented DVDs
- Working on professional DVD projects

Linking Media and Menus

DVDs are interactive. Clicking a button takes you somewhere, displays something, or plays a clip. Consumer-level products, such as MyDVD, automatically create those interactive links. With higher-level products, however, you play a much larger role.

That's not to say DVDit! does nothing for you. In the case of the current DVD project, DVDit! already has created nearly a dozen button links for you. Each time you add a media element—video or still—DVDit! makes a thumbnail button with a link to that media file.

But a few buttons do need links; these typically are graphic buttons you add to menus. We'll start by linking a video to a button and then link menus to buttons.

Task: Create Links

▼ TASK

In the previous hour, you added media to menus, creating buttons in the process. In this task, you simply create a link between an existing button and a video or menu. Here's how to do that:

1. Open the main menu.
2. As shown in Figure 20.1, right-click anywhere on the menu and select Show Links. This removes the button graphics and displays only their outlines, button numbers, and any links (there are none in the main menu).

Each button in this menu has two rectangles: one for the button graphic and the other for the text label. As you create links from these particular buttons to media and menus, you'll need to link only the graphic portion (you can link the same media to both the button and text, but it complicates matters). This is why you should turn on Show Links to see exactly where to drag media to create links.

FIGURE 20.1
Right-click any menu and select Show Links to display to which media and menus buttons and text are connected.

3. Open the drop-down menu below the menu preview window and select Main Movie. As shown in Figures 20.2 and 20.3, click and drag its Chapter 1 triangle to the Main Movie Button 1 (taking care to not drag it to the smaller Button 7).

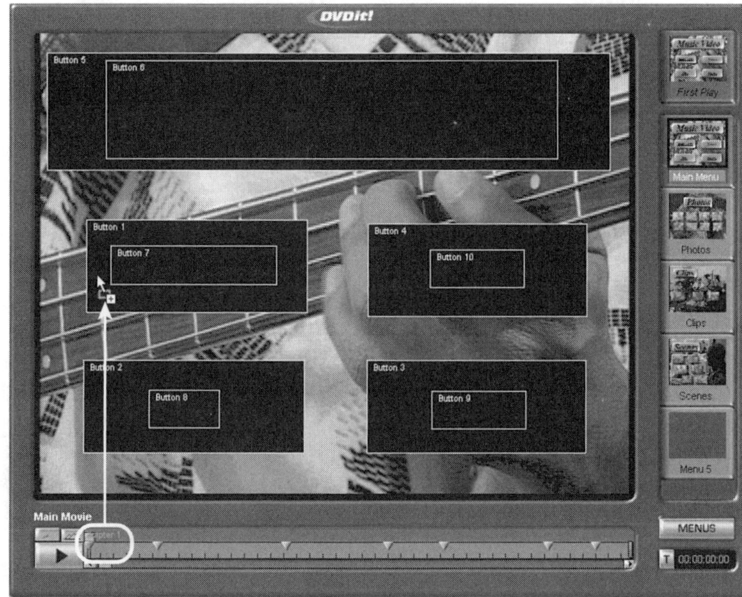

FIGURE 20.2
Use the Show Links display to ensure you drag your media to the correct button.

FIGURE 20.3
A close-up look at how you position the cursor when dragging a menu to a button.

4. Take a close look at the newly created button link text. It should have the same text and link description as shown in Figure 20.4 (I moved Button 7 out of the way to expand the view): Button 1, Chapter 1, `Music video.mpeg`, `Music video.wav`. (The Movie number might be different.)

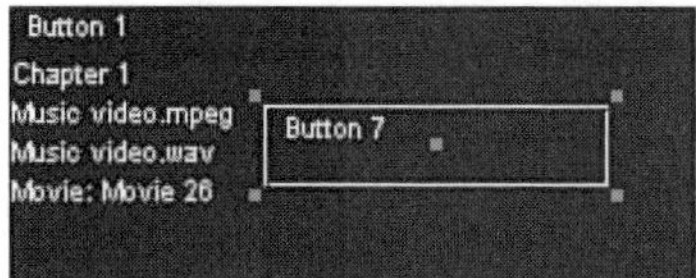

FIGURE 20.4
How your button links will look after dragging a video and audio file to a button.

5. Now you'll use a similar drag-and-drop method to link buttons and menus. Remind yourself which buttons are which by right-clicking anywhere in the main menu and selecting Hide Button Links.
6. Right-click again and select Show Button Links.
7. As shown in Figure 20.5, drag the Photos menu placeholder thumbnail and drop it in the Photos button box outline (not the smaller Photos *text* box). As shown in Figure 20.6, you'll see a hand cursor with a tiny thumbnail of the entire menu showing you that this drag-and-drop link will work.

As happened when you dragged the Music Video to the Main Movie button, DVDit! displays the linked object's name in the button box outline. In this case, it displays `Menu: Photos`.

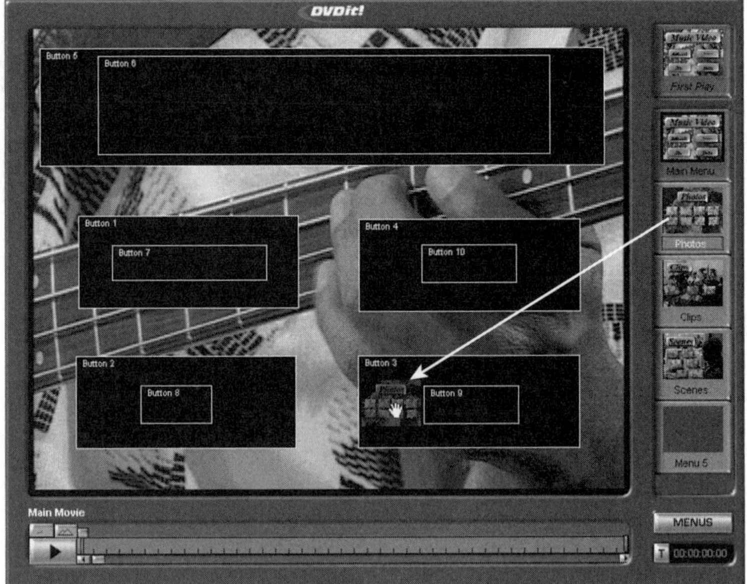

FIGURE 20.5
Drag and drop a menu to a button to create a link from that button to that menu.

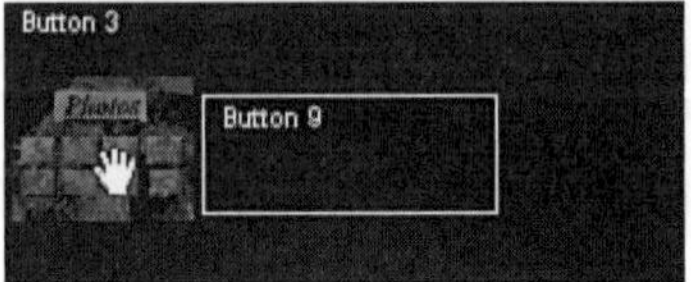

FIGURE 20.6
DVDit! gives you visual feedback letting you know that this drag-and-drop link will be effective.

8. In the same fashion, drag and drop the other two menu thumbnails to their respective buttons on the main menu. Now, as shown in Figure 20.7, all four buttons should have links (the main menu text button at the top should remain link-free).

> If you were to drag a menu placeholder thumbnail to an empty place in a menu, DVDit! would add a thumbnail button using the menu's appearance as its image.

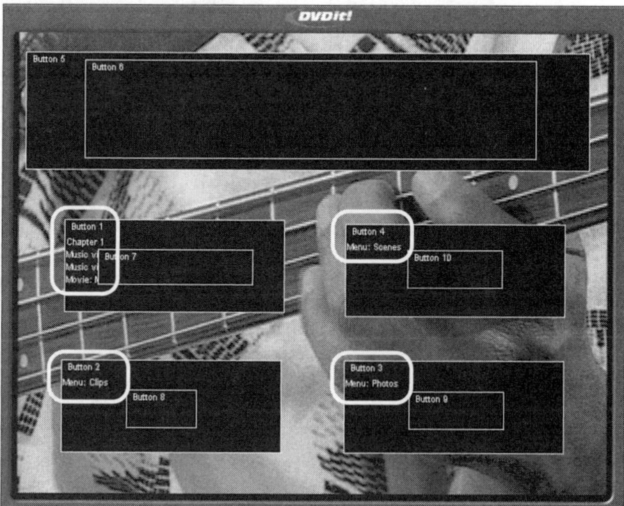

FIGURE 20.7
All four main menu buttons should have links associated with them.

9. Because you dragged and dropped media to create all the buttons in the other three menus, those buttons already have links associated with them. Confirm that by clicking each menu, right-clicking, selecting Show Button Links, and noting the link information in each button box. Take a look at the Clips menu in particular because its buttons link to both video and audio files (see Figure 20.8).

10. Save your project.

FIGURE 20.8
Checking the other menus (such as this Clips menu) shows that all their buttons already have links.

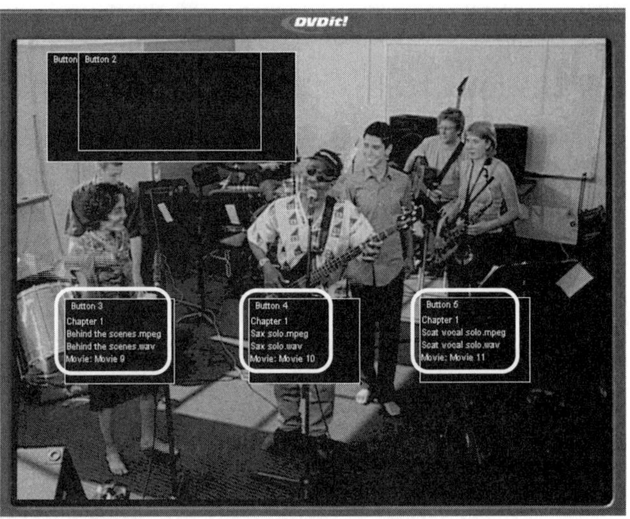

Adjusting Menu and Media Properties

The final authoring step is to make some small but critical adjustments to some of your media and menus. The goal is to define how your menus and media will behave when someone is watching your DVD. For example, you need to define what happens when a video ends or what displays when a viewer presses the Next or Menu button on the remote control.

You'll use the Movie and Menu Properties dialog boxes to tell the DVD what to do. It's a time-consuming but necessary process.

Task: Edit Movie Properties

TASK

You'll start by adjusting how long the still images display. Then, if you choose, you can adjust the settings for the videos. In general, accepting the default settings means your DVD will work in a predictable fashion. But I want to at least let you know how the movie properties work. Here's how to do all that:

1. Open the Photos menu and display the button links by right-clicking anywhere and selecting Show Button Links.
2. Right-click in one of the photo button's boxes and, as shown in Figure 20.9, select Link Properties.

FIGURE 20.9
Use Link Properties to access the properties dialog boxes.

3. Look at the Movie Properties dialog box shown in Figure 20.10. It displays these items:
   - The name DVDit! gave to this image (`Movie #`)
   - The actual image filename (`Music video-#.bmp`)
   - Information that states that no audio is associated with this clip
   - Its duration (5 seconds is the default)
   - Three Remote Mapping drop-down lists (I explain them in step 5)

FIGURE 20.10
Tie up loose ends in the Movie Properties dialog box.

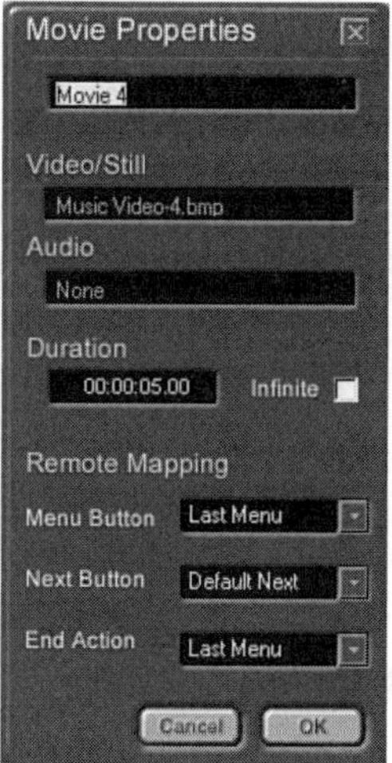

4. You can change how long the image will display by clicking the Duration time code to highlight it and typing in a new figure. If you want to change the duration for other images, you must close the Properties dialog box by clicking OK or Cancel, right-clicking a different button, and selecting Link Properties.

If you click the Infinite check box, the image will display until the viewer presses a remote control button. If your movie is a video (instead of a still image as in this case), selecting Infinite means the video will play to its conclusion and display the last frame until the viewer presses a remote control button.

Because many videos end with a black frame, checking Infinite for them could be disconcerting for your viewers—they'll end up staring at a blank screen, wondering what went wrong. Therefore, I suggest using Infinite only when you know your video's last frame is something other than black.

5. This dialog box has three Remote Mapping drop-down lists (the default values in this case are fine). Basically, they are set such that when an image is finished playing, the DVD returns to the Photos menu. They have the following functions:
   - **Menu Button**—This tells the DVD player which menu to display when the viewer presses the remote control's Menu button. The options are Do Nothing, Select One of Your Menus, and Last Menu. Selecting the Last Menu option moves the viewer back to menu she used to arrive at the current position. Selecting the Do Nothing option does just that—nothing. The viewer must press the Next button to go to another movie menu.
   - **Next Button**—This tells the DVD player which menu or movie to display when the viewer presses the remote control's Next button. The options are Do Nothing, Select One of Your Menus or Movies, and Default Next.
   - **End Action**—This tells the DVD player which menu or movie to display when this movie finishes playing. The options are Same As Next, Select One of Your Menus or Movies, Last Menu, and Loop.
6. Close the Movie Properties dialog box by clicking either OK to accept any changes or Cancel to accept the original defaults.

Task: Edit Menu Properties and Add Audio

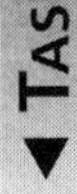

Menus have some slightly different options, but again the defaults should work fine. In this case, you'll also add a music clip to all your menus just to see how that works. Follow these steps:

1. Open the main menu.
2. Add music to that menu by dragging the `Music Video Menu.wav` file from the Media section of the Theme palette to some place on the main menu outside the

button boxes. As shown in Figure 20.11, this adds a small speaker icon below the lower-right corner of the menu preview window.

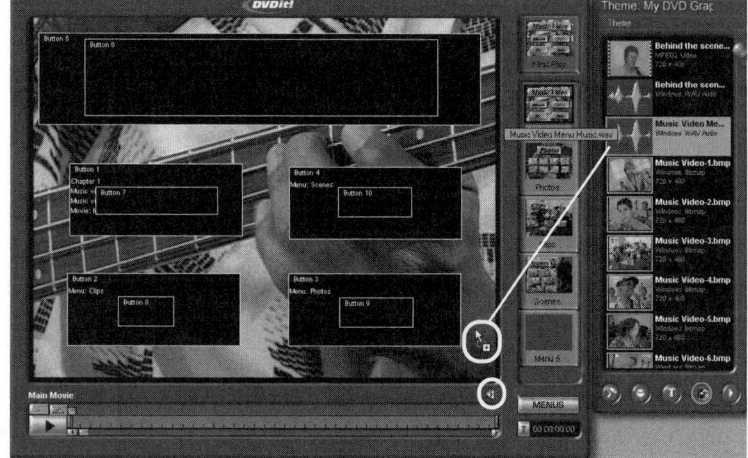

FIGURE 20.11
Dragging and dropping an audio file onto a menu adds a speaker icon below the menu preview screen.

3. Add the same audio file to the other three menus using a slightly different method. As shown in Figure 20.12, drag the `Music Video Menu.wav` file directly onto each of the other menus' placeholder thumbnails.
4. Right-click anywhere in the main menu outside the button boxes and select Show Properties. As shown in Figure 20.13, the Menu Properties dialog box opens.
5. The Menu Properties dialog box has three drop-down menus with some minor differences from the menus in the Movie Properties dialog box:
   - **Default Button**—This is the menu button that is automatically highlighted (the button gets a bit brighter) when this menu first displays. As your viewer moves the cursor around the screen, other buttons become highlighted, one at a time.
   - **Return Button**—Select the action that occurs when the viewer presses the Return button on the remote.
   - **End Action**—If you set a Duration value (other than Infinite), use this menu to tell the DVD what to do when time is up. Typically, you'll select a video clip or submenu.

FIGURE 20.12
Another way to add audio to a menu is to drag an audio file to a menu's placeholder thumbnail.

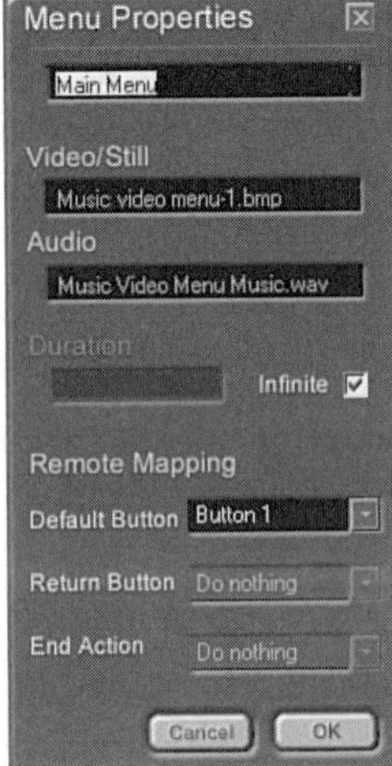

FIGURE 20.13
The Menu Properties dialog box has slightly different options from the Movie Properties dialog box.

The default settings should work fine. In particular, leaving the menu set to Infinite means the music will *loop*—play repeatedly until the viewer presses a button on the remote. Unchecking Infinite lets you select an End Action option, one of which is Loop.

Making all these selections is labor-intensive work. But failing to tie up all these loose ends could leave your viewers stranded in the middle of your DVD. Therefore, when it's completed, you should give your project a thorough test. I cover how to do that in Hour 21. That's when you'll record this project and take it for a real spin on your set-top player.

Creating Personal Projects

The tasks these past few hours have all been about showing you how you can use DVDit! 2.5 to make any type of DVD project you have in mind.

In Hour 1, "Discovering What DVDs and DVD Authoring Software Can Do for You," I presented a few examples of DVD project ideas—from family projects to corporate and professional. Perhaps back then it wasn't clear exactly how DVDs and DVD authoring software could help you create such projects. Now that you have a feel for how it works, I'm guessing you are chomping at the bit to create some of those projects.

For the remainder of this hour, I'll remind you of the myriad possibilities and give you some real-world examples of DVD authoring in action.

Family Stories

Your first DVD projects will probably start at home. I suggest getting your feet wet creating some of the following DVD projects before moving on to more demanding efforts:

- Storing video archives
- Reliving vacations with videos and slideshows
- Family tree history
- Sports team season highlights
- Holiday greetings

A First-Time DVD Authoring Experience

"Exciting" is how Leonard Broz describes burning his first DVD. "I waited my whole life for this moment," he adds with a smile. When he shows family video albums and "mini-travelogues" to relations and friends, they are "elated" with the results.

The former technology and industrial arts teacher from suburban Chicago winters in Arizona where he teaches computer-based video- and photo-editing techniques to other seniors in his community.

Broz uses Sonic Solutions' MyDVD to convert his still photos and digital video into crowd-pleasing DVDs.

He edits 90 minutes of raw video down to 10 minutes or so, converts his digital images into a slideshow with transitions, and then adds a narration and music. He uses MyDVD's custom styles to personalize attributes such as the font, text size, and background images of his menus; then he saves that updated style for use later.

FIGURE 20.14
Broz creates travelogues combining videos and digital photos with a narration and music.

Besides travelogues, Broz produced a DVD of his son's wedding. He used photos from the pre-wedding showers, the ceremony, and the honeymoon along with a video from the reception and put them all together on a DVD.

FIGURE 20.15
Broz's son's wedding DVD was a big hit.

His next task is creating "the story of our life on DVD" from the many family photos and the 300 or so videotapes he has shot over the years, starting with his wedding day.

Using Prosumer Techniques on Personal Projects

This next story demonstrates the power of DVDs. Not only are they a media repository, but they also preserve memories. Plus, their interactivity and ease of use mean your DVDs' viewers will want to share those memories again and again.

The Peoria, Arizona Hockey Mites on DVD

Ed Loeffler had a surprise for the Peoria, Arizona, Junior Polar Bears hockey team. The group of eight-to-ten year olds and their families had gathered for their season-ending party at a home with a large-screen TV.

Loeffler, the team historian, popped in a DVD and for the next two hours the gathered families clicked back and forth through menus and videos, reliving the highlights of their travels all around the Southwest. What they saw "amazed them," says Loeffler. "They loved it."

FIGURE 20.16
Ed Loeffler and his eight-year-old hockey-playing son, Taylor.

When completed, Loeffler handed out 25 copies of that DVD (he burned them one at a time on his PC), giving each family "a lifetime of memories," he says. "I saw joy on each person's face."

FIGURE 20.17
Loeffler's opening menu organizes the season by each road trip.

Those DVDs are the distillation of a lot of digital video and 3,000 (!) digital photos. Using Sonic's DVDit!, Loeffler created nested menus to ease access to the individual clips. The opening menu has links to each road trip, and each of those menus has links to the game plus other family fun. He used photos and graphics software to create custom menu backgrounds and then added text links.

FIGURE 20.18
Loeffler used photos to create menus.

Loeffler's already making plans for next year. He wants to step up to a full-featured authoring product such as Sonic DVD Producer that lets him create animated menus and buttons, work in Dolby audio, and use wide-screen features. "It's addictive," he says. "I want to do it even better next time."

Designing Business-Oriented DVDs

DVDs' interactivity, excellent video quality, and capability to include data files make them the perfect media for corporate training, marketing, and meeting presentation tools. That latter use is particularly powerful.

Consider the following MGM Mirage story as a case in point. Instead of thinking of DVDs as a collection of media with some data assets thrown on for extra measure, the Mirage corporate executives make presentations using Web browsers peppered with HTML links. At any point in their presentations, they can click a link that starts a software DVD player and then quickly access the DVD menu and its many high-quality videos—a very slick tool.

MGM Mirage Has DVD Vision

Randy Dearborn was there at the beginning. As multimedia director for MGM Mirage's massive casino empire, he has led the charge to DVD. Five years ago, he began replacing the company's unwieldy, expensive, and unreliable laser discs with DVD players. Now, all the laser discs are gone and DVD and MPEG-2 videos are on display throughout the MGM Mirage resorts nationwide.

FIGURE 20.19
MGM Mirage uses DVD-based kiosk displays to promote its resort offerings.

Dearborn and his Mirage Resorts team use professional-level Sonic Solutions DVD authoring applications to create point-of-sale video displays, exterior signage, corporate archives, and "back of the house" communications with the company's 10,000 employees.

FIGURE 20.20
Mirage retail store customers use touch screens connected to DVD players to sample video and audio products.

The resorts use Pioneer 7200 drives (the precursor to the current 7400 drives) that have built-in software that allows playback by clips, directly to the giant reader boards on the casinos' towering marquees on Las Vegas Boulevard and through interaction with touch screens. Some interactive touch-screen kiosks let customers select music CDs and listen to specific tracks.

FIGURE 20.21
Corporate executives eschew PowerPoint and use DVDs linked to HTML pages as the AV tool for their road show presentations.

MGM Mirage corporate executives now supplement their investor road shows with DVDs. They click a button in an HTML page to bring up the DVD menu and they jump instantly to an MPEG-2 video to further amplify a point.

Dearborn's latest efforts are to transfer the entire video archives of the company to DVD, categorizing clips and creating menus that let users outside the department easily access the material.

Working on Professional DVD Projects

You might consider turning your growing DVD authoring expertise into a business. Or you might have an ongoing video production business and want to give it extra cachet by offering clients DVDs in addition to videocassettes.

Typical professional DVD projects entail the following:

- Using DVDs to enhance event videos, such as weddings, meetings, and concerts.
- Letting clients view multiple editing options in the comfort of their offices or homes.

- Creating great demo reels on DVD. DVD's ubiquity means not having to worry about clients not having compatible playback devices.

High-end production firms take DVDs to new levels, using high-end DVD authoring applications to explore the arcane realms of the DVD specification to find new ways to entertain. The following story is a case in point.

Creating Interactive DVD Fun for Children

The Chicago Recording Company, one of Chicago's leading music and advertising recording studios, is a DVD pioneer. In 1999, the company started using Sonic DVD Creator to create DVDs for music groups—repackaging archival audio tapes and videotapes of concerts and television appearances.

The company's latest DVD products are light years beyond those early pioneering efforts says CRC's DVD authoring specialist, Sean Sutton. "We are pushing the envelope of the DVD specs."

FIGURE 20.22
CRC creates Big Idea Productions' 3-2-1 Penguins! *DVDs as well as* Veggie Tales *and* Larryboy.

CRC's latest efforts have focused on creating innovative and interactive DVDs for children. Working with Big Idea Productions, developers of *Veggie Tales* and other children's products, CRC has produced DVD movies with kid-friendly mazes, trivia quizzes, and other fun activities.

The challenge is to use "a technology designed only to play nice-looking pictures and good sound," says Sutton, "to somehow make games that are fun and accessible to children who may not yet be able to read." Navigating the maze in Figure 20.23, for instance, requires only the use of the arrow keys on a DVD remote control. Each button click seamlessly displays a new screen showing the changed location. It takes 300 such images and some horribly tedious coding of `if-then` statements to create this maze.

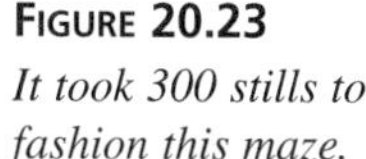

FIGURE 20.23
It took 300 stills to fashion this maze.

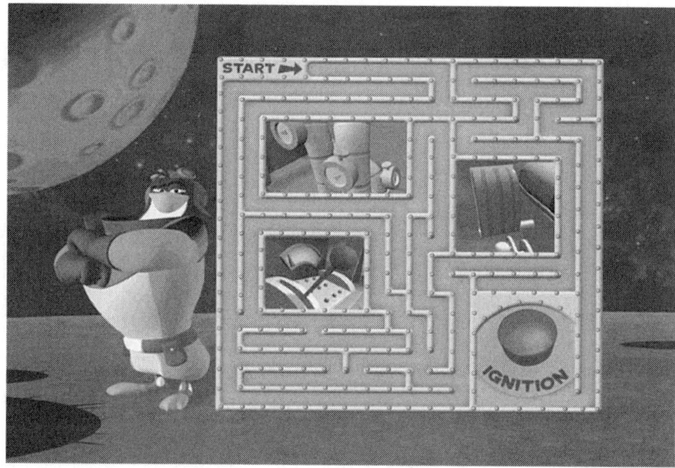

The deceptively straightforward-looking Penguin quiz in Figure 20.24 takes advantage of part of the standard DVD specs that allow randomizing—a bit like hitting random play on your music CD player. Sutton exploited that function to display questions in different sequences each time children play the DVD.

Other elements create additional DVD design complexity, including

- Fashioning a way to get the DVD to respond to wrong answers by returning to the original question
- Having it not repeat previously asked questions
- Getting it to keep track of a child's score

FIGURE 20.24
Official DVD specifications include a randomizing feature used to keep this quiz fresh.

Creating the Veggie Tales Voice Swap—it lets kids give characters the voices of other characters in a variety of settings—required 96 separate video clips.

FIGURE 20.25
This mix-and-match character and voice activity required 96 video clips.

Clearly, it's no longer enough to simply slap a 30-minute video on a DVD. CRC fills each Big Idea product to the limit.

Summary

I wrapped up DVD authoring with DVDit! 2.5 by showing you how to make the final links between media and menus. DVDit! automatically creates most of those links when you drag and drop media into menus, but you do need to connect a few dots.

You then checked out the many menu and media property options—most of which were fine in their default modes—but you saw that you exercise a lot more control over your DVD's functionality using DVDit! I saved the last step—DVD testing and burning—for the next hour.

Finally, I reminded you of the myriad possibilities that DVDs present and gave you four real-world examples of how DVD authors used DVD authoring applications to create some exciting projects.

Workshop

Review the questions and answers in this section to reinforce the final few DVD authoring steps covered in this hour. Also, take a few moments to take the short quiz and perform the exercises.

Q&A

Q I dragged a menu placeholder thumbnail to another menu to create a link to one of that first menu's buttons, but a new button (that looks just like the menu I'm dragging) appears in the Menu preview screen and no link appears in any of the menu buttons. Why is that?

A To create a link with a specific button, you need to drag the menu thumbnail from the placeholder to that button. If you drop that menu thumbnail outside a button, DVDit! thinks you want to make a new button. By default, when dragging a menu to another menu, DVDit! creates a rectangular, beveled button with the linked menu as its thumbnail image.

Q I dragged media onto a menu and created buttons and links in the process. I want the center button to be the one that highlights first when the viewer gets to that menu, but the upper-left button is listed as Button 1. How do I fix that?

A Button number one (or in other menus where there is no linked button 1, the lowest-numbered button with a link) is the one highlighted when a viewer first arrives at a menu. To change that, go to that menu's properties—right-click the menu and select Show Properties—and then select the correct button number from the Default Button drop-down list.

Quiz

1. You can change a still image's duration in two ways. What are they?
2. You automatically create links to menus and movies by dragging them from different locations to existing buttons. What are those two locations?
3. You place an image in first play and give it a specific duration. When it finishes playing, you want to jump automatically to a menu. How do you do that?

Quiz Answers

1. Right-click that image's button or placeholder screen, select Show Properties, and change the Duration setting. Alternatively, you can drag an audio file to that image placeholder screen. The image will display for the length of the audio clip.
2. To make a link to a movie, select it from the Movie drop-down list below the lower-left corner of the preview window; then drag its chapter one triangle to its associated button. To make a link to a menu, drag its placeholder thumbnail to its associated button.
3. Right-click the still image's movie placeholder screen and select Show Properties. In the End Action drop-down menu, select the action you want to take place after the still image reaches the end of its duration. Your choice usually is the opening menu, but you could have the DVD play through a sequence of stills and video clips (like previews of coming attractions).

Exercises

1. Take a look at some Hollywood DVDs and note how they behave. Note which button highlights first when a screen opens, the order in which buttons highlight, what happens at the end of a short clip, how scene selection menus work, and what happens when you press the Title or Menu buttons on your remote. The DVD developers defined all those behaviors, as will you if you work with prosumer- or professional-quality DVD authoring applications.
2. Use DVDit! to build a project using your own media. Remember that to add media, buttons, or menu backgrounds to DVDit!, you need to import them from the Theme palette rather than simply pasting them into the DVDit! Theme file folder. Otherwise, DVDit! won't be capable of working with them.

HOUR 21

Burning DVDs and Dealing with Mass Replicators

Now it's time to see the fruits of your labor, record your project to a DVD, and pop it in your TV's set-top player for your viewing pleasure.

In this hour, I show you how to use DVDit! to preview your DVD and check for broken links or other unexpected behavior. I explain DVDit!'s output options, briefly cover how to select recordable media, and then have you burn and test your DVD.

I wrap up this hour with a quick look at MyDVD's Direct-to-DVD module, showing you the steps to take if you want to mass-replicate your DVD. I also tell you about one way to market your product.

The highlights of this hour include the following:

- Checking menu and media links and project flow
- Selecting recordable media
- Burning and testing your DVD
- Using MyDVD's Direct-to-DVD option
- Going the mass-replication route
- Using Sonic Solutions' Publishing Showcase

Checking Menu and Media Links and Project Flow

Before you use DVDit! to burn your DVD, you should test all the links and settings. DVDit! lets you preview the project, mimicking how the finished DVD will play back in the player.

Task: Test Your Project

▼ TASK

In this exercise, you simulate what it's like for a viewer to sit down with a remote control and start clicking through your DVD. Follow these steps:

1. Start DVDit!, open your project, and click the Play (or Preview) button in the lower-right corner of the Theme palette (see Figure 21.1). This opens the Remote Control interface shown in Figure 21.2.

The first thing you'll notice is that music starts to play—that's the audio file you dragged to the main menu (as well as to the three other menus). If you do nothing, it will loop, playing to its conclusion and then starting over.

2. If you look at the remote control buttons, you'll see that most of them are self-explanatory. For instance, the four triangles navigate around the menu buttons and Menu takes you to whichever menu you were last at (although you can specify a different action in the Properties dialog box). Four functions are highlighted in Figure 21.2 that might not be that intuitive. Moving clockwise from the upper-left corner, they are

▼

- **Title**—Clicking this takes you to the menu at the top of your Menu placeholder list.
- **Return**—This performs whichever function you specified in the Movie/Menu Properties dialog box (you probably did not specify a return function).
- **Next**—This jumps to the next button's menu or movie.
- **Previous**—This jumps back to the previous button's menu or movie.

FIGURE 21.1
Click the Play (or Preview) button to access the Remote Control interface.

FIGURE 21.2
Four buttons in the Remote Control interface might not be self-explanatory: Title, Return, Next, and Previous.

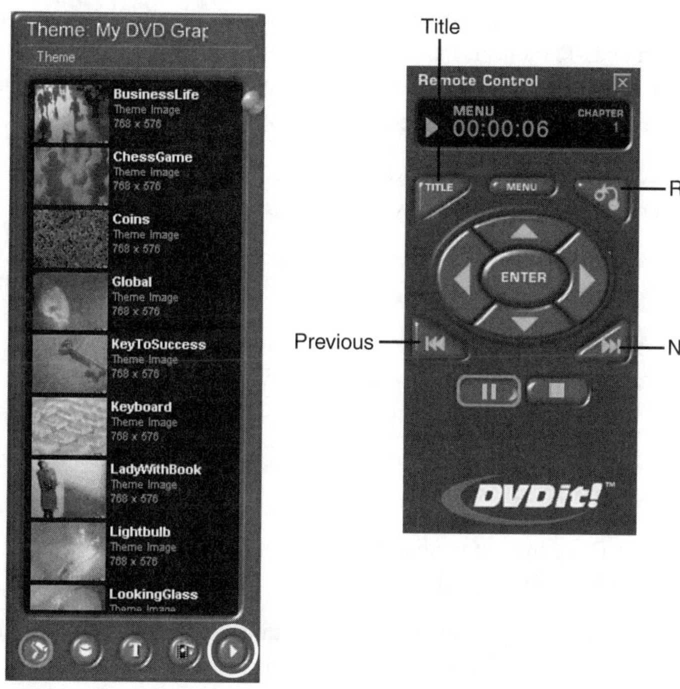

3. Use the remote control to navigate through your project. Try everything, making sure that all the menus do what you expect. View the still images, clips, or scenes. After they finish playing, you should automatically return to their respective menus. If any menu buttons take you nowhere or take you to the wrong media or menu, return to your project's menu and movie Properties windows to fix them.
4. To exit the Remote Control interface, click the × in the upper-right corner.

Selecting DVD Recordable Media

DVD recordable media are swiftly approaching commodity status. That is, soon there might be no discernible difference between one brand and another. For the moment, though, it still appears that you get what you pay for.

Selecting "house brand" media remains a hit-or-miss proposition. One unsubstantiated test I saw showed some store-brand recordable DVDs with 40% failure rates. This means four out of ten discs became proverbial cocktail coasters.

Obviously, companies that sell branded drives along with their own branded media—such as H-P, Pioneer, Panasonic, and Sony—want you to think that you can't have one without the other. But selecting from that top tier of the price chart can be overkill (although, their prices are coming down).

I recommend taking the middle ground: Select name-brand, generic media from firms such as Verbatim, Memorex, Maxell, Mitsui, and TDK.

Because prices continue to drop, it makes little sense to include specific prices here. In general, though, the following are true:

- DVD-R is the least expensive option.
- DVD+R, DVD-RW, and DVD+RW all cost about 50% more than DVD-R. Because DVD-RW and DVD+RW are rewritable, you don't need very many of them anyway.
- DVD-RAM remains the most expensive media; it's nearly three times as expensive as DVD-R.

As DVD recordable media approach commodity status, there are still a few minor things to watch out for, including

- **Authoring versus general**—Authoring media are designed specifically for authoring DVD recordable drives such as Pioneer's DVR-S201. You probably don't own such a drive. Early DVD-R drives were capable of making bit-for-bit copies of other DVDs. By early 2001, the movie industry responded by creating the Content Scrambling System (CSS) encryption scheme and getting DVD manufacturers to change their drives to make it physically impossible to copy CSS discs. So, if you have a DVD-R–capable drive, it probably uses DVD-R general recording media.

- **Media quality and your drive recording speed**—If you want to take advantage of faster DVD recording speeds, you need to buy media rated for those speeds. For instance, look for 4X in the product name or specs if that's the speed of your drive.
- **Capacity**—You might never need to buy anything other than 4.7GB DVD recordable media. Such media is standard single-side, single-layer media at full DVD capacity. Several other capacities exist, so be sure you get the one that suits your project size and drive specifications.

Burning and Testing Your DVD

This last step, in what has been a lengthy process, is relatively simple. Primarily, it comes down to selecting a few settings, noting whether you're going to add DVD-ROM data to the disc, and choosing the correct DVD recorder.

Task: Make a DVD Disc

TASK

If all goes smoothly, in a few minutes you'll have a DVD in hand ready to play on your set-top DVD player or PC. Follow these steps:

1. Start DVDit! and open your project.
2. Put a blank recordable disc (DVD-R, DVD-RW, DVD+R, DVD+RW, or DVD-RAM) in your recorder. You can also record to CD-R or CD-RW. For more on that, see step 6.
3. As illustrated in Figure 21.3, select File, Project Settings and select an output size from the drop-down list to match your recordable disc. Check the Current Project Size box to ensure that your project does not exceed your disc capacity. Finally, name your project.

If you followed the tasks in the previous four hours, your project will be about 450MB—about one tenth the capacity of a standard DVD-R.

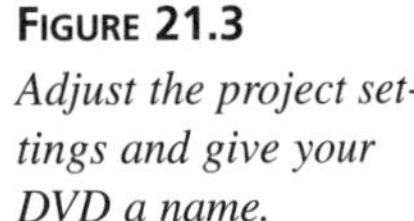

FIGURE 21.3
Adjust the project settings and give your DVD a name.

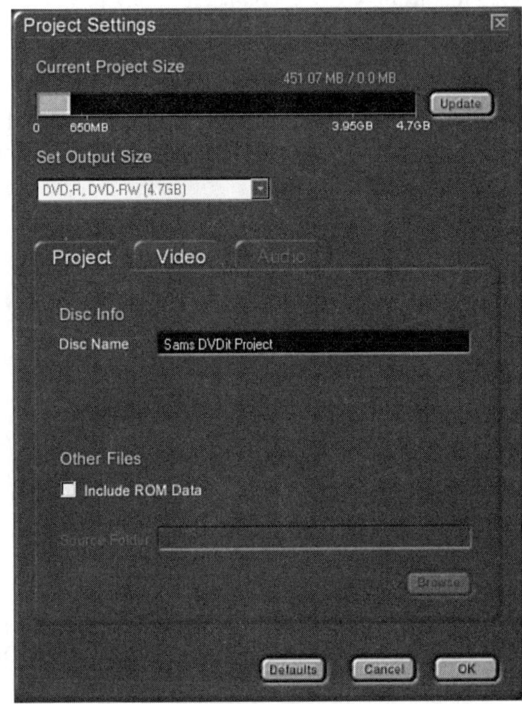

4. Click the Video tab and select either Set Best Quality for AVI and QT Files or Manually Set Quality for AVI and QT Files (see Figure 21.4). If you choose the latter, which is necessary only for large projects that might exceed a DVD's capacity, use the slider to select from the five settings. Click OK.

The Project Settings Audio tab is disabled for this SE version of DVDit! It's activated in the PE version, though, allowing you to convert audio to Dolby Digital (as opposed to the uncompressed PCM audio standard for the SE version) and letting you select from a dozen bit-rate quality levels.

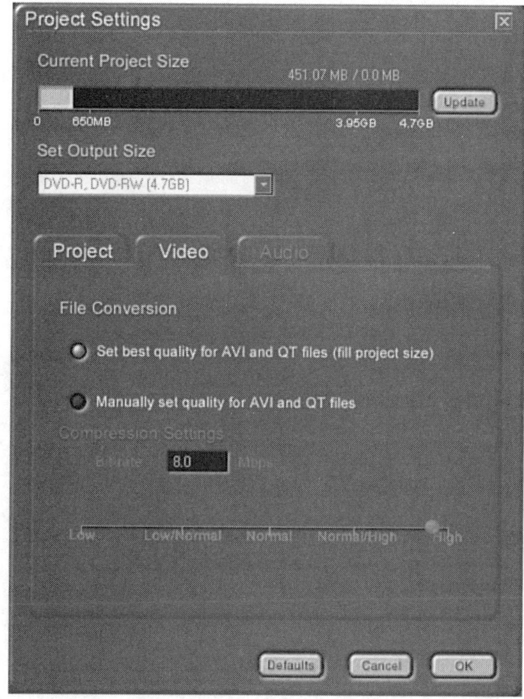

FIGURE 21.4
In the Video tab, you can opt to reduce the bit rate of your compressed MPEG files, saving space on your DVD but losing quality in the process.

5. Select Build, Make DVD Disc to open the Make a DVD Disc dialog box, shown in Figure 21.5.

After selecting Build in the main menu, you also can choose to make a DVD *folder*. This is a hard drive storage folder that contains all your transcoded (converted to MPEG) project media. Its purpose is to let you play your MPEG files to ensure that they work well before actually creating your DVD. It's an extra step few will use. It's also a bit confusing because, when you later create a DVD from that folder, DVDit! calls it a DVD *volume*.

6. Select the source. In this case, select the default setting, Current Project.

From the Make a DVD Disc Source drop-down list, you also can select DVD Volume ("Folder") or Disc Image. A *disc image* is a file produced by higher-end authoring products such as Sonic's Scenarist. It has no directory structure and contains all the DVD project data in the exact position it will appear on the final DVD.

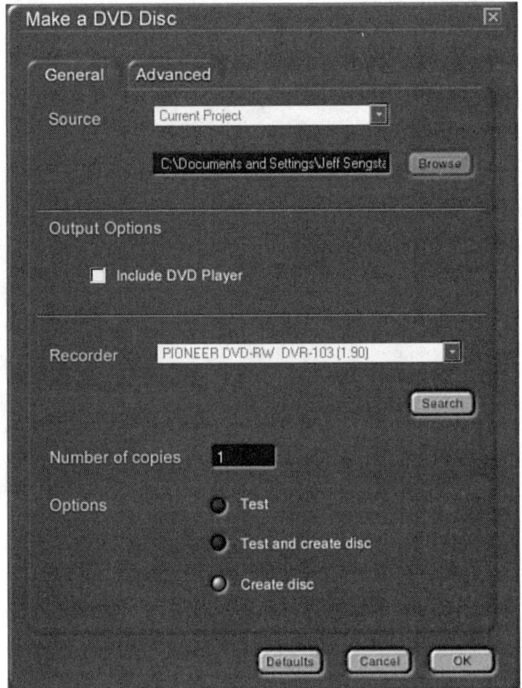

FIGURE 21.5
Set a few parameters in this interface and then start recording your DVD.

7. If you're writing to a CD-R or CD-RW, select the Include DVD Player check box. Doing so adds Sonic's proprietary cDVD player to the CD so it can play your DVD project from a PC CD player. The finished CD will *not* play in a set-top, standalone DVD player, however.
8. Select your DVD recorder from the drop-down list.

The PE version of DVDit! lets you output your project to a digital linear tape for use in replication. I cover this in some detail later this hour in the "Going the Mass-Replication Route" section.

9. If you want to make more than one copy, change the value in the Number of Copies box.
10. Your choice in the Options section depends on whether you have burned a DVD before with no errors. If so, select Create Disc. If not, select Test and Create Disc.

If you select Test or Test and Create Disc, DVDit! simulates project legalization and verification, multiplexing (combining) of the MPEG video and audio, formatting the DVD, imaging, and passing the data to the drive. After the virtual write has been deemed successful, the actual write to DVD media can occur.

If your DVD media is rated at a slower speed than your recorder's top speed, select the speed that matches your media.

11. Click the Advanced tab to switch to the interface shown in Figure 21.6. You have the following two general options:
    - **File System**—Use one or both of these two options only if you are adding DVD-ROM files to your project or are including the cDVD player on a CD-recordable disc. In the latter case, accept the default settings—Use Joliet and Use Long File Names.
    - **Temporary Storage**—By default, DVDit! places your media and project files in a temporary file folder before recording them to your DVD. Only if you have limited hard drive space should you deselect this option. In that case, DVDit! writes all files directly to the DVD, bypassing caching. You can opt to save the cached files, but there is little reason to do so.
12. Click OK. DVDit! immediately starts transcoding your files. As shown in Figure 21.7, it keeps you informed of its progress each step along the way.
13. Within a few minutes DVDit! will have completed recording your project to DVD. It will eject your DVD and display this message: `DVD created successfully.`

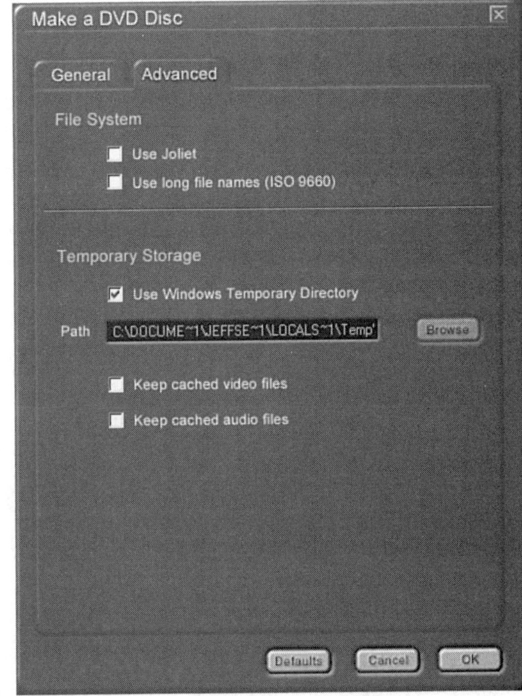

FIGURE 21.6
The Advanced page lets you select filename types and whether you will use a temporary directory while burning your DVD.

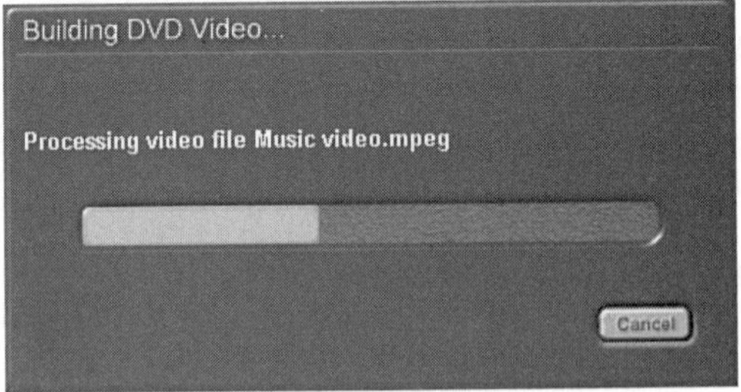

FIGURE 21.7
As DVDit! prepares your project to record to a DVD, it lets you know about each step in the process.

Take your DVD to a set-top standalone DVD player and enjoy this musical interlude you've created.

If that DVD player is reasonably new and therefore fully compatible with all DVD specifications, your DVD should work fine. If not, try it on your PC. Use whatever DVD software player you used in Hour 2, "Getting Your Gear in Order—DVD Recorders and Media."

If compatibility problems exist with your set-top DVD player, visit one of several online sites that list DVD players and whether they play back DVD-Rs or other recordable DVD media. Apple has an exhaustive list at `http://www.apple.com/dvd/compatibility/`. At the very least, take your DVD to a local consumer electronics store and try it in several of their DVD players.

Using EZ/CD Print for Professional-Looking DVDs

You've seen Hollywood DVDs with artwork printed directly on the DVD. You, too, can give your DVDs a professional look.

The standard ways to label DVDs fall into two camps: marking pens and paper labels. Now, thanks to a collection of affordable printers from EZ/CD Print (`http://www.ezcdprint.com/`), you can give your DVDs that Hollywood look by printing directly onto your burned DVDs (see Figure 21.8).

EZ/CD Print, using Epson printer engines, manufactures three CD/DVD printers ranging in price from $400 to $500 (they also function as regular, full-featured paper printers).

You simply use a CD/DVD labeling template available with most CD/DVD recording software. Instead of printing to a gummed paper label, you insert a special inkjet-printable, recordable DVD (or CD) into the EZ/CD printer and in about a minute (or less for the $500 model), out slides a DVD with a full-color image on it. One source for ink jet printable media is EZ/CD Print's distributor: `http://www.atdiscount.com`.

If you are in the DVD authoring business, this is a great way to impress clients.

FIGURE 21.8
The $400 EZ/CD 4200 prints full-color labels directly onto your DVDs.

Using MyDVD's Direct-to-DVD Option

You know what a pain it is to use your camcorder to view a video on your TV. You have to connect the cables and then fast forward or rewind to find a clip that interests you.

A more elegant viewing approach is MyDVD's Direct-to-DVD option. This feature lets you capture video and transfer it directly to a DVD. You can add an opening menu and a chapter menu, and you can have MyDVD automatically select chapter points or select them manually.

After the DVD is burned, it's simply a matter of viewing your DVD with your set-top DVD player. You don't have to hassle with extra cables, and you can skip quickly to your favorite scenes. Plus, this is an easy way to archive videos.

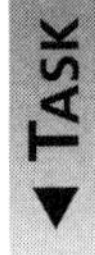

Task: Take the Direct-to-DVD Route

If you've used MyDVD's video capture module, this DVD burning procedure will be remarkably easy. If not, I think you'll still find that it's pretty effortless. Here's how to do it:

1. Start MyDVD and in the opening screen, shown in Figure 21.9, select Transfer Video Direct-to-DVD.

FIGURE 21.9
Using MyDVD's Direct-to-DVD option is a quick and easy way to archive and play back your videos.

2. The MyDVD Direct-to-DVD Wizard, shown in Figure 21.10, offers several options. As you did in Hours 4–6, you can select a menu style to suit your taste. In addition, you must name your DVD; otherwise, MyDVD won't let you go to the next step. Finally, be sure to select the correct DVD recorder; then click Next.
3. The final dialog box, shown in Figure 21.11, lets you set output video quality, chapter features, and a couple of other options. Here's a brief rundown:
    - **Settings**—Accept the default, highest-quality video compression setting or use the drop-down list to reduce the video quality (as you adjust the quality, MyDVD updates the Record Time Available in the Details window at the bottom of the screen).

- **Capture Length**—Normally, you'll stop recording manually. If you have a specific duration in mind, you can have MyDVD stop recording automatically by setting a time limit here.
- **Scene Detection**—MyDVD can manually detect scene changes. You can opt to do that manually, or you can do nothing. If you do mark scenes—automatically or manually—you can have MyDVD create a chapter menu (or menus, depending on the number of scenes).
- **Chapter Settings**—This sets the scene change detection sensitivity.

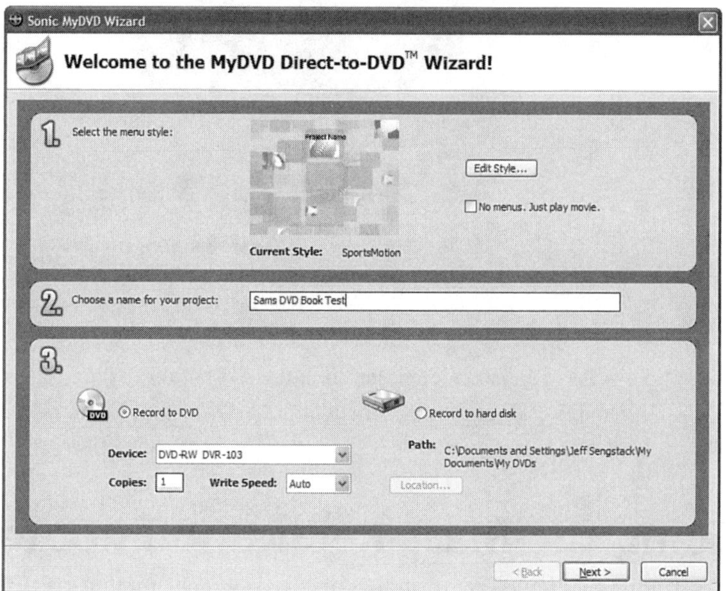

FIGURE 21.10
Select a menu style before capturing video.

4. When you're ready, use the VCR-like controls to cue your tape. Then, click the red Record button.

> If you are recording from an analog camcorder/VCR, rewind to a few seconds before your record start point, start playing the video, and then click the red record button when you want to start recording.

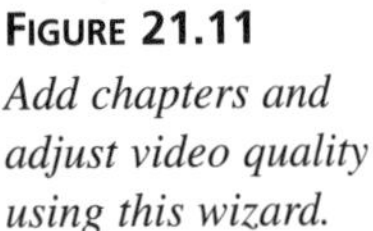

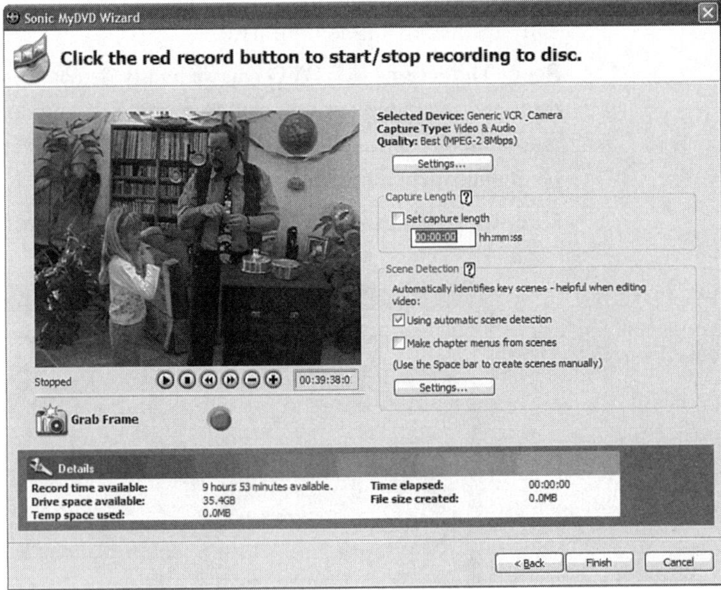

FIGURE 21.11
Add chapters and adjust video quality using this wizard.

5. When you're done, either click the red button to stop recording or watch as MyDVD does that automatically. MyDVD then transcodes the video and burns the disc.

Going the Mass-Replication Route

Burning multiple DVDs of the same project, one at a time, is tedious. Mass-replication can therefore be an appealing alternative.

These days, working with a replicator is not very difficult. Your finished project recorded to a DVD-R and some liner and DVD disc artwork is about all you need.

To ensure a smooth DVD replication process, follow these tips:

- Start by visiting the DVD Association Web site: `http://www.dvda.org`. It has a list of member replication firms. Many of these companies are *mass* replicators, meaning they deal with film studios and publishers and handle multiple orders annually for millions of copies. Few will touch an order of fewer than 1,000 discs. Don't let that discourage you, though. Contact a replicator from the list and ask whom they recommend for smaller orders.

- If you want to make a single-sided, single-layer DVD disc (referred to in the industry as a DVD-5), all you need is a DVD-R (or a digital linear tape—DLT) to serve as a master.

Digital Linear Tape Versus DVD-R

If you plan to have a replication firm mass-produce your DVD project, you have two primary options for mastering media: DVD-R and digital linear tape.

It used to be that DLT was the only option, meaning you had to buy a DLT machine and upgrade your DVD-authoring software to one that offered DLT as an output option.

Now replicators recognize the ubiquity of DVD-R and have developed ways to use that medium to create duplication masters.

You should remember these two caveats:

- Most mass-replicators do not accept DVD-RW, DVD-RAM, DVD+R, or DVD+RW masters. In particular, rewritable discs are not suitable because the data might not be contiguous on the disc.
- DVD-Rs are not 100% reliable when used as replication masters. Some desktop DVD recorders create DVD-Rs that have information that is illegal per the industry book standard. Those issues sometimes prevent the disc from being mastered for DVD replication or cause playability issues with the resulting replicated DVDs.

My recommendation is for infrequent users of replicators to use DVD-R masters. Just make sure that you try a pre-replication test disc on your set-top player to ensure that it works properly.

For those of you who will use replicators more frequently, buy a DLT recorder. The de facto industry-standard format was used in the now-discontinued Quantum 4000. You can still get that format at a value-oriented price ($1,200) with the Quantum DLT VS80, shown in Figure 21.12. It has twice the capacity of the DLT 4000 and uses the same DLT tape IV media.

The bottom line is that DLT is a very stable format and is strongly encouraged for those who want to make a business out of producing DVDs for mass distribution.

FIGURE 21.12
AT $1,2000, the Quantum DLT VS80 is the digital tape recorder of choice for value-oriented DVD producers.

- For a double-sided, single-layer so called DVD-10, you'll need two DVD-R discs.
- DVD-9 (single-sided, double-layer) and DVDs encrypted with the CSS copy-protection scheme need DLT masters.
- Be sure you've tested your DVD-R master on a set-top DVD player. Click every menu and press all the remote buttons to ensure that your DVD does what it's supposed to do before you send it off for duplication.
- Your most difficult hurdle might be artwork to create your liner and DVD label art. Most DVD replicators provide artwork templates you can use in various graphics programs.
- Be sure you proofread everything and check your colors. Your labels can look dramatically different from one replicator to another.
- Allow enough time for your project. From delivery of your master along with label artwork, expect to wait two weeks for completion of your order.

Walking Through the Replication Process

Creating DVDs is a multistep process that uses high-pressure injection molds, lasers, and electroplating. Following is a brief overview of the process:

1. Technicians coat a glass master with a very thin layer of light-sensitive material.
2. A sharply focused blue or ultraviolet laser converts digital data into a series of pulses, burning holes into that thin chemical layer along a spiral track (not a series of concentric rings) from the center to the outside.
3. A chemical bath washes away the film at each burnt pulse location leaving tiny pits. A thin layer of metal (usually nickel) is applied as a means to conduct electricity.

FIGURE 21.13
Application of the adhesive to the bottom layer of the DVD prior to the placement of the second layer.

4. This nickel-coated glass master is placed in another bath with an electric current flowing through it. More metal ions fill the tiny pits and then cover them with a thin wafer.
5. That thin layer (now covered with tiny bumps), called the *father*, is a reverse image of the glass master. It becomes the master mold for the actual stampers used to create the DVDs.

FIGURE 21.14
DVDs that have just been through the bonding process.

6. The stampers are placed in molding machines into which hot, molten polycarbonate is injected under extreme pressure (several tons) to create reverse images of the stampers.
7. Each layer is "metalized," typically with aluminum to create a reflective surface.
8. Each metalized surface is bonded to a non-metalized substrate.
9. So-called single-layer DVDs actually have two such metalized layers (CD-ROMs have only one). Double-layer DVDs have four.
10. Lacquering is applied to protect the reflective upper layer and serve as a surface on which to print labels.

Figure 21.15
Molded DVDs ready for printing.

Using Sonic Solutions' Publishing Showcase

Sonic Solutions, in an alliance with CustomFlix, presents a relatively painless and potentially profitable means to market your DVDs.

SONIC PUBLISHING SHOWCASE

For a one-time set-up fee (as we went to press it ranged from $29 for Sonic Solutions customers to $50), CustomFlix will help you create an e-store Web page, complete with a 30-second video promo, and will fill and ship all your orders (see Figure 21.16).

You can have CustomFlix manufacture your DVD in small quantities as orders come in (for which it receives $10 plus 5% of the sales price per item) or have it replicate your DVD in much larger quantities. If you choose the latter option, CustomFlix charges a reduced per-transaction fee but adds a monthly inventory charge.

You set the DVD price and do your own advertising while CustomFlix offers your DVD for sale on at least five Web sites, including its main store, Amazon zShop, Froogle, and its Yahoo! and eBay stores.

FIGURE 21.16
CustomFlix offers a relatively simple way to publish your DVDs.

If you choose the one-at-a-time approach, you can make a profit with sales of only 10 copies (with a list price of $20). If you try the replication model (1,000 minimum print run), you'll see profits after you sell about 250 copies within a year. But with the greater risk comes greater rewards. After your DVD has annual sales of 500 copies, you start making more per transaction than if you opted for the one-at-a-time scheme.

Visit `http://www.customflix.com/Special/SonicShowcase.html` to check out its offerings. When I last looked, it offered a wide assortment of DVDs from a how-to on porcelain doll making to a feature-length, off-color comedy about Bigfoot.

Summary

Burning DVDs is a three-step process: You check whether your project's links and menus function as you expected, tweak the output options within DVDit! 5, and burn your DVD. After it's burned, you should take the acid test of viewing your DVD in a set-top player.

For those straight-to-DVD projects—single VCR copying or archiving—MyDVD's Direct-to-DVD module is an easy-to-use alternative to the full authoring approach.

As you gain experience in DVD production, you might think in terms of marketing your projects. Using a replicator has become a reasonably routine process, as has turning to the Web as a publishing tool. Sonic Solutions and CustomFlix have teamed up to give Sonic Solutions authoring application users a simple and inexpensive means to market and distribute DVDs.

Workshop

Review the questions and answers in this section to reinforce your DVD finalizing knowledge. Also, take a few moments to take the short quiz and perform the exercises.

Q&A

Q I noticed that the transcoding presets have a maximum of 8,000kbps (8Mbps) for video. That's not the maximum allowable MPEG-2 video quality of 9,800kbps (9.8Mbps). Why not?

A Sonic Solutions limits it to 8Mbps so enough room exists for the audio. PCM audio takes up 1.5Mbps, so add that to the 8Mbps and you get 9.5, which is scarily close to the 9.8Mbps maximum for DVD. If you use the PE version of DVDit! 2.5, you can select Dolby Digital audio, which takes up less bandwidth and could allow you to increase the video quality. However, that option is not available in either version of DVDit!

Q I want to make 50 copies of a DVD to send to all the families who attended my daughter's wedding. Should I use a replicator? I don't have a DLT tape recorder. Do I need to get one?

A Fifty copies falls into the category of needing to ask how much manual labor you are willing to do. Burning 50 copies one at a time will take several days, and few major replicators will handle such a small run. Those that do typically burn DVD-Rs rather than build a master and replicate copies. Therefore, you don't need to use DLT. Obviously, prices are moving quickly in this field, but my guess is that using a service bureau to burn 50 DVDs will cost about $250.

Quiz

1. When must you use a digital linear tape when replicating a DVD?
2. DVDit! PE offers PCM and Dolby Digital audio. Why select one over the other?
3. When you test your DVD—either using DVDit!'s Preview interface or playing an actual burned DVD of your project—you notice that the button that automatically is highlighted when a menu opens is not the one you wanted to have highlighted. How do you fix that?

Quiz Answers

1. Actually, you personally never need to use a DLT. You always can use DVD-R discs and, if necessary, the replicator will put them on DLT for DVD mastering. Generally, this occurs only if you want to create a DVD-9 disc, a double-layer disc

that can play a movie longer than 2 hours without skipping a beat. The caveat here is that DVD-R is not an absolutely reliable medium for a master, but DLT is.

2. PCM is uncompressed digital audio that is similar to CD audio but with higher sampling frequencies. At its best quality setting (the only setting available in DVDit!), it's as good as audio gets on a DVD. But it requires extra bandwidth, which limits the bit rate allowable if you want to transcode your video at the best rate permitted by the DVD specification. So, to get great video with what is still excellent audio, use Dolby Digital audio.
3. In DVDit!, go to that menu and display its links. Note the button number of the button you want to highlight. Right-click the menu and open its properties. In the Menu Properties window, change the default button to the one you want to be highlighted when the menu first displays.

Exercises

1. Go to `http://www.dvda.org`, select a DVD replicator, and talk to them—or better yet, pay them a visit. Check on minimum order quantities, prices, label/liner graphics templates, and whether the company accepts DVD-R masters. If their minimum quantities are too large, ask whether they can recommend smaller replicators.
2. Go online to check out the latest recordable DVD drives and other developments in this very dynamic industry. I'd suggest using `http://news.com.com/` as your starting point. Navigate to the Hardware/Storage section (use the Site Map link at the upper-right corner of the page as an aide); then select DVDs and go from there.
3. Visit `http://www.customflix.com/` and take the online tour. It's an excellent introduction to the company's marketing program. When you're done, check out the latest offerings and note how they're marketed. Because this site is in its infancy, a wide range of promotional ideas and production values is available. Ask yourself what works well and what doesn't; then adjust your marketing plans accordingly.

HOUR 22

Creating Custom MyDVD Templates with Style Creator

MyDVD ships with numerous menu templates that give you a wide range of DVD project options, from a family-friendly look to several corporate styles.

At some point, you might want to create a project with a more customized look and feel. Sonic Solutions offers Style Creator, a menu template creation tool that helps you craft entire menus that work in MyDVD. In effect, Style Creator turns a consumer-level DVD authoring application—MyDVD—into a prosumer product.

In this hour I introduce Style Creator's powerful functionality and show you how to make changes to an existing template. In a sidebar, I also explain how to use graphics software such as Adobe Photoshop to create buttons and backgrounds for DVDit! and other authoring applications.

The highlights of this hour include the following:

- Introducing Sonic Solutions' Style Creator
- Changing a template background
- Editing buttons and button elements
- Testing your template edits in MyDVD

Introducing Sonic Solutions' Style Creator

I mentioned in Hour 6, "Authoring Your First DVD Project Using MyDVD: Part III," that Sonic Solutions offers a Photoshop plug-in you can use to create menu templates or styles for MyDVD.

Style Creator is a *macro*. Those who have worked with macros in word processors, for instance, know they duplicate some number of user-defined steps. The same holds true for this plug-in. When you save a template using it, you'll see it perform various functions (very quickly) as it converts your graphics into a template file that will work within MyDVD.

Templates are a great way to give DVDs with similar themes a consistent look. For instance, a production company that specializes in weddings can use a template to simplify its workload while giving clients a product that matches any demos they've viewed.

Those templates can have static or animated menus and buttons as well as customized button highlights. The highlights can change color or shading as viewers navigate their remotes to them or select them.

How Style Creator Works

Style Creator takes a carefully formatted set of graphics you've created in Photoshop (or Photoshop Elements) and converts it to a file type that works within MyDVD.

For the tasks in this hour, you'll install the Style Creator plug-in into the trial version of Photoshop Elements included on this book's companion DVD. Sonic Solutions created Style Creator for the full $600 version of Photoshop. But by the time this book is published, Sonic expects to have qualified Style Creator for the much less expensive $99 Photoshop Elements. In any event, it works fine with Photoshop Elements, qualified or not, and using it in the upcoming tasks will be a practical introduction to the concept of layered graphics for which Photoshop is so well known.

Style Creator started its life as a tool for Sonic Solutions software engineers and artists to create menu templates or styles to ship with MyDVD. Each template has very specific rules about where buttons appear on MyDVD menus and how large they can be. In the meantime, Sonic Solutions realized the value that Style Creator gave its users, so it released it to the public on its Web site.

I've provided a copy of version 2 of the plug-in on this book's companion DVD. By the time this book is published, Sonic Solutions might have released an update—check `http://styles.mydvd.com/`. Even if a newer version is not available, you should visit this Web site because you can download, at no charge, a collection of custom styles created by MyDVD users.

Taking a Close Look at the Plug-in

Style Creator works only if you create a layered Photoshop graphic that follows specific rules. Figure 22.1 shows a listing of Photoshop layers with some of the naming conventions used by Style Creator.

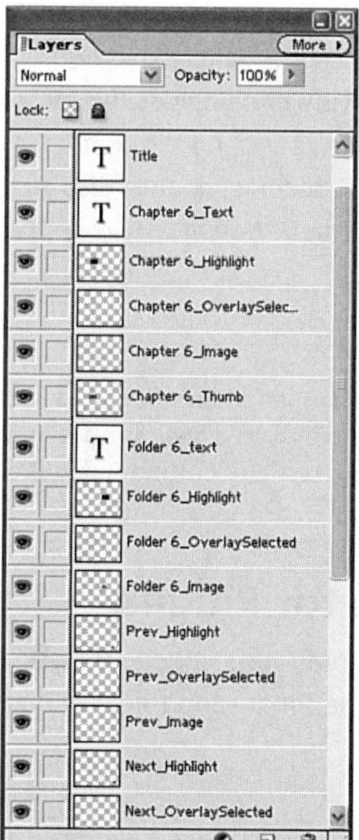

FIGURE 22.1
You must follow the Style Creator layer naming conventions exactly for it to work.

A full explanation of all that Style Creator offers along with all its rules would consume much more than one hour. I've limited this hour's discussion to a basic explanation and an introduction. For more details, read through the detailed Style Creator Tutorial. You'll find it in the Style Creator Plug-in file folder on this book's companion DVD. You'll need the industry-standard document viewer Adobe Acrobat Reader to view it; it's available for free download at `http://www.adobe.com`.

Style Creator recognizes each layer name and treats whatever graphic it finds in each layer according to its intended use. For instance, the `Home_Image` layer is the button that returns viewers to the DVD's main menu. When you import a Style Creator Style template into MyDVD, the DVD authoring application automatically places whatever graphic is in the `Home_Image` Photoshop layer in the lower-left corner of a menu and links that graphic to the main menu.

Each image layer can have up to four other elements associated with it: `Highlight`, `Selected`, `Activated`, and `Text`. The `Highlight` layer defines the *hotspot zone*, which is the area in which the button resides (hotspots cannot overlap). The `Selected` and `Activated` graphics display when the viewer navigates the remote control to a button (`Selected`) and then presses Enter (`Activated`) on the remote. These typically are simple highlights. In MyDVD, they are usually rectangles or circles around the perimeter of a button or frame. Finally, the `Text` layer defines where text appears relative to its associated button.

After you've created or changed a template, you use the Style Creator Photoshop plug-in to convert it to a style for use in MyDVD. I explain that later in the chapter, in the section "Editing Buttons and Button Elements."

Installing the Style Creator Plug-in

Normally, installing the Style Creator plug-in is done automatically using an executable file that searches your hard drive for the Photoshop folder and adds two files to two Photoshop subfolders.

If you own a copy of Photoshop (version 6 or higher), feel free to use it for the upcoming tasks. Simply run the plug-in executable—`SonicStyleCreator.exe`—that you'll find on this book's companion DVD in the Sonic `Style Creator plug-in` folder.

The executable won't work with Photoshop Elements, so you need to do some manual copy-and-paste work instead. If you haven't installed Photoshop Elements, do so now. Simply go to the `Photoshop Elements` folder on this book's companion DVD and double-click `Setup.exe`. To simplify things, accept the default location: `C:\Program Files\Adobe\Photoshop Elements 2`.

This is a trial version of Photoshop Elements. It's equivalent to the full version, but it expires 30 days after you install it.

Install the Style Creator plug-in by navigating to its file folder on the DVD. Copy `SonicStyleCreator.8LI`, and then paste it to this file folder: `C:\Program Files\Adobe\Photoshop Elements 2\Plug-Ins\Automate` (if you installed Photoshop Elements somewhere else, paste accordingly). Return to the Sonic `Style Creator plug-in` file folder on the DVD, copy `SonicHidden.8BF`, and paste it into `C:\Program Files\Adobe\Photoshop Elements 2\Plug-Ins\Filters`.

To ensure that all went well, open Photoshop Elements. You'll see the splash screen shown in Figure 22.2. Note that it tracks how many days you have left in your trial period.

FIGURE 22.2
The trial version of Photoshop Elements times out in 30 days.

Click Try to open the user interface. Check to see that the plug-in loaded properly by selecting File, Automation Tools and then noting whether Create Style for Sonic MyDVD is listed as an option (see Figure 22.3). If it's not there, you need to check the two sub-file folders—`Photoshop Elements 2\Plug-Ins\Automate` and `Photoshop Elements 2\Plug-Ins\Filters`—and make sure the proper files are in their respective locations.

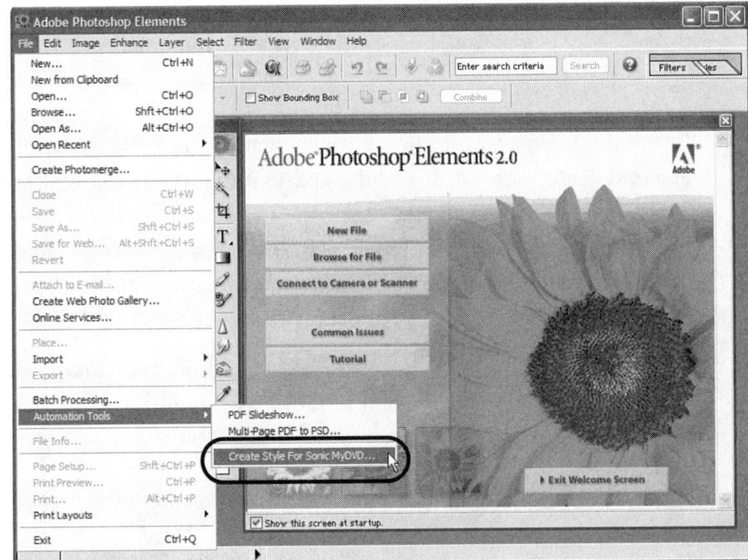

FIGURE 22.3
Check Photoshop Elements to see whether the Style Creator plug-in loaded properly.

Changing a Template Background

In the next two sections, you'll use Photoshop Elements to edit an existing menu. Then you'll run the Style Creator plug-in to convert that menu to a MyDVD Style pack. This is an easy way to get an overview of the template creation process, to take a look at the precise nomenclature required for the plug-in to work properly, and to see how button highlight layers work.

The purpose of the coming tasks is to introduce you to what Photoshop can do to help you create customized menus and buttons. Because this is not a book about Photoshop, sometimes I'll combine several steps into one task item and minimize the explanations of the associated features.

Task: Replace a Background

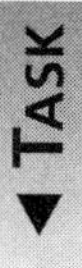

Even if you've never worked with Photoshop, you can make Style Creator menu style templates. The template requires very specific layer names, so you should take your first stab at creating a menu style template by editing an existing, conforming layered graphic. We'll start simple by changing the graphic menu's background. Here's how to do that:

1. With Photoshop Elements open, select File, Open and navigate to the Style Creator Plug-in file folder on this book's companion DVD. Select `ExampleSonicStyle.psd` and click Open.

> `.psd` is the standard file extension for Photoshop files. `ExampleSonicStyle.psd` is a straightforward, layered Photoshop file. After you've made some changes to it, you'll convert it to a style using the Sonic Solutions Style Creator Photoshop plug-in.

2. The goal now is to make your workspace to look similar to Figure 22.4. You do that by minimizing (or closing) the How to and Hints folders. Then click and drag the Layers tab from the upper-right corner to the workspace to open it (or select Window, Layers). Drag its lower edge to make it taller and display more layers.

> As a means to create clearer illustrations for this book, I generally ran most software in 800×600 resolution (the lowest resolution Windows XP will allow). I suggest you run in 1024×768 or higher if your monitor supports it, and if it doesn't create too much eyestrain. If you do opt for a higher resolution, your workspace will look different from the images here. For instance, at 1024×768, you'll see more layers than are displayed in Figure 22.4.

3. Each of the boxes in the Layers palette represents one layer of the image. Turn off the eyeball for every layer except the last one, Background. You can click them one at a time to turn them off individually or simply click the top eyeball—Safe Area—and hold down the mouse button as you drag it down the palette, leaving the last eyeball (Background) turned on. Your image and Layer palette should look similar to Figure 22.5.

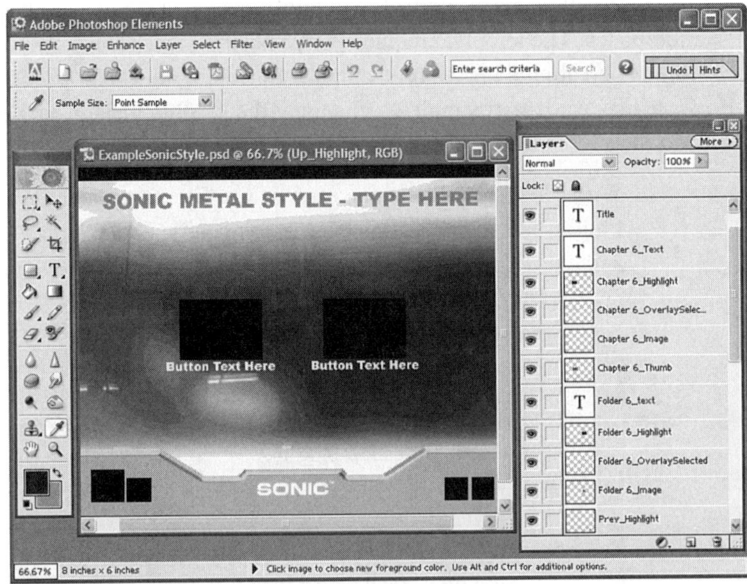

FIGURE 22.4
Arrange your Photoshop Elements workspace along these lines.

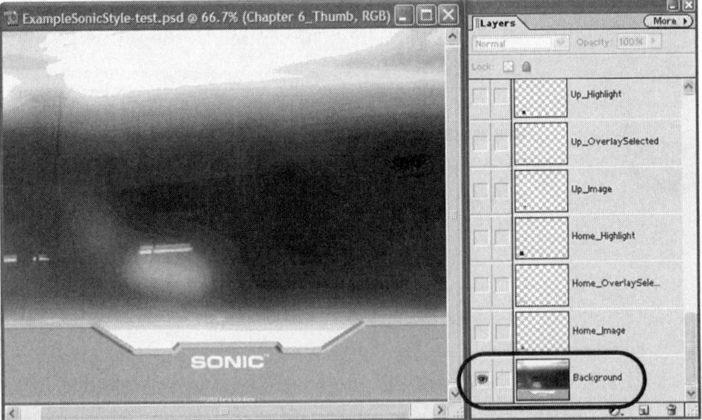

FIGURE 22.5
Turn off all layers (except Background) by clicking their eyeballs.

4. You will replace the Sonic Solutions background with a simple pattern. Start this process by opening a new project (select File, New). Then, using Figure 22.6 as a reference, type in 768 and 576 pixels, select RGB Color (any resolution will work), and make sure the Transparent radio button is selected. Then click OK.

When making menus using the Style Creator plug-in, you should use background graphics that are 768 pixels × 576 pixels. The Style Creator plug-in scales all non-768×576 background images to 768×576. This shrinking, expanding, or skewing of graphics that don't match this resolution or its 4:3 aspect ratio results in distorted or lower-quality graphics. Photoshop notes that 768×576 is a standard PAL preset, but it works fine in NTSC as well.

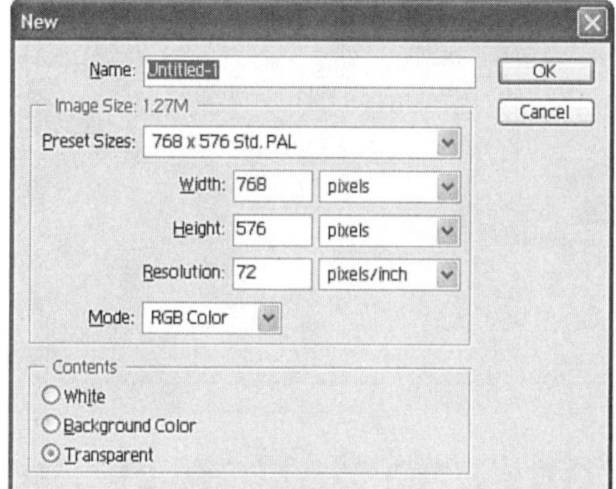

FIGURE 22.6
Set the dimensions of your new Photoshop graphic to 768×576 pixels.

5. Create a new background by selecting the Paint Can icon shown in Figure 22.7, clicking the Fill drop-down list at the top, selecting Pattern, and selecting a Pattern from the palette to the right of the word `Pattern` (I chose Wood). Now move your cursor over the Untitled-1 window (it turns into a paint can) and click. The window fills with the selected pattern.
6. In the main menu, select Select, All to put a dashed line around the pattern; then select Edit, Copy to copy that entire 768×576 pattern.
7. Minimize the Untitled-1 window, and then select the `ExampleSonicStyle.psd` graphic. Activate the Background layer by clicking the word `Background` in the Layer palette. From the main menu, select Select, All; then press Delete to remove the Sonic Solutions background graphic.
8. Paste the pattern from Untitled-1 onto the Background layer by selecting Edit, Paste from the main menu (you can also use the keyboard shortcut for paste, which is Ctrl+V).
9. Save your project by selecting File, Save As; then navigate to a file folder of your choice (the Photoshop Elements file folder is a logical location), give your project a new name such as `ExampleSonicStyle-test`, and then click Save.

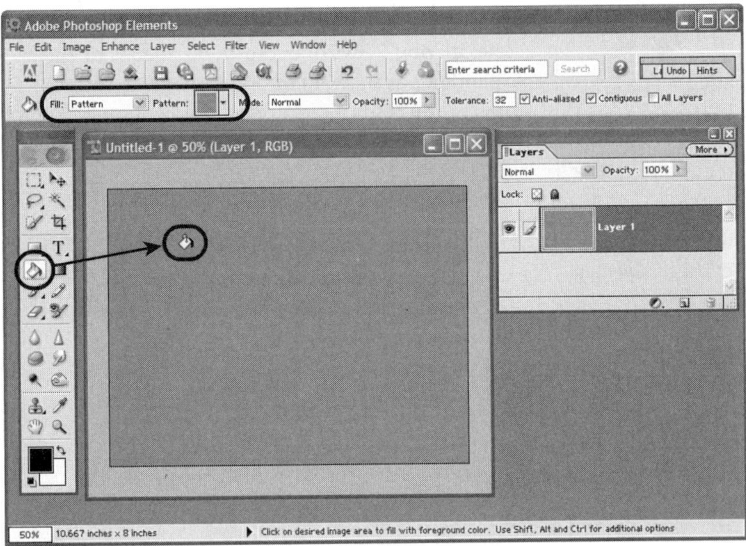

FIGURE 22.7
Use a Photoshop pattern and the Paint Can tool to create a new background for your style template.

Editing Buttons and Button Elements

What really sets customized menus apart from the crowd is their buttons and button highlights. Each button has an associated hotspot zone—called `Highlight` in Style Creator. In addition, it might have a thumbnail video still locator—called `Thumb`—and up to two associated highlight graphics: `Selected` and `Activated`. Style Creator lets you give your buttons and their associated graphics virtually any look you want.

Task: Examine Button Characteristics

Before making changes to the buttons in the sample template, you should look at the various button types and see how their associated elements work together. Here's how you do this:

1. Turn off all the eyeballs in your `ExampleSonicStyle` layer listing. Using Figure 22.8 as a reference, scroll to the Chapter layers and switch on the `Chapter 6_Image` and `Chapter 6_Thumb` eyeballs.

When MyDVD sees the word `Chapter` in a style template, it knows it's a button that will both take viewers to a video or slideshow and have a thumbnail image inside a frame. `Chapter 6_Image` is the frame that surrounds that thumbnail, and `Chapter 6_Thumb` is a blue rectangle that defines where the video still image thumbnail will appear. The number 6 refers to the maximum number of buttons that can appear on a MyDVD menu that uses this particular template. If you add more than six menu buttons (in this case), MyDVD creates another menu.

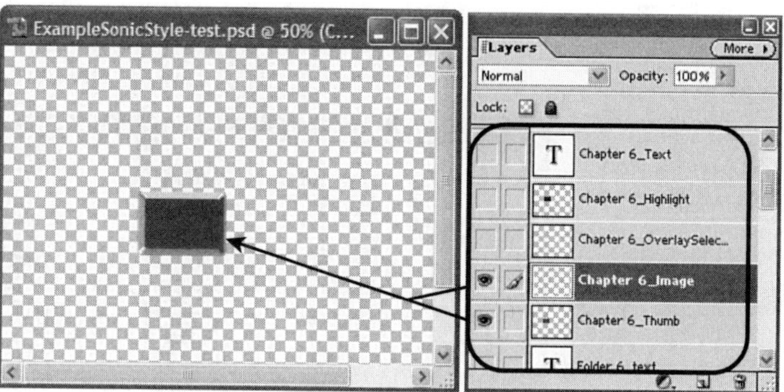

FIGURE 22.8
The `Chapter 6_Image` *and* `Chapter 6_Thumb` *graphic layers define the appearance and functionality of this button.*

2. Turn on the other three associated eyeballs in turn, working up the list starting with `Chapter 6_OverlaySelected`, then `Chapter 6_Highlight`, and finally `Chapter 6_Text`.

As you step through the other three chapter button elements, you'll see how they work. `Chapter 6_OverlaySelected` is a red box that displays when the viewer navigates the remote control to that button, whereas `Chapter 6_Highlight` is the hotspot zone that defines the button's boundaries (button highlights or hotspot zones cannot overlap). The `Chapter 6_Text` defines a placeholder for whatever text you want to add there later in MyDVD (it's white and hard to see on the checkerboard background).

3. Turn off all those eyeballs. Using Figure 22.9 as a reference, turn on the eyeballs for all the remaining buttons: `Folder 6_Image`, `Prev_Image`, `Next_Image`, `Up_Image`, and `Home_Image`.

Each of these buttons has a specific function in MyDVD. The Folder button automatically creates a link to a subfolder, Prev and Next take viewers to the previous or next menu in a chain of menus (when the number of buttons exceeds the maximum allowed by Style template), and Up takes viewers from a submenu to its parent menu. Home takes viewers to the main opening menu.

All menu templates that ship with MyDVD use the same graphics—triangles in circles, swooping arrows, and the small house icon—for these standard DVD functions. The cool thing about Style Creator is that you can use it to completely change those, and any other, buttons that will appear in projects created using MyDVD. Style Creator lets you create customized, professional-looking DVDs with a consumer-priced product.

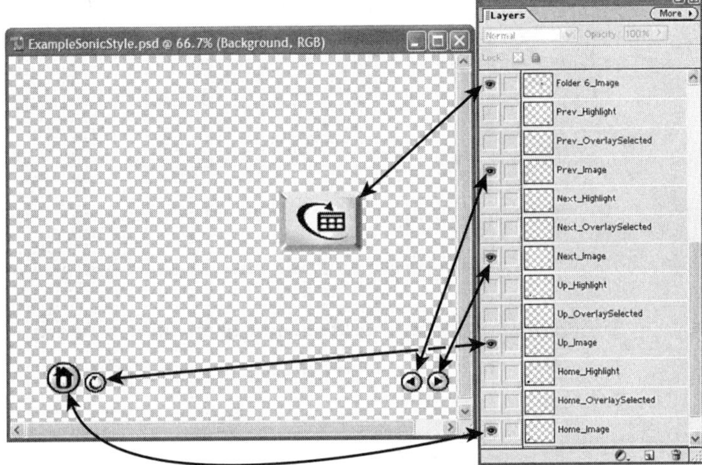

FIGURE 22.9
Each button type in the Style Creator template has a different function.

4. Experiment with these button types by clicking their associated elements on and off. You'll see that the highlight hotspot zones vary in size depending on the size of the button and that the `OverlaySelected` graphic is always a red box or circle.

You might note that one button element is missing from all the layers in the `ExampleSonicStyle` file: `OverlayActivated`. None of the templates provided with MyDVD includes this element, although MyDVD can handle activated overlays. Therefore, the next task explains how to change all the existing button elements and shows you how to add a layer for the `OverlayActivated` element.

The purpose, simply, is to introduce you to this process, not create a completely revamped template. To do that, you'd need to duplicate these steps several times. I do suggest you try to make a complete, personalized template in this hour's exercises.

Task: Edit a Button

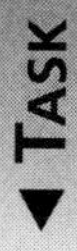

Changing a button's characteristics means changing graphics on several layers. You'll start with the button graphic itself and then tackle its hotspot and overlays in the next task. Do the following:

1. Turn off the `Background` layer by clicking its eyeball; then select the `Home_Image` layer above it by clicking its name. As shown in Figure 22.10, it displays the return-to-main-menu house icon used in all menu templates that ship with MyDVD.

> Clicking a layer's eyeball displays that layer but does not select the layer. If you want to edit a layer, you must click its name to select it as well as display it.

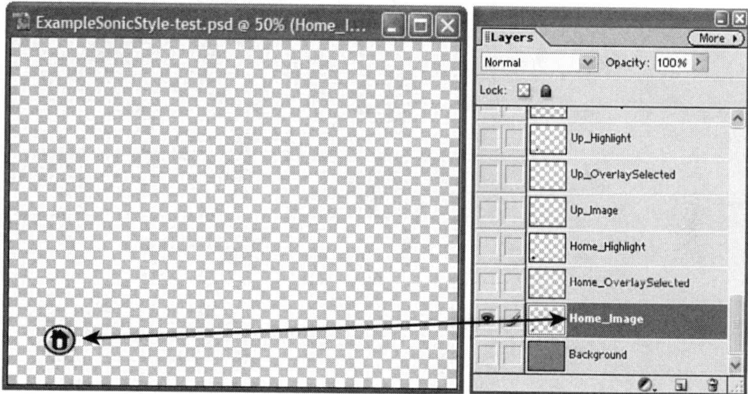

FIGURE 22.10
The `Home_Image` layer is the familiar MyDVD return-to-main-menu house icon button.

2. In the Photoshop Elements main menu, select Select, All and then delete the house icon graphic by pressing Delete.
3. Return to the Untitled-1 window by clicking it. Then add a layer by clicking the Create a New Layer icon at the bottom of the Layer palette (see Figure 22.11).

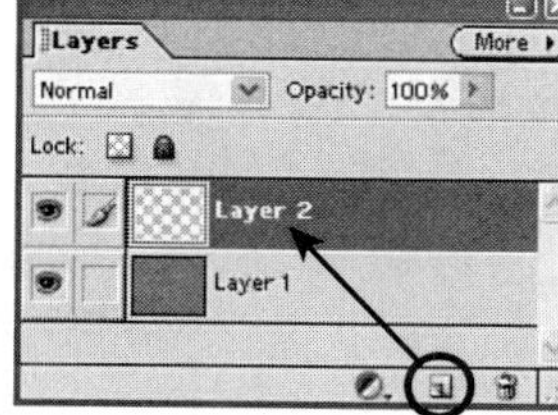

FIGURE 22.11
Use the Create a New Layer icon to add a layer to make a new menu button.

4. Using Figure 22.12 as a guide, create a new button graphic by selecting the Shape Drawing tool from the toolbar and selecting the Rounded Rectangle (or a shape of your choice). Then click and drag a shape in the bottom-left corner of the screen, give it a style by using the Style drop-down menu (I chose Simple Inner), and select a color from the Color swatch.

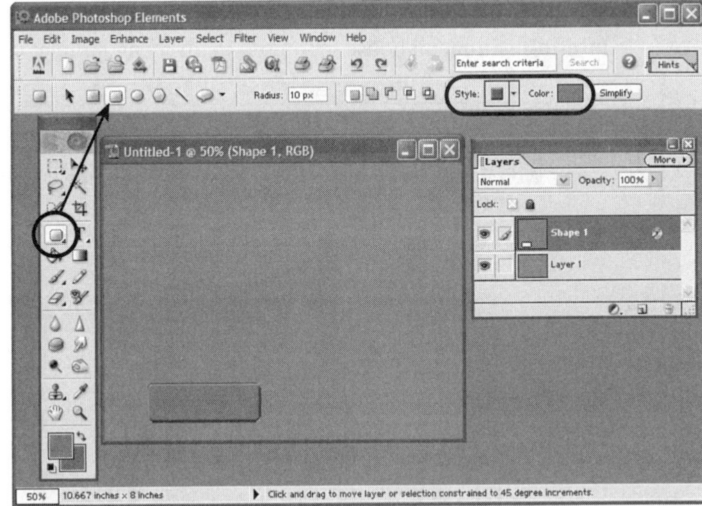

FIGURE 22.12
Use the Photoshop Elements Drawing tool and its associated styles and colors to create a new menu button.

5. To add text to the button, click the Text tool (see Figure 22.13). Then select a font and size (I chose Times New Roman, 36 Point, Bold), position the cursor over the left side of the button to ensure your text fits on the button, and type **`Main Menu`**.

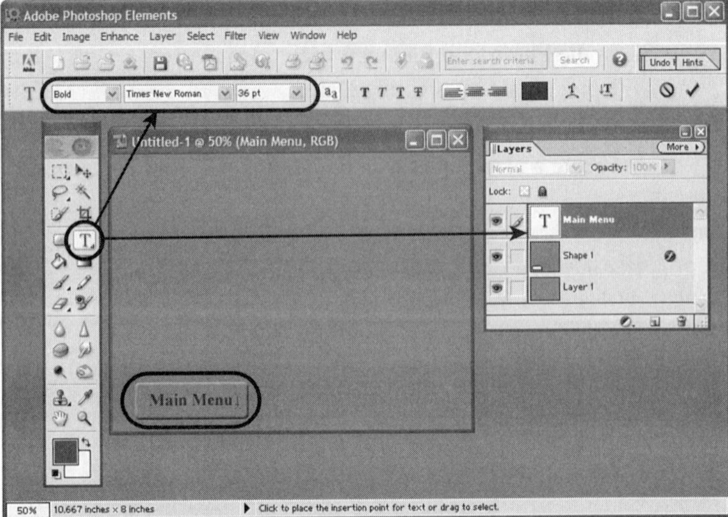

FIGURE 22.13
The Text tool is a simple way to clarify your button's function.

6. Adding text to a Photoshop graphic automatically creates a new layer called, in this case, **Main Menu**. If you didn't properly position the text on the button graphic, you can use the Move tool (shown in Figure 22.14) to change the text's location.

FIGURE 22.14
Use the Move tool to adjust the location of your newly created text.

7. You need to merge the text and shape layers into one layer before adding the button to your menu template. As shown in Figure 22.15, you do this by deselecting the Layer 1 eyeball and leaving the other two eyeballs turned on. From the main menu, select Layer, Merge Visible.

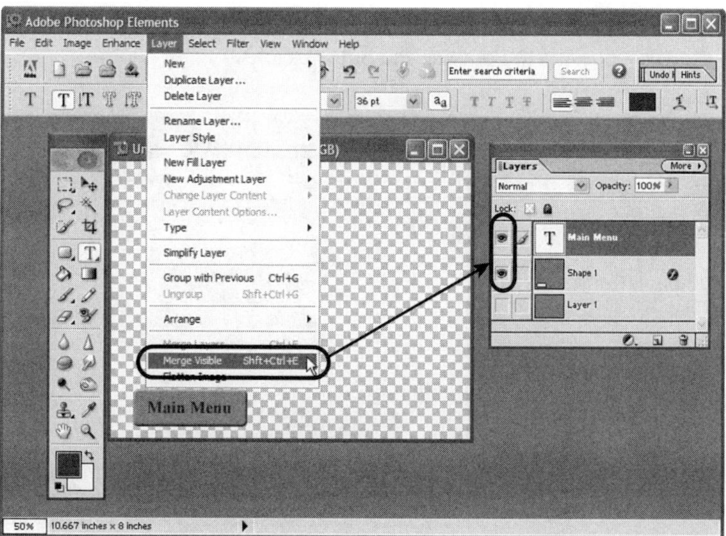

FIGURE 22.15
Merge layers by clicking their respective eyeballs and then selecting Merge Visible.

8. Add the main menu button to your template by selecting Select, All from the main menu. Select Edit, Copy (or press Ctrl+C) to copy the merged layer, and return to `ExampleSonicStyle`. Make sure you've selected the `Home_Image` layer by clicking its name in the Layers palette. From the main menu, select Edit, Paste (or press Ctrl+V) to drop the button in the middle of the workspace. You now can close Untitled-1.

> To fully exploit the benefits of the Style Creator, you need to go beyond simply changing this single button. One likely scenario would be for you to create a set of duplicate buttons to perform the four principal DVD menu functions: main, previous, next, and up. The only difference among the buttons would be the text on each one. To create those four similar buttons, undo the merge by using Photoshop Elements' Step Backward command—select Edit, Step Backward or press Ctrl+Z. Then either highlight the text to replace it or click the text tool to create additional text layers. Add each new text object to your existing button and merge and copy each one in turn to the proper layer in the `ExampleSonicStyle` graphic.

9. To properly position this button and any subsequent buttons, you must do two things: From the Photoshop main menu, select View, Grid to add grid guidelines to the workspace. Then you must scroll to the top of the `ExampleSonicStyle` layer palette and turn on the `Safe area` layer eyeball. In Figure 22.16, I dragged that layer name from the top line to directly above the `Home_Image` layer for ease of reference. Use the Move tool to adjust the location of your new button so that it doesn't fall outside the Safe Area.

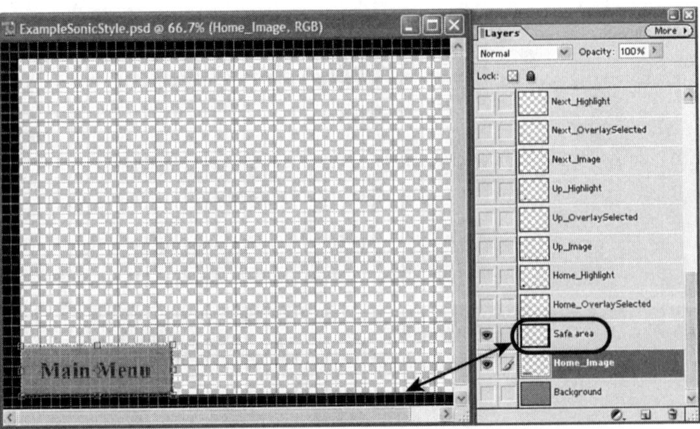

FIGURE 22.16
Use the Photoshop elements grid and the Style Creator Safe Area to align your buttons.

22

As mentioned earlier in the book, the black Safe Area box defines the portion of the NTSC TV signal that might fall outside the edges of most TV sets. You want your graphics to fall inside that perimeter to ensure viewers can see them.

10. Finally, you should do a bit of housekeeping. This step is not necessary, but it will avoid confusion later when you view your template in MyDVD. Using Figure 22.17 as a reference, turn off the grid, select the three up layers in turn, and use the Move tool on each graphic to slide them on top of each other (just to the right of the main menu button).
11. Save your graphic file by selecting File, Save.

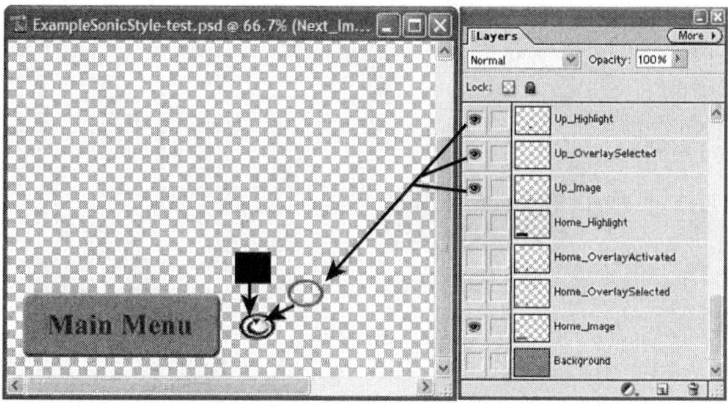

FIGURE 22.17
To avoid having the Up button show up on top of your newly created Main Menu button, move it and its elements off to one side.

Task: Change Button Elements

TASK

After you change the shape or size of a button, you need to change its associated elements to ensure they all work together properly. In this task you change the highlight hotspot zone, edit the `OverlaySelected` graphic, add a new `OverlayActivated` layer, and add a graphic to it. Here's how:

1. Click the `Home_Image` eyeball and select the `Home_Highlight` layer by clicking its layer name. Your screen should look similar to Figure 22.18.
2. Click the Move tool and then grab the handles on the black highlight square to move and expand it to completely cover the Main Menu button. You should end up with a rectangle just like the one in Figure 22.19.
3. To simplify this next step, you'll use a graphic from the `ExampleSonicStyle` file. Unclick the `Home_Highlight` eyeball and select the `Home_OverlaySelected` layer; a red circle appears.

FIGURE 22.18
Here's how your screen should look with the `Home_Image` *graphic displayed and the* `Home_Highlight` *graphic layer selected.*

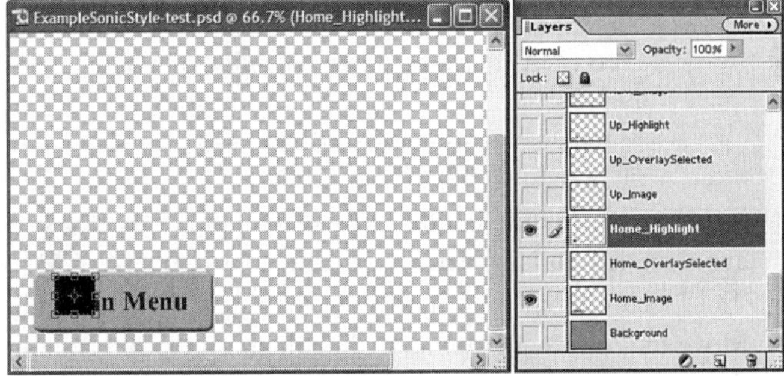

FIGURE 22.19
Here's how your `Home_Highlight` *hotspot zone should look after moving and expanding it to cover the Main Menu button.*

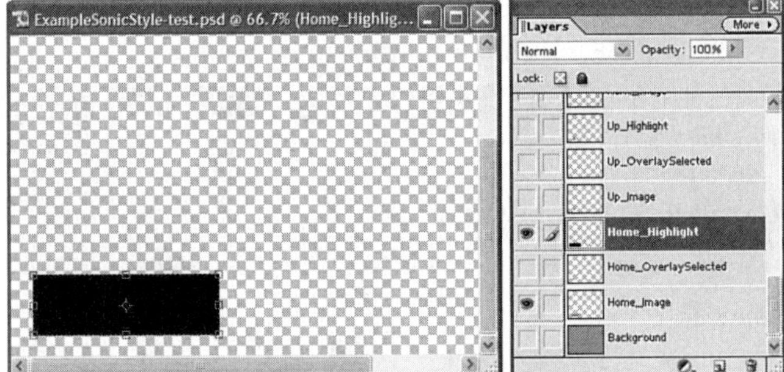

4. From the main menu, select Select, All; then press Delete to remove the `Home_OverlaySelected` circle.
5. Select the `Folder 6_OverlaySelected` layer. As shown in Figure 22.20, it's a rectangle. From the main menu, select Select, All; then select Edit, Copy or press Ctrl+C.

FIGURE 22.20
Use a graphic from a different layer to serve as the `Home_Image` *selected overlay.*

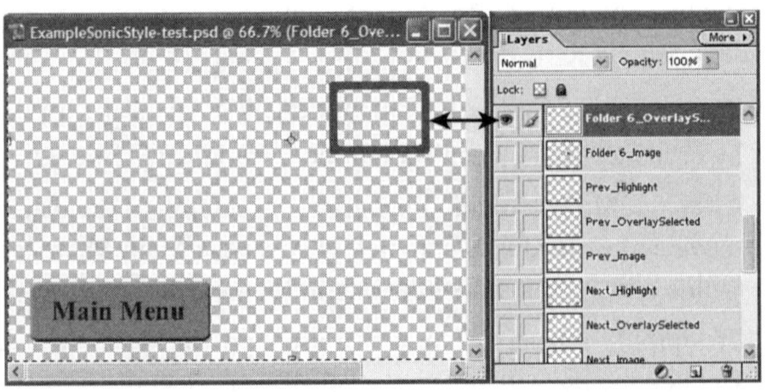

6. Unclick the `Folder 6_OverlaySelected` eyeball and select the `Home_OverlaySelected` layer. Select Edit, Paste (Ctrl+V) to drop the red rectangle into that layer.
7. Use the Move tool to slide the rectangle over the Main Menu button, and then use its handles to change its size and proportions to cover the perimeter of the Main Menu button. Your graphic should look similar to Figure 22.21.

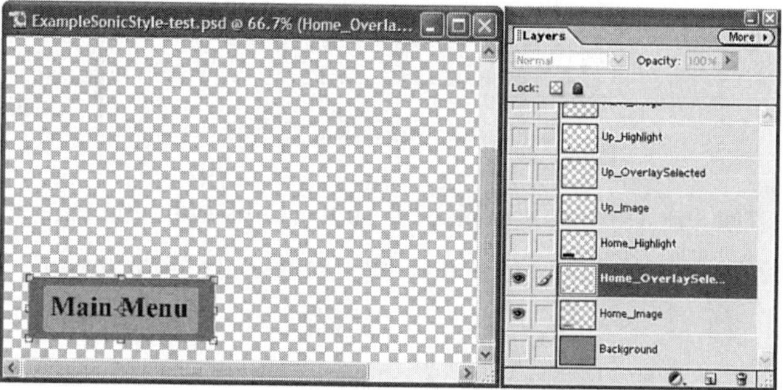

FIGURE 22.21
Use the Move tool to fit the red rectangle around the outside edges of the Main Menu button.

8. Add a layer for the `Home_OverlayActivated` graphic by clicking the Create a New Layer icon at the bottom of the Layer palette (see Figure 22.22). Right-click, select Rename Layer, and type **`Home_OverlayActivated`**. Be sure you use the exact spelling, capitalization, and underscore.

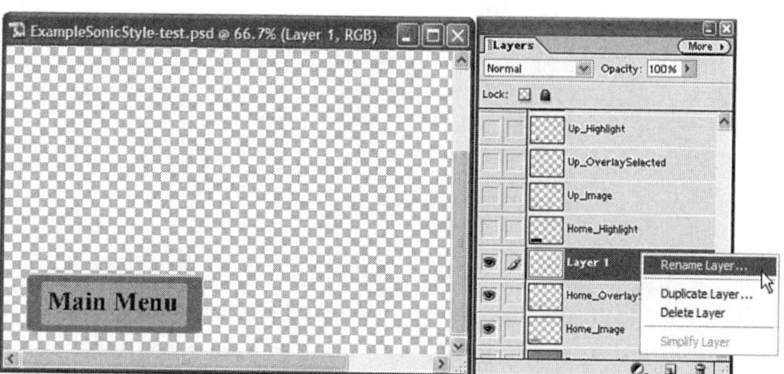

FIGURE 22.22
To give your buttons an activation highlight, add a new layer named `Home_OverlayActivated`.

9. Select the `Home_OverlaySelected` layer; select Select, All; and select Edit, Copy. Select the newly created `Home_OverlayActivated` layer and then select Edit, Paste. As shown in Figure 22.23, doing so adds the newly stretched rectangle to this newly created layer.

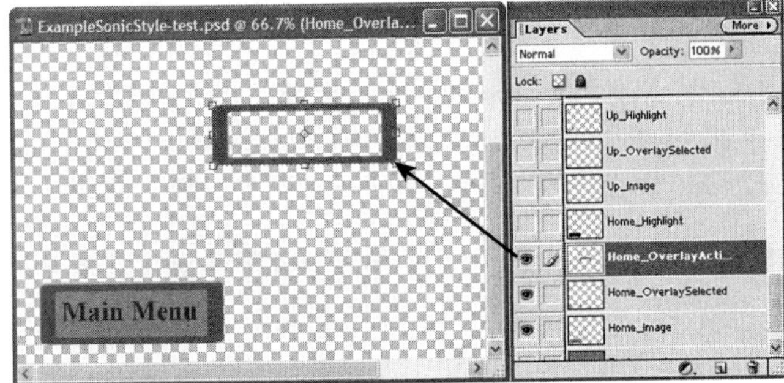

FIGURE 22.23
Paste the newly stretched rectangle into the newly created `Home_OverlayActivated` *layer.*

10. Use the Move tool to place it on top of its twin rectangle.
11. Click the Set Foreground Color swatch to open the color selector (see Figure 22.24).

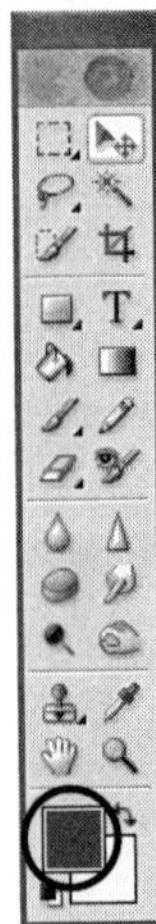

FIGURE 22.24
Use the Color swatch to open the Color Picker.

12. Using Figure 22.25 as a guide, use the Color Picker to change the rectangle color from red to any color of your choice. I chose pure blue by setting the blue value to 255 (its maximum) and red and green to 0. Click OK.

The plug-in won't let you use more than three colors for overlays. In this case, all selected overlays are red and all activated overlays are blue. If you choose more than one color for the same overlay layer (selected or activated), MyDVD selects the most commonly used color for all the elements in that category.

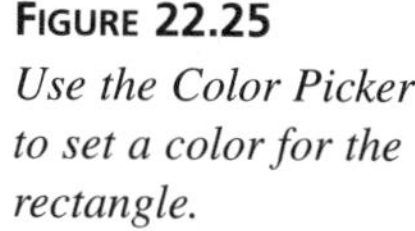

FIGURE 22.25
Use the Color Picker to set a color for the rectangle.

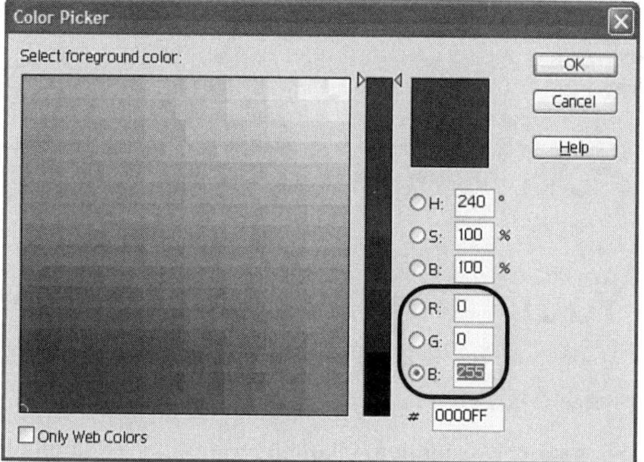

13. Complete the color change process by clicking the Paint Can icon and clicking its cursor on the rectangle.
14. Select File, Save.

Creating Graphics for Use in DVDit!

DVDit! offers a nice variety of backgrounds and buttons, but its developers assume most users will opt to create their own graphics in Photoshop or some other graphic program that works with Photoshop layers. If you do, here are three tips:

- **DVDit! resizes any graphic or still to a 4:3 aspect ratio**—To avoid distorted images that are flattened or elongated, here are two points to keep in mind:
    - Images for menu backgrounds should have a 640×480 resolution or a 4:3 aspect ratio. If you use less than 640×480, DVDit! will expand your image to fit, and it might not look as sharp as you want.
    - Images that you'll use as stills should have a 720×540 resolution or a 4:3 aspect ratio. For images that don't match that resolution or aspect ratio, add borders to properly size them. Again, lower resolutions will lead to less crisp-looking images.
- **You must compensate for square pixels**—When using Photoshop (or other PC graphics software), to compensate for its square pixels and NTSC-DV and PAL's non-square pixels, create full-screen images (typically backgrounds) with Photoshop's opening palette set to 720×540, its Mode set to RGB color, and its Contents set to Transparent. When completed, resize your graphics to 720×480. They'll look squashed in Photoshop, but when you import them to DVDit!, they'll look as they did when you created them.

- **DVDit! supports Photoshop layers and alpha channel transparencies**—Using layers you can create a button in Photoshop on one layer and then add slightly altered versions of that button in several layers within the same graphic. After you import this button to DVDit! (by selecting Theme, Add Files to Theme), DVDit! displays each layer as a separate button in the Theme window. And you can use Photoshop's alpha channel to make part of the button transparent. That means you can create buttons with holes in them that let the DVD menu show through.

Testing Your Template Edits in MyDVD

The purpose all along has been to show you how to use Style Creator to create templates that will work in MyDVD.

I limited this hour's tasks only to changing the background and a button and its elements, so when you test this edited style in MyDVD, it won't look very impressive. But I think you'll see how it all works and will have a clearer idea of how these elements all work together.

MyDVD has rigid rules about button locations, and you can't separate out button elements. But if you want to create a simplified and consistent template for use by several employees or students, for example, Style Creator and MyDVD make a great partnership.

Sonic developers have told me that the new version of Style Creator, which should be released by the time this book publishes, will have a small check box that tells MyDVD whether to adhere to the positions of buttons as they're laid out in the style or whether it should position the buttons according to the layout rules in MyDVD. This is a powerful tool that will make the consumer-friendly and reasonably priced MyDVD behave even more like a prosumer-priced authoring application.

Task: Test Your Style in MyDVD

TASK

Because you worked with MyDVD in Hours 4–6, much of this task will be routine. What's of interest here is how MyDVD handles the changes you made to the `ExampleSonicStyle` file. Here's how to view those changes:

1. In Photoshop Elements, convert your graphic to a Sonic Solutions style by selecting File, Automation Tools, Create Style for Sonic MyDVD. This opens the window shown in Figure 22.26.

FIGURE 22.26
The Style Creator plug-in interface.

2. Set the Source to Current Document, select Create New Style Pack in the Destination window, and select a file folder. Then give your style a name, click Save, and click OK. The plug-in will turn on its macro, perform a few dozen functions (it might take about 30 seconds), and then let you know the export has been completed successfully.

As the plug-in steps through its various functions, it might display one or more error messages. Most of these are inconsequential, and if you followed the previously mentioned steps, none of the messages should cause the plug-in to not complete its task. One message notes the presence of the Safe Area layer, which is not part of the style format; therefore, the plug-in merely alerts you about it and ignores it after you click OK. Other messages typically refer to possible overlay color conflicts, but MyDVD usually resolves them automatically. Simply click OK if any error messages appear.

3. Open MyDVD. In the opening interface, select Create or Modify a DVD-Video Project. In the main user interface menu bar, select Edit Style.
4. In the Edit Style window shown in Figure 22.27, click Import Style.
5. In the Import Style interface, select Create New Style Category.

In the Import Style interface shown in Figure 22.28, the Import Style into Current Category radio button might be grayed-out (disabled). The specific reason is that MyDVD does not let you import a style into its two default style collections, Default Styles and DefaultMotion Styles. This forced choice occurs if you have not yet imported a style into MyDVD or have one of the two default style packs selected.

FIGURE 22.27
Import your style by clicking that button in the Edit Style window.

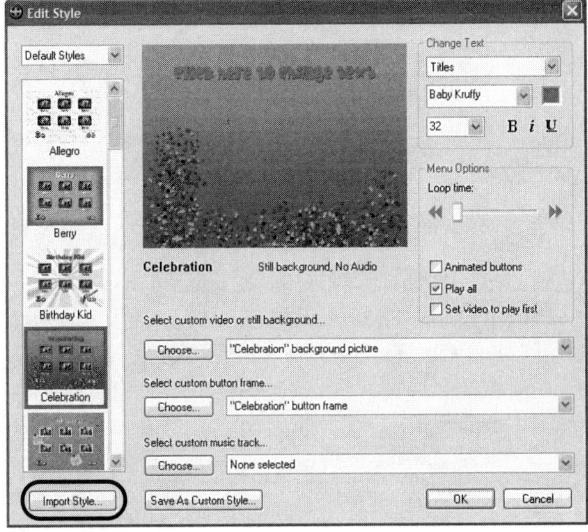

FIGURE 22.28
You might have only one option in the Import Style interface, depending on whether you've already imported a style into MyDVD.

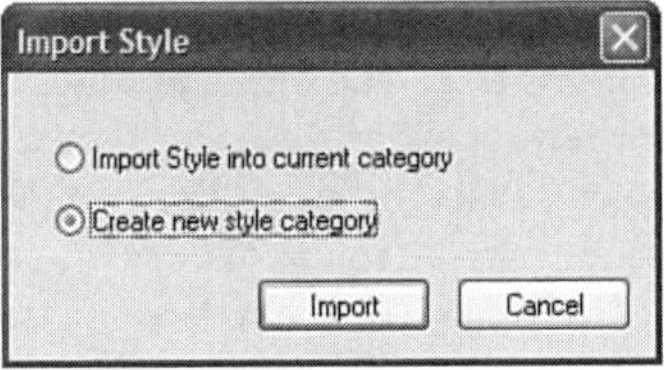

6. Navigate to the style you created in the previous tasks, highlight that file, and click Open. It will load into the Edit Style window. As shown in Figure 22.29, click your edited style to display the background and the default menu text. Click OK in the bottom-right corner to return to the main MyDVD interface.
7. Click OK to return to the main interface, and then click Add Sub-menu two times to place two menu buttons on the main menu screen.
8. Switch to Preview mode by clicking the Preview button.
9. In the Preview window, shown in Figure 22.30, check out the button behavior by clicking one of them. It should turn blue (or whatever color you selected for the activated layer) and take you to a submenu.

FIGURE 22.29

After importing your edited style, click its thumbnail image to display it.

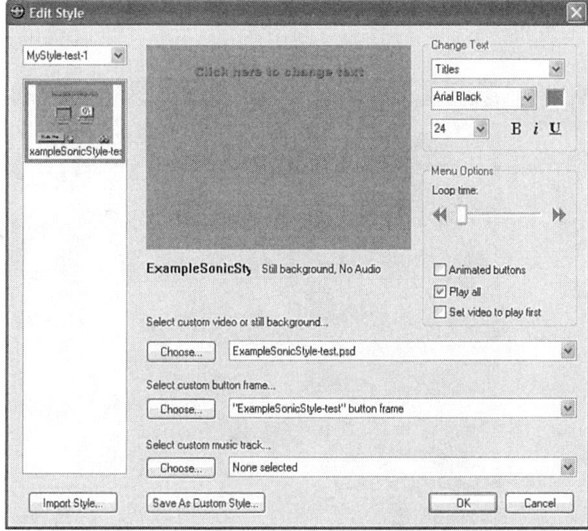

Because you used Style Creator to add an `OverlayActivated` layer to this style (giving it a color other than red), all other buttons in the template gain that activated feature automatically.

FIGURE 22.30

In Preview mode, note that your buttons now have an activated highlight color.

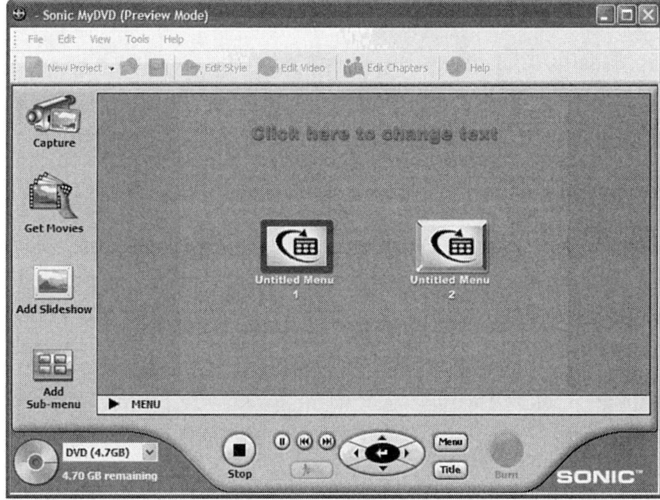

10. In the submenu shown in Figure 22.31, move the cursor back and forth between the Main Menu button and the Up button to see their respective red selected highlights display.

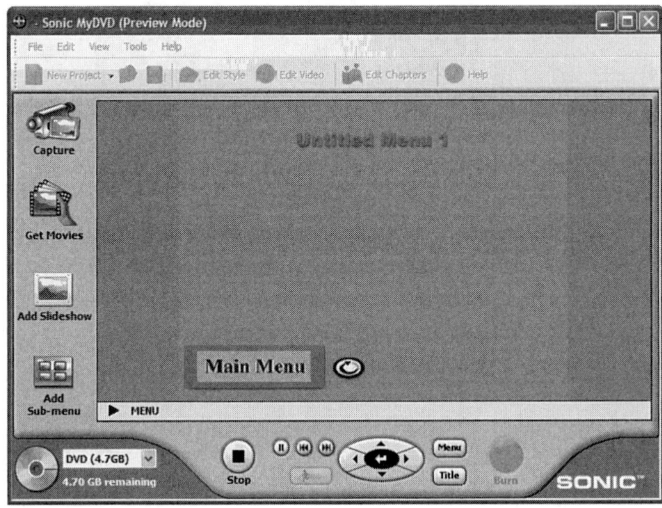

FIGURE 22.31
The Main Menu button has rectangular red and blue selected and activated boxes.

The Main Menu button will display the red rectangle you swapped for the red circle, and the Up button will be in the new location you specified in Style Creator.

11. Finally, click either button and note that the blue activated highlight displays.
12. Click the Stop button and then close MyDVD.

Summary

As you create more DVDs, you might want to create customized menus and buttons. Sonic Solutions' Style Creator is an excellent way to accomplish this task. It uses specifically named Photoshop layers that, when imported to MyDVD or DVDit!, ensure that your graphics match the appearance and functionality you envisioned for them.

Although it takes a while to get up to speed with Style Creator, after you get the hang of it, you might come to rely on it as a the primary means to personalize your DVDs.

Workshop

Review the questions and answers in this section to reinforce your Style Creator knowledge. Also, take a few moments to take the short quiz and perform the exercises.

Q&A

Q I prefer the drag-and-drop method to move graphics from one place to another versus the copy-and-paste method. I tried that method in Photoshop to take a newly created button from one workspace to the `ExampleSonicSample` file, but it adds a new layer and won't put the graphic in the proper layer.

A That is a characteristic of Photoshop. Just as adding text to a workspace automatically adds a layer for that text, dragging a graphic to a workspace does the same thing. So, use the copy-and-paste method in this case to add a graphic directly to the layer you've selected.

Q I want to create style templates from scratch. What do I need to know?

A There are several rules you must follow for the Style Creator plug-in to convert layered Photoshop graphics into styles. Those rules are too detailed to list here. In general, though, you must follow the Style Creator layer naming conventions and include some specific layers—background, title, and all button types. Read the Style Creator tutorial included on this book's companion DVD for more information.

Quiz

1. Describe overlays and explain the difference between selected and activated.
2. You created a button using a drawing tool. Now you want to give it a 3D look and a new color. How do you do that?
3. You've created an `OverlaySelected` graphic for the Chapter button and want to use it for that button's `OverlayActivated` layer. How do you do that?

Quiz Answers

1. Overlays are button graphic elements that switch on when a viewer navigates the remote to them and presses Enter. The selected overlay displays when the viewer navigates to a button, and the activated overlay displays for a moment after the viewer presses Enter on the remote.
2. Select that button's layer and click the Style drop-down list. The default Bevels collection will appear. Click several styles to note how they change your button's

appearance. Several more styles are available by clicking the arrow in the upper-right corner. To change the color, use the Set Foreground Color box in the toolbar and select a color from the Color Picker.

3. Create a new layer and name it **`Chapter #_OverlayActivated`** (where # is whatever number of buttons you want to appear on a MyDVD menu before making a new menu). Copy the graphic from the `OverlaySelected` layer and paste it into the newly created `OverlayActivated` layer. Give it a new color and position it on top of the button.

Exercises

1. Make a style template. The easiest way to do this is to revisit this hour's tasks and simply repeat some of the steps. If you're really motivated, build one from scratch. To do that, read the Style Creator tutorial included on this book's companion DVD; it explains all the rules referred to in the "Q&A" section earlier. You can create a wide variety of buttons and overlays in a single template (see page 52 of the tutorial). This works well when creating buttons for DVDit! projects or for MyDVD if you want some number of a button type on one menu and different buttons on the next menu. If your style is something you want to share, post it to the Sonic Solutions Web site (`http://styles.mydvd.com/`).
2. Overlays don't have to look like their associated buttons. For instance, a button could be shaped like a crescent moon and the selected overlay could look like a star. The only constraint is that overlays must be one color and can't go beyond the edges of the highlight hotspot zone (you can adjust the highlight borders to accommodate a larger overlay). With that in mind, use Photoshop's Custom Shape tool and its extensive palette of shapes to create some buttons and overlays with distinct differences that will add some extra interest—or surprises—to your menus.

HOUR 23

Professional DVD Authoring with ReelDVD: Part I

As you improve your DVD production skills, you might find that prosumer products such as DVDit! don't offer the full range of customizability and control that you need.

You might want more subtitle or audio tracks, more button and menu navigation features, or greater flexibility in the creation of menus.

Only Sonic Solutions makes such professional-level DVD authoring products for the PC. Its three high-end products—ReelDVD ($699), DVD Producer ($2,499), and Scenarist (starting at $7,999)—offer those features and more, as well as improved user interfaces and productivity tools for the professional user.

In this hour, I'll give you part one of a two-part, hands-on tour of ReelDVD (a trial copy is on this book's companion DVD). I'll cover part two and offer a brief feature demonstration of DVD Producer in Hour 24, "ReelDVD Part II: DVD Producer and DVD Trends."

The highlights of this hour include the following:

- Professional versus prosumer: explaining the differences
- Taking a hands-on tour of ReelDVD
- Installing and setting up ReelDVD
- Adjusting ReelDVD's sub-picture color and opacity scheme
- Importing project elements

Professional Versus Prosumer: Explaining the Differences

Stepping up to ReelDVD or DVD Producer has advantages and disadvantages. Using professional DVD authoring software such as these two products increases your creative potential, but taking that upward step frequently means leaving some nifty prosumer convenience behind.

Typically, professional DVD authoring products do *not* offer menu templates, button creation tools, or text layout options. And they generally don't accept a wide range of video, audio, or graphic file types.

At this higher level, the working assumption is that you have a suite of graphics creation and MPEG encoding software/hardware tools and want to do all that preparatory work outside the authoring environment.

That makes sense if every DVD you produce needs very detailed menu and button navigation, multiple audio or subtitle tracks, or menus that meet exacting visual standards.

But most DVD producers don't work at that level all the time. The best scenario, from my point of view, is one in which, as you go up in price and quality, you get all the options of the lower-priced products with all the functionality and flexibility of the higher-priced line.

That's generally not the case at this moment. The one exception is DVD Producer, as it offers text formatting, menu design features such as automatic motion button creation, and accepts any graphics file and several video file types. With that in mind, you can expect that most professional DVD authoring applications will have most of the following attributes:

- User interfaces that better represent the interactive nature of your finished project
- Precise control over all button, chapter, still image, and menu navigation
- Greater control over colors for button *states*—how they change appearance when the viewer highlights or selects them
- A subtitle editor that simplifies that tedious process
- Allowing both 4:3 and 16:9 aspect ratio videos in the same project
- Setting language attributes for audio and subtitle tracks (I'll explain this later in the hour)
- OpenDVD compliance—that is, the capability to take any DVD created with another OpenDVD authoring product, reauthor it, and burn a new disc
- Improved *proofing*, meaning simulation/emulation modes that pinpoint timing issues and missing links between buttons, media, and menus
- Permitting the maximum number of audio (32) and subtitle (8) tracks allowed by the official DVD specification
- Enhanced control over file layout when burning your DVD to ensure shorter seek times during playback
- Support for some of the DVD specifications' lesser-known features such as Jacket Picture and Text Data

A *Jacket Picture* is a user-selected still image that displays whenever the disc is loaded and the DVD player is in stop mode. *Text Data* is information in text form that is shown on a DVD Player's front panel display. Not all DVD players support Text Data. Text Data typically is chapter names or other information relevant to the DVD title currently being played.

Taking a Hands-on Tour of ReelDVD

I think ReelDVD's user interface is elegant. As shown in Figure 23.1, it has a look and feel unlike any other DVD authoring product covered so far in this book. It integrates a timeline and an object-oriented project layout that uses a flowchart-like authoring approach.

FIGURE 23.1
ReelDVD's user interface at first might seem somewhat technical for a novice, but I think you'll come to appreciate its elegant organization.

Each rectangular filmstrip icon represents a media or menu object. The lines and directional arrows indicate the menu/chapter navigation flow, and the timeline displays the attributes for a selected media object, including audio, subtitles, and chapters.

After you get your feet wet with ReelDVD, I think you'll wonder why all DVD authoring products don't take this approach.

You build your project by bringing your prebuilt menus and media into the flowchart, or *storyboard*, area. Then you drag and drop links among menus, movies, slideshows, and chapters. It's a simple and quick way to create precise navigation.

This is the first of all the DVD authoring applications covered so far in this book that offers a subtitle editor, shown in Figure 23.2. It lets you use standard font and text options and adjust the color and opacity of the text's inside and outside edges, as well as add a transparent or opaque background color.

ReelDVD's subtitle text editor is a real selling point when compared with the only other DVD authoring product covered so far in this book that works with subtitles, Pinnacle Impression DVD. As opposed to Sonic ReelDVD's integrated subtitle creator, Pinnacle's authoring application requires that you create all subtitles in a separate graphics program (Windows Paint, for example). With that approach, you need to create individual, 1MB BMP files for each phrase and then import them into the authoring application.

Because a typical feature film might have thousands of such subtitle files those 1MB files can consume a fair amount of hard drive space. However, it's important to note that all DVD authoring applications (including Impression and ReelDVD) convert their subtitle data into DVD-standard .04 Mbps data streams during the final output to disc.

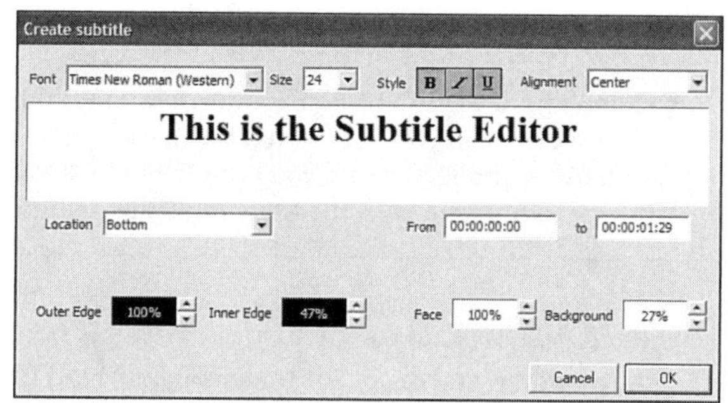

FIGURE 23.2
ReelDVD's subtitle editor is a real productivity enhancement.

Because professional users typically create elements such as menus, buttons, and text in separate graphics programs such as Photoshop, ReelDVD does not include those DVD authoring features.

Most of Sonic's products are moving toward implementing the new OpenDVD specification that allows finished DVD titles created with OpenDVD-compliant authoring systems to be revised after they have been burned to disc. That means, without access to the source media files originally used to create the title, you can open a finished DVD to add or delete video, audio, menus and navigation to create a new DVD. Choosing to create an OpenDVD title is optional, of course since you might not want to grant such access to your DVD, but for an industry that thrives on revision, OpenDVD technology really makes it easy to update, fix, or change content on your DVD. DVD Producer, the professional-level product covered in Hour 24 is OpenDVD compliant.

Installing and Setting Up ReelDVD

In the next few tasks in this hour and the next, I'll take you through a simplified ReelDVD authoring process to give you an idea of how this product works. The purpose is to introduce you to ReelDVD, not to offer complete explanations.

Task: Prepare ReelDVD for Your Project

▼ TASK

In this task, you'll install ReelDVD, make a few adjustments to some files to ensure ease-of-access, and change some project settings. Here's how:

1. Locate the ReelDVD folder on this book's companion DVD. Double-click `ReelDVDSetup.exe` to install it. To simplify the next step, accept the default location on your `C:` drive.

> You don't have to install Cineplayer. By this time, you probably have a DVD software player on your hard drive. Cineplayer's primary role is to emulate your finished product before you burn the DVD. (Emulation is a more accurate way to test your DVD than the simulation feature in ReelDVD.)

2. Locate the ReelDVD Tutorial Assets file folder on this book's companion DVD. Copy and paste or drag and drop that folder to your hard drive. I'd suggest using the default ReelDVD directory: `C:\Program Files\Sonic\ReelDVD`.
3. All the tutorial assets are set to read-only. As shown in Figure 23.3, switch that attribute off by right-clicking the ReelDVD Tutorial Assets file folder, selecting Properties, deselecting Read-only, and clicking OK. Then select Apply Changes to This Folder, Subfolders and Files from the next window and click OK.

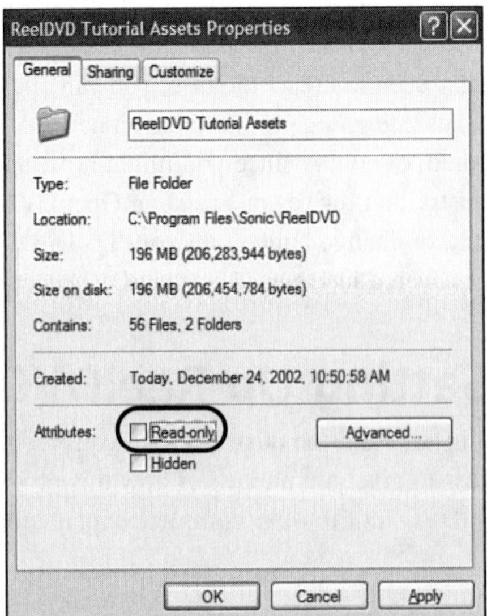

FIGURE 23.3
Switch off the tutorial assets' read-only attribute to make them accessible to ReelDVD.

▼

4. Change a line in the subtitle text file to ensure ReelDVD knows where to find the subtitle files. Do that by locating that file: `C:\Program Files\Sonic\ReelDVD\ReelDVD Tutorial Assets\Subtitles\04_Subtitle_Script.sst`. Right-click it and select Open with, Choose Program. Then select Notepad to open the text file.
5. Change the line highlighted in Figure 23.4 to note the location of the ReelDVD Tutorial Assets Subtitle folder on your hard drive. If you installed ReelDVD to the default location, insert this line in that text file: `C:\Program Files\Sonic\ReelDVD\ReelDVD Tutorial Assets\Subtitles\`. If you chose a different file folder, note its path in this text file. After you've changed the line, select File, Save; then close Notepad.

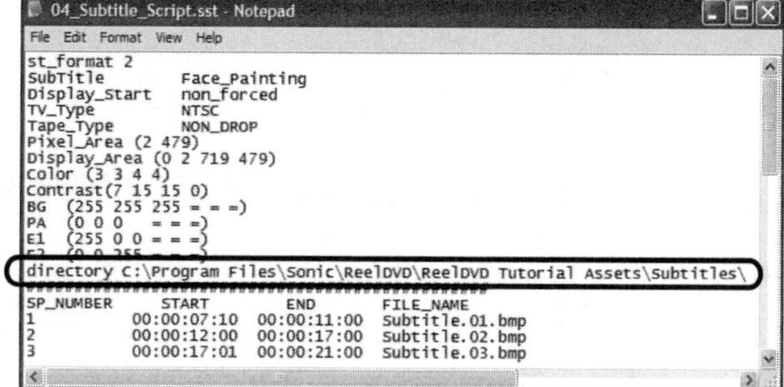

FIGURE 23.4
Change the highlighted line in this text file to tell ReelDVD the location of the subtitle file folder.

6. Launch ReelDVD either by double-clicking its icon on the desktop or selecting Start, Programs, Sonic, ReelDVD, Sonic ReelDVD.
7. Change the Project Settings by selecting Edit, Project Settings and then clicking the Project tab if necessary. Change the disc name to `Facepainting` (it defaults to all caps, as shown in Figure 23.5). Ensure the default settings match Figure 23.5: NTSC (29.97 fps), Non-drop Frame, MPEG 2, Standard - 4:3, 720×480 – Full D1, As is, and 8.0 Mbps.
8. Click the Languages tab and set the following items to English: Audio Languages Stream #1, Subtitle Languages Stream #1, Simulation Defaults Audio, and Simulation Defaults Subtitle. Use Figure 23.6 as a guide.

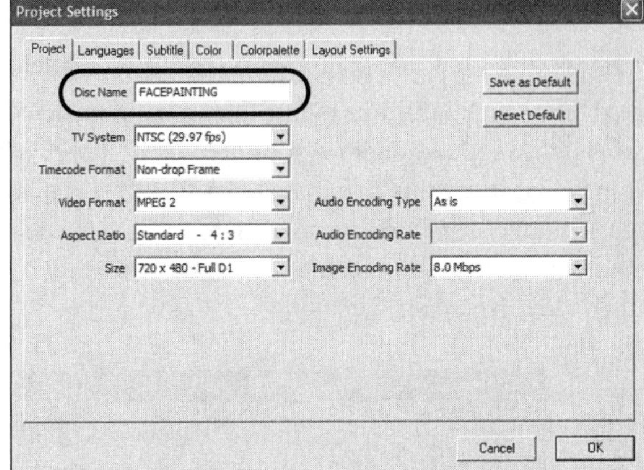

FIGURE 23.5
Name your project and then accept the default project settings.

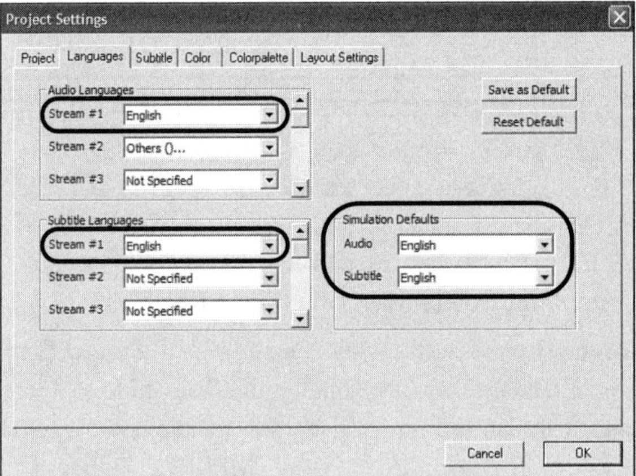

FIGURE 23.6
Check to see that all the language options are English.

Setting default languages lets you *localize* your DVD, ensuring it displays initially in the language of your target audience. You can select from dozens of languages. When a viewer presses the Language button on the remote, it will display whichever languages have been selected for the subtitle/audio streams in your project.

Adjusting the Sub-Picture Color and Opacity Scheme

This next task takes some explanation. In it you will set the menu button *state* characteristics—their highlight colors and opacity levels. These determine how the buttons change color or shading as viewers navigate their remotes over them or select them.

You need to apply some changes globally to your entire project before you work on any menus. If you wait and do this *after* adding menus to your project, the program will not automatically apply these color and opacity attributes to the existing menus. This is a purposeful program design to ensure that the user experience is consistent. If ReelDVD permitted global attribute updates after-the-fact, undesirable changes to the DVD menus might occur.

With ReelDVD, you need to create menu buttons and other graphics outside the program. After they're created, you need to tell ReelDVD how to add highlights to those buttons depending on the viewer's choices. For instance, scrolling the remote over a button can cause its border to lighten, or selecting it (pressing the Enter key on the remote) can turn the entire button red.

Depending on how you create the original button graphics, you might choose to create the highlights as separate graphics files or on a transparent layer using Photoshop or some other graphics program that handles Photoshop-layered files. That highlight graphic file or transparent layer—the so-called *sub-picture* overlay (see the following note)—uses only a single color, creating something similar to a silhouette of the portion of the button you want to highlight.

For that silhouette-like representation, the DVD specification lets you use one of only four colors: blue, red, black, or white (few DVD producers use white for a variety of reasons).

Those colors don't necessarily show up in your final project; they simply define the button highlight area. Later, you tell ReelDVD which color and opacity level to apply, or *map*, to each button highlight. You need those three or four colors because you might want to create several highlight colors and opacities for different buttons in a single menu.

Sub-picture is a standard, but confusing, DVD term. A more accurate descriptor is *super-picture* because sub-pictures are actually transparent overlays with graphic elements on them. The DVD specification enables the layering of graphics on top of other graphics—typically, for button states (different colors/opacities depending on whether the viewer has navigated over or selected a button) and subtitles.

Less-expensive DVD authoring products handle the sub-picture process automatically, which is easy on the user but offers virtually no control over the highlight size, shape, color, or opacity. For instance, these products don't allow you to highlight only a portion of the button or outside the button boundaries. In ReelDVD and other higher-end products, however, you take charge of these overlaid sub-pictures. Most of that work is in their creation (usually in Photoshop) and selecting colors and opacities for the various button states.

You worked with layered Photoshop files in Hour 22, "Creating Custom MyDVD Templates with Style Creator." Those were menu templates that fit specific criteria. To see how a basic Photoshop menu or button file looks, open the `06_DVD_Credits.psd` file in Photoshop, Photoshop Elements, or any other graphic program that works with PSD-layered files. As shown in Figure 23.7, you can see the transparent layer (as indicated by the checkerboard display). The only button graphic element on it is the highlighted word `back`.

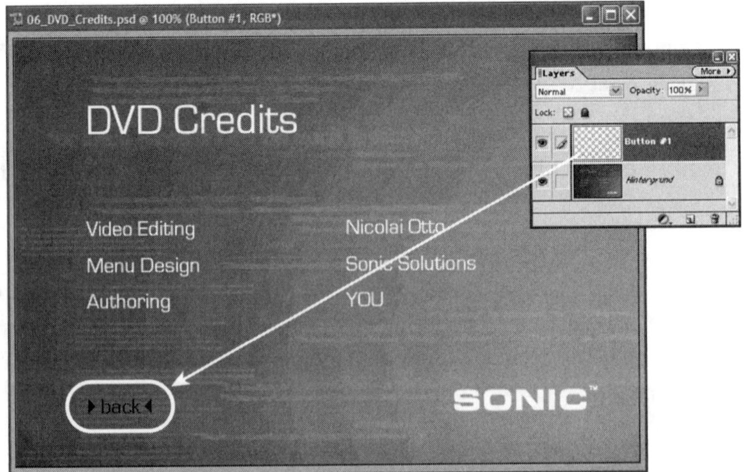

FIGURE 23.7
Opening the DVD Credits PSD file in Photoshop shows its extra layer—in this case, a simple back button.

To get a better idea of how the sub-picture color/opacity scheme works, open Project Settings by selecting Edit, Project Settings. Then, select the Color tab. Figure 23.8 shows the four colors—blue, red, black, and white—and the default colors and opacity levels ReelDVD associates with them. They are as follows:

- **Display**—The color/opacity that displays when the viewer takes no action with a button, neither navigating to it nor pressing the Enter button on the remote. Usually, you'll set this to 0% opacity to let the original button color and other characteristics show through.
- **Selection**—The color/opacity that displays when the viewer uses the remote to navigate to this button. Usually, you'll use a color different from the original button and reduce that new color's opacity so the original button still shows through.
- **Action**—The color/opacity that displays when the viewer presses Enter on the remote. This highlight displays for just a moment before the DVD moves to the selected media or menu.

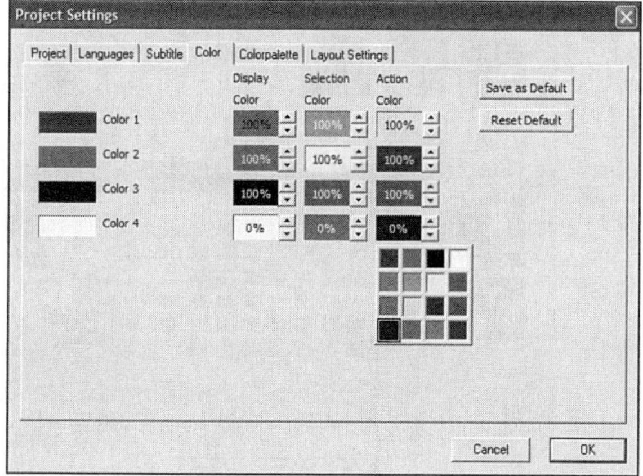

FIGURE 23.8
Use the Project Settings Color tab to set the default button highlight color and opacity characteristics.

ReelDVD has default color and opacity values for each of the four sub-picture colors. You can make global changes to those settings, either making new default settings or on a project-by-project basis. You also can make adjustments to individual buttons. I'll explain how to do the former in a moment and the latter in Hour 24.

You need to make these global changes before starting your project. If you wait until after you've added some menus to a project, the new global settings will have no effect on those menus.

The DVD specification allows for only 16 colors in any single DVD project's subpictures, but they can be any 16 colors you choose. As shown previously in Figure 23.8, if you click a color box under the Display, Selection, or Action column, a 16-color palette pops up from which you can make your selection.

As shown in Figure 23.9, you can customize that 16-color selection. Simply click the Project Settings Colorpalette tab, select the color you want to change, and use the Define Custom Colors palette to create a new color.

> You'll use that 16-color palette when you set up your subtitle appearance settings. If you click the Subtitle tab in the Project Settings window, you'll note that you can adjust the color and opacity settings for the text's outer edge, inner edge, face, and background.

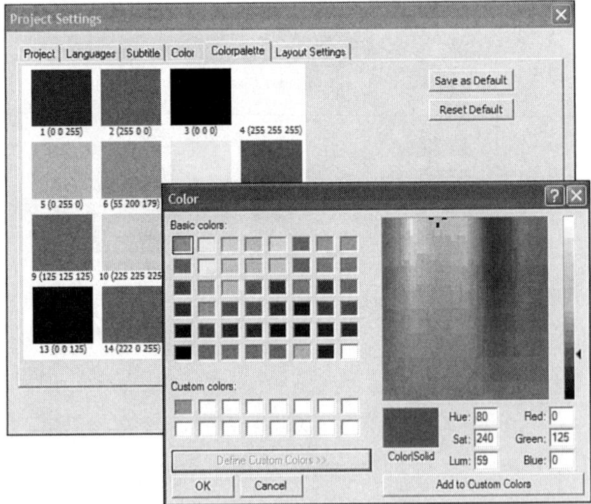

FIGURE 23.9
Use the Colorpalette tab to define your project's 16-color button highlight palette.

Task: Adjust Button Highlight—State—Colors and Opacities

▼ TASK

In this task, you'll take the global approach to adjusting button highlight characteristics; you'll make some individual button adjustments in Hour 24. In this task, you'll set all the display button highlights to black, all the selection button highlights to light gray, and all the action button highlights to red. You'll also make some opacity adjustments. Here's how:

1. Return to the Project Settings Color tab.
2. Under the Display column, set Color 2 and Color 3 to black. As shown in Figure 23.10, do that by clicking their respective boxes and selecting Black from the

▼

pop-up color palette. Leave the opacity at 100%. Later, when working with individual menus, you'll drop the opacity to 0% for two specific buttons.

The reason you change only Colors 2 and 3 is because the menu sub-picture button highlights created for this tutorial use red (Color 2) and black (Color 3). This is an example of using two of the four available sub-picture colors to give them different highlight characteristics within the same menu.

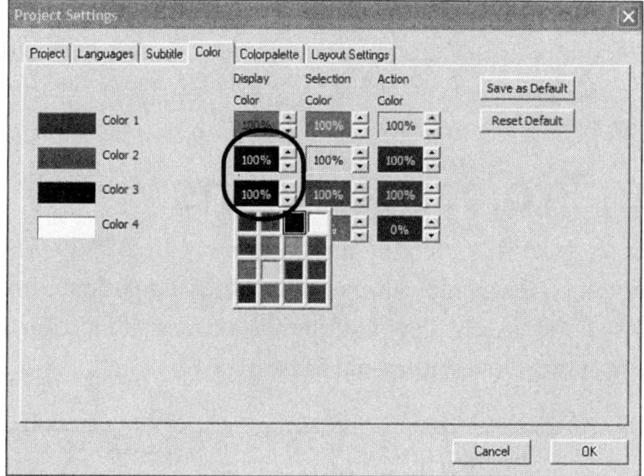

FIGURE 23.10
Click a color box to display the pop-up palette from which to select a color.

3. Under the Selection column, change Colors 2 and 3 to Light Gray.
4. Reduce the opacity setting on these two colors to 80% by clicking the down arrows next to the colors three times (each click reduces the opacity by 6% or 7%).

For this task, you can choose any color and opacity settings you want. The colors and opacities settings selected for this task simply work well with the menus and buttons in this tutorial project.

5. Under the Action column, change Colors 2 and 3 to Red. Then, give them each a 60% opacity. When you're done, your project settings should look similar to Figure 23.11. Click Save As Default, click OK, and then click OK again to close the window.

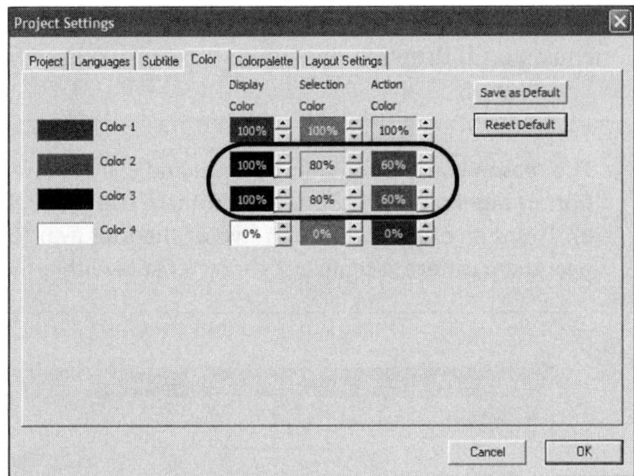

FIGURE 23.11
How your updated button highlight color settings should look.

Importing Project Elements

You are now ready to start creating a project using ReelDVD. The project will have a brief opening video that jumps automatically to a main menu. From there, you will let viewers access a movie, chapters within that movie, a slideshow with audio, and closing credits. The program flow will match Figure 23.12.

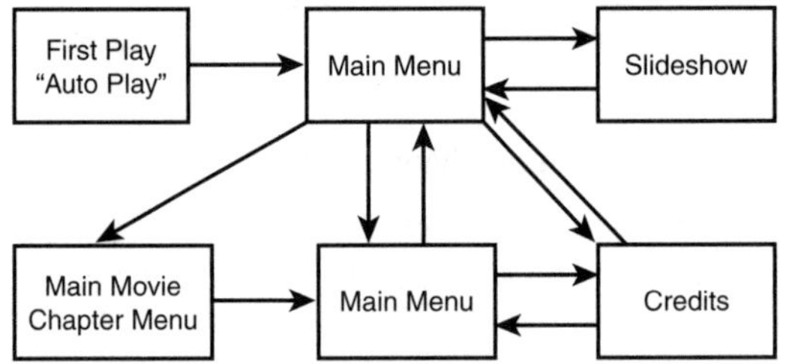

FIGURE 23.12
The project connects a number of typical DVD elements.

Before tackling the next task, take a look at ReelDVD's user interface. As shown in Figure 23.13, it has four workplaces: the Storyboard Area, the Timeline or Track Window, the Preview window, and an Explorer window.

You can easily adjust the relative size of any of the four work areas by dragging their borders.

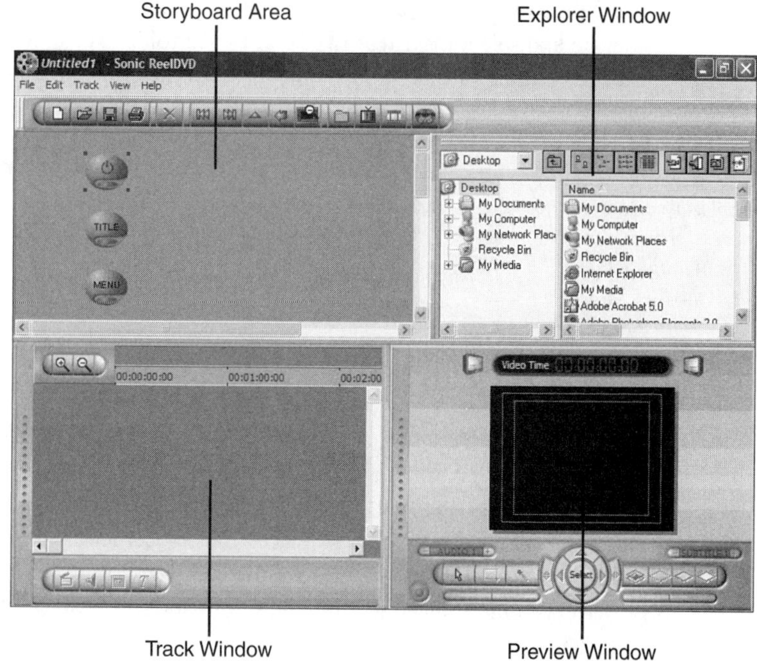

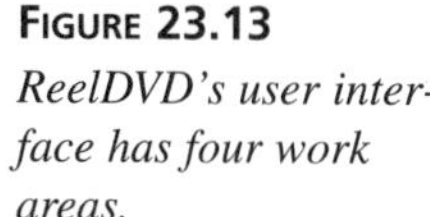
FIGURE 23.13
ReelDVD's user interface has four work areas.

You'll drag and drop media and menus from the Explorer window onto the Storyboard. You'll then add sub-picture overlays, subtitles, and audio to the media and menu icons in the Storyboard window, set movie chapter points in the Track window, and then edit menus and buttons. Finally, you'll preview, or *simulate*, your project in the Preview window.

> ReelDVD uses some terms that don't quite match those used in other Sonic Solutions products. For instance, each media/menu/subtitle object is called a *track*, First Play is called Auto Play, and the Timeline window is called the Track window.

Task: Import Media, Menus, and Subtitles

TASK

This is where you'll begin to see the power and utility of ReelDVD's combo flowchart/timeline user interface. You'll drag and drop items onto the Storyboard and later make adjustments using the timeline (Track window). Here's how you begin this process:

1. In ReelDVD's Explorer window (in the upper-right corner of the user interface), locate and select the `Tutorial Assets` file folder. As shown in Figure 23.14, this displays some of this project's media as well as two additional file folders.

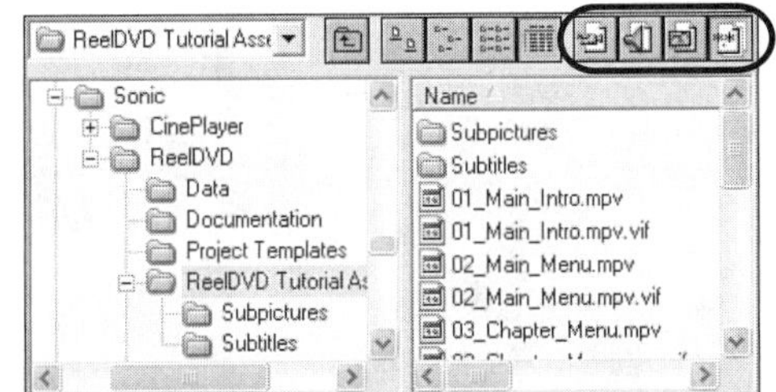

FIGURE 23.14
Use ReelDVD's Explorer window to access project assets. The highlighted icons limit the files shown to the specified media types.

2. Select the camera icon (the left icon shown in Figure 23.14) to limit the display to only the video assets. Select the four MPV files (click the top video file and then Shift+click the fourth to select them all at once), and drag them onto the Storyboard window. Doing so places four object icons in the Storyboard window; drag and drop those icons into a more compact arrangement, as shown in Figure 23.15 (you might need to slide the scrollbar at the bottom of the window to locate all four icons).

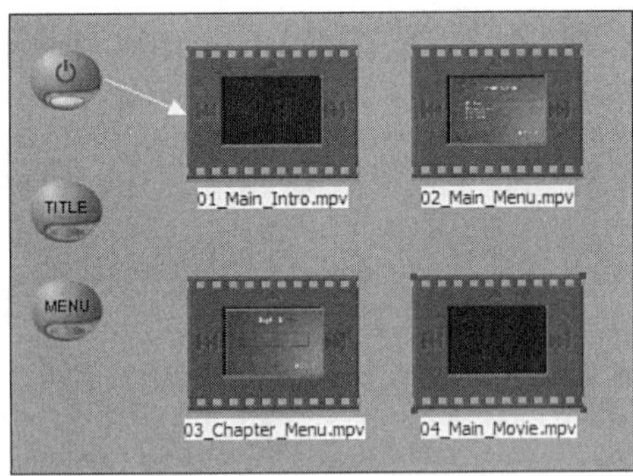

FIGURE 23.15
Arrange the media objects in the Storyboard window to make them more manageable.

ReelDVD automatically connects the first object dropped onto the Storyboard to the Auto Play (First Play) icon. In this case, that's `01_Main_Intro.mpv`. It's a video of the Sonic logo and will in fact be the Auto Play video for this project.

Two of the four movie files are menus: `02_Main_Menu.mpv` and `03_Chapter_Menu.mpv`. ReelDVD handles animated menu backgrounds with ease.

3. In the Explorer window, click the still image icon (third from the left in the group shown earlier in Figure 23.14). This displays four still images you'll use in a slideshow, as well as three images that comprise the credits menu.
4. Drag the first slide (`05_Slide_A.bmp`) to an empty space on the Storyboard. When prompted, select Slide Show and click OK.

The single image creates a slideshow *track* in the Timeline/Track window. You'll build out that slideshow by adding the remaining three images to the Track window, not the Storyboard. You could place them individually on the Storyboard and connect them to play sequentially, but you could not associate a single audio track with all four. Instead, the audio would stop and restart as each new slide appeared—not a very elegant process.

5. Select the remaining three slide images (use the Shift+click method explained earlier), and drag them to the Track window, next to the first slide. Your Track window should look similar to Figure 23.16.

You can improve your view of the still image thumbnails in the Track window by expanding the time divisions on the Timeline. To do that, click the plus sign (+) shown in Figure 23.16.

When you play your DVD, each slide will display for the default time of 10 seconds. You can change the duration for any slide by dragging its edges to lengthen or shorten it.

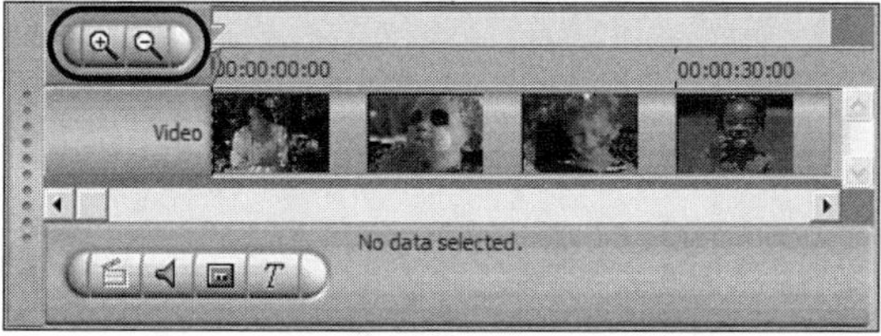

FIGURE 23.16
You create your slideshow in the Track (Timeline) window. Use the plus sign (+) to expand the view.

6. Import the Credits menu by dragging its Photoshop file (`06_DVD_Credits.psd`) to an empty space on the Storyboard. When prompted, select Still Menu with Sub-Picture and click OK.
7. The window shown in Figure 23.17 opens and displays any layers in the Photoshop file. In this case, the single extra layer is simply the back button. Click OK.

When adding a layered still menu to a project, ReelDVD automatically notes the *hotspot*, the nontransparent graphic(s) in the extra transparent layer(s). In addition, ReelDVD automatically creates a sub-picture overlay file; in this case, it's called `06_DVD_Credits.sp.bmp`. ReelDVD then places it in the `Tutorial Assets` file folder. It matches the appearance of the transparent Photoshop layer.

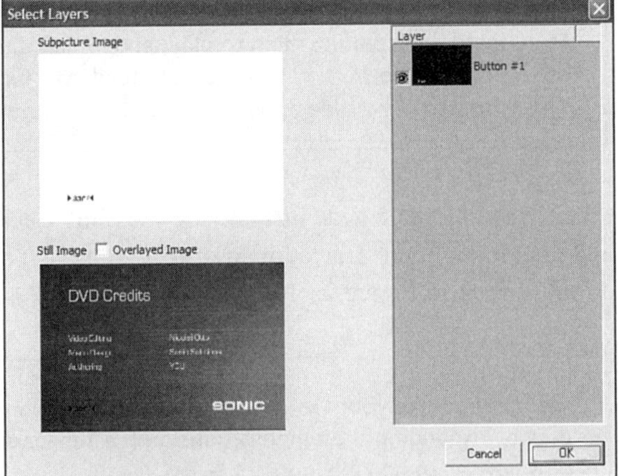

FIGURE 23.17
ReelDVD displays Photoshop graphic layers when you import a layered PSD.

8. In the Explorer window, click the Audio Files icon (second from the left in the group shown earlier in Figure 23.14). Drag the file `04_Main_Movie.ac3` to the `04_Main_Movie.mpv` icon in the Storyboard window.

After you've dragged the audio file to the movie icon, look at the Track window. As Figure 23.18 shows, ReelDVD automatically has added an audio track in the Timeline.

FIGURE 23.18
Dragging an audio file to a movie icon in the Storyboard window adds an audio track to the Track/Timeline window.

In higher-level DVD authoring, you'll usually work with MPEG audio and video elementary streams—that is, separate files for audio and video. Typically, you'll use MPEG-2 for video and AC-3 (Dolby Digital) or PCM (AIFF or WAVE) audio.

9. From the Explorer window, drag the `05_Slide_Show.ac3` file to the `05_Slide_A.bmp` icon (track) in the Storyboard window. This also automatically adds an audio track to the Timeline.
10. In the Explorer window, navigate to the Subpictures folder by double-clicking it. Select the All Files icon (the one on the right in the group shown earlier in Figure 23.14) to show the two BMP files.
11. Drag the `02_Main_Menu_Sub.bmp` file to the `02_Main_Menu.mpv` file icon (Track) in the Storyboard window. Then, drag the `03_Chapter_Menu_Sub.bmp` file to its associated Storyboard icon (Track)—`03_Chapter_Menu.mpv file`.

Both of these sub-picture files are overlays that will serve to highlight menu button icons (in this case, triangles) or box borders. If you want to look at them, simply open them in any graphics software that can view BMP files. Earlier in this hour, you defined the colors for the various button states by making global changes to the sub-picture color and opacity settings. You'll fine-tune two of them in the next hour.

12. Navigate to the Subtitles file folder by double-clicking it; make sure the All Files icon is active to view all 38 files. You'll import only the script file that tells ReelDVD where to find the 37 other subtitle files and where to place them in the movie. Drag `04_Subtitle_Script.sst` to the `04_Main_Movie.mpv` icon (Track) in the Storyboard window.

After you've added the subtitle script file to the main movie, look at the Timeline in the Track window. ReelDVD has automatically added a subtitle track called `Sp 1` (Sub-picture Track #1). The script file has all the file and subtitle timing information. If you expand the Timeline view by clicking the plus sign, you'll get a better idea of how this works. As shown in Figure 23.19, as you drag the green timeline cursor, the subtitles display in the Preview window.

FIGURE 23.19
You can view the subtitles by dragging the highlighted green timeline cursor and viewing the video in the Preview window.

Summary

The more you work with entry-level or prosumer DVD authoring software, the more you will come to recognize its limitations.

If you start tackling larger DVD projects intended for wider audiences, you might need subtitles or additional audio tracks for foreign language dubs. To give your menus extra pizzazz, you might want to add specialized button highlights that don't exactly match the size and shape of the button. Or you might want more button highlight color and opacity choices.

The solution is to spend more money on higher-end DVD authoring products that give you this extra level of control.

As of fall 2003, the upper-end choices on the PC side all come from Sonic Solutions: ReelDVD, DVD Producer, and Scenarist. While some of the automation features that are included in entry-level DVD creation packages are not present in these more advanced applications, using any of these products gives you more control over your DVD's look and functionality.

In ReelDVD's case, users rely on separate graphics or video editing applications for menu and button creation. And button navigation will no longer default to the order in which you place the buttons on a menu; you must handle all the detail work.

Higher-level authoring products require greater diligence on the part of the user but yield a greater flexibility that adds distinction to your DVD title.

In this hour, you took your first stab at working with the low end of this high-end market: ReelDVD. You installed it (noting some issues about subtitle script files in the process), toured its interface, delved into the arcane world of sub-pictures and highlight color and opacity schemes, and imported your project elements.

In the next hour you'll complete this ReelDVD production process.

Workshop

Review the questions and answers in this section to reinforce your newly acquired ReelDVD background knowledge. Also, take a few moments to take the short quiz and perform the exercises.

Q&A

Q When looking at a layered graphic, I see that the button on the transparent layer is solid black. Won't that cover up any button beneath it, not highlight it?

A That solid black graphic merely defines the size and shape of the highlight. ReelDVD digitally replaces that opaque black (or red, blue, or white) color with the color and opacity you choose in the Project Settings Color menu.

Q When I dragged the four movie files to the Storyboard window one at a time, the wrong one connected to the Auto Play button. What happened, and how do I fix that?

A When dragging them one at a time, the first one you imported must not have been `01_Main_Intro.mpv`. By default, ReelDVD sets the first imported media file as the Auto or First Play. To fix that, click the oval icon at the base of the Auto Play button and drag its cursor to the correct icon: `01_Main_Intro.mpv`.

Quiz

1. You want to make all selection button highlights for your project light green with 60% opacity. How do you do that?
2. Working with this hour's project, how would you change the opening sequence of events, making the DVD Credits the Auto/First Play file, then playing the Sonic Logo `01_Main_Intro.mpv` file, and finally displaying the `02_Main_Menu.mpv` file?
3. You have a client who uses a specific color in her graphics. She wants you to make your button highlights match that color. How do you do that?

Quiz Answers

1. Select Edit, Project Settings and then select the Color tab. Click each color box in turn in the Selection column (if you used fewer than all four available sub-picture graphics colors—red, blue, black, and white—you need click only the applicable boxes). Clicking a box pops up the 16-color palette. Select the light green color swatch. After you've fixed all four colors, change all the opacity levels to 60 (click the down arrow button six times). Make these and any other global sub-picture color changes before you start importing menus into the Storyboard. Otherwise, the changes will have no effect on the current project.
2. All it takes is a few drags and drops. Click the Auto/First Play button's oval icon and drag it to the DVD Credits menu. Click the DVD Credits Next icon (the double-arrow on right side) and drag it to the `01_Main_Intro.mpv` file icon. Finally, click the Main Intro Next icon and drag it to the `02_Main_Menu.mpv` icon. This is an easy and much faster process than you'd have if you were working with an entry-level or prosumer DVD authoring product.
3. Select Edit, Project Settings and click the Colorpalette tab. Select one of the 16 color swatches—a color you don't anticipate using for this project—and open the Define Custom Colors window. Your client probably has a numerical representation of that color (0–255 values for red, green, and blue), so type them in the respective boxes, click Add to Custom Colors, and click OK. The new color will show up in the 16-color palette and be available under the Color tab or later in the Preview window to use as a button highlight.

Exercises

1. Get a feel for subtitle creation. Do that by first customizing the subtitle look and screen placement in the Project Settings/Subtitle window. Select the `05_Slide_A.bmp` icon in the Storyboard window to open it in the Track/Timeline window. Click the Subtitle icon (it's the large, italic *T* at the bottom of the screen). Type in a subtitle and click OK. It will show up in the newly created subtitle track and display in the Preview window.
2. If you want learn how to create a layered Photoshop graphic to import into ReelDVD or any other Photoshop-compliant DVD authoring product, select Help to open the ReelDVD Help file. Go to Contents and select Planning and Preparation, Preparing Assets. Scroll down that long page to Layered Photoshop Image Files for an explanation about how to proceed. As an aside, the folks at Sonic Solutions take great pains to create informative and practical help files. Feel free to check out other Help topics.

HOUR 24

ReelDVD: Part II, DVD Producer and DVD Trends

In this book's concluding hour, you'll perform the actual authoring of your ReelDVD project. You'll set movie chapter points, use links, create button hotspots and highlights, link those buttons to your media and menus, and finally simulate (or *proof*) your project.

I'll take you through DVD Producer, the next step up in DVD authoring price and performance. I'll also explain what makes it different from DVDit! and ReelDVD.

And I'll gaze into my crystal ball and take a guess as to what the future holds for DVD production and technology.

The highlights of this hour include the following:

- Working with chapters
- Linking media and menus

- Editing buttons
- Proofing your project
- Touring DVD Producer
- DVD trends

Working with Chapters

You'll use a simple drag-and-drop method to connect all your media and menu objects. Each of the objects, or Track icons, in the Storyboard window has three or four means to make those connections. Figure 24.1 shows those links.

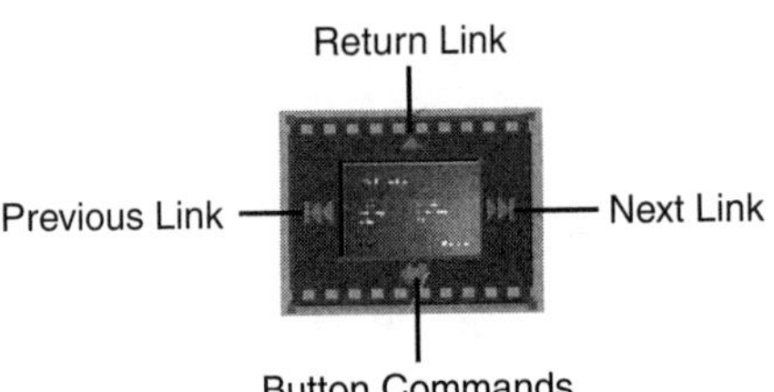

FIGURE 24.1
The four links available for media and menu Track icons.

Previous, Next, and Return correspond respectively to the Skip Backward, Skip Forward, and Return keys on a typical DVD remote control. The bottom link—Button Commands—is available only for menus with buttons. In the following task, you'll click those links and drag connecting lines to the appropriate icons. This connect-the-dots process is a remarkably simple yet elegant authoring tool.

Task: Set Chapter Points and Link Movies and Menus

▼ TASK

This is a two-step task: organize the Storyboard window and mark chapter points in the main video. Here's how you do that:

1. Organize the Storyboard window by dragging and dropping all the Track icons and the Menu, Title, and Auto Play buttons into a manageable area, something along the lines of Figure 24.2.
2. Next, you add chapters to the main movie in the Track window by clicking `04_Main_Movie.mpv` in the Storyboard window to display it in the Track window. Drag the green timeline marker to about `1:04:20` (use the Preview window time display—the Chapter Time will show up in that window only after clicking the Chapter icon).

▼

FIGURE 24.2
Use this figure as a reference when organizing your Storyboard window icon layout.

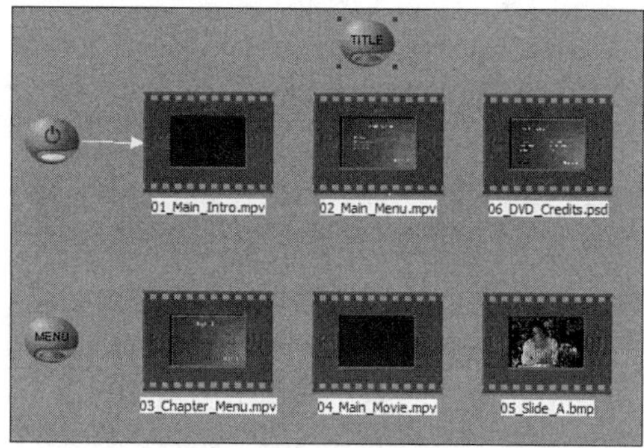

3. Expand the Timeline view by clicking the plus sign (+). As shown in Figure 24.3, make sure the marker is over an open space between two subtitles (the DVD specification does not allow chapters within subtitles).

> Setting chapter points is an inexact process for two reasons: Dragging the timeline cursor to an exact timecode is a hit-or-miss proposition, and the DVD specification limits available chapter point locations to about two per second of video. To overcome the cursor issue, set a frame-specific chapter break point by typing in `1:04:20` in the Chapter Time box at the bottom of the Track window. As for the DVD specification, ReelDVD automatically finds the closest allowable chapter break.

FIGURE 24.3
Set chapter points by moving the green timeline marker and then clicking the Chapter icon.

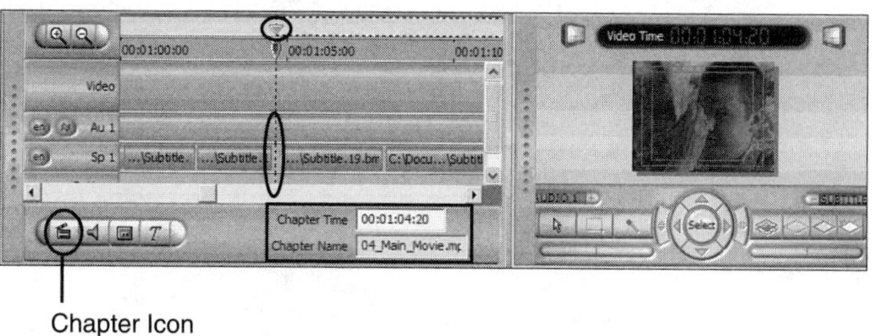

4. Click the Chapter icon shown in Figure 24.3 to mark a chapter. It shows up as a yellow triangle above the green timeline marker.

5. Set another chapter point at 1:58:00.

Linking Media and Menus

With your project elements in place, it's time to create the DVD navigation scheme by creating links among the media and menus. In MyDVD this occured automatically (more or less), and in DVDit! you used separate menus. Here, though, you'll rely on a simple but elegant drag-and-drop method.

Task: Link Media and Menus

TASK

Now you'll link items by doing the following:

1. ReelDVD already should have connected Auto Play (First Play) to `01_Main_Intro.mpv`. If not, as shown in Figure 24.4, click the oval at the bottom of the Auto Play icon and drag your cursor to `01_Main_Intro.mpv`.

FIGURE 24.4
Use the Auto Play Command button to connect it to your First Play movie: `01_Main_Intro.mpv`.

2. Set `02_Main_Menu.mpv` as the DVD's Title menu by clicking the small oval in the Title icon and dragging the cursor to `02_Main_Menu.mpv`. Do the same with the Menu icon's oval button and `03_Chapter_Menu.mpv`. Your Storyboard should look similar to Figure 24.5.

The Title and Menu icons in ReelDVD's storyboard represent the Title and Menu buttons on a standard DVD remote control. Generally, when viewers press the Title remote control button you want to return them to the DVD's main menu, and when they press the Menu button on the remote you want to return them to a submenu on the DVD (usually the previous menu visited).

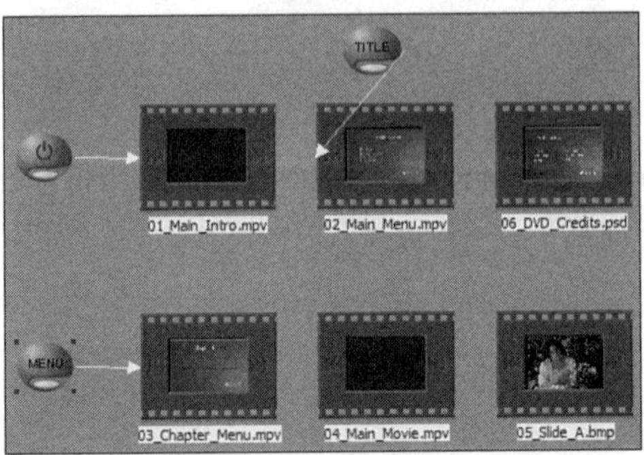

FIGURE 24.5
How your Storyboard should look after connecting the three icons that represent the main buttons on your DVD remote control.

3. Set the Next links using Table 24.1 as a guide. Click each Storyboard icon's Next button (the double triangle on the right) and drag it to the destination icon (for the two menus, you'll drag the cursor inside them so they will loop until the viewer presses a menu button or the remote control).

ReelDVD does not have an undo command, but if you dragged a link to the wrong icon (track), a work-around is available. In this task's case, simply click the Next/Return button again and drag the cursor to the correct icon. If you made a link when you intended to have no link, erase that link's connecting line by clicking the Next button and dragging the cursor to an empty space in the Storyboard.

TABLE 24.1 Setting the Next and Return Links

Source Track	*Next Track*	*Return Track*
`01_Main_Intro.mpv`	`02_Main_Menu.mpv`	No Return Link
`02_Main_Menu.mpv`	`02_Main_Menu.mpv`	No Return Link
`03_Chapter_Menu.mpv`	`03_Chapter_Menu.mpv`	`02_Main_Menu.mpv`
`04_Main_Movie.mpv`	`02_Main_Menu.mpv`	`02_Main_Menu.mpv`
`05_Slide_A.bmp`	`02_Main_Menu.mpv`	`02_Main_Menu.mpv`
`06_DVD_Credits.psd`	`02_Main_Menu.mpv`	`02_Main_Menu.mpv`

4. Using Table 24.1, do the same for the return links (the button at the top of each icon). This lets the viewer return to the main menu simply by clicking the Return button on the remote. When you've completed steps 3 and 4, your Storyboard window should look similar to Figure 24.6.

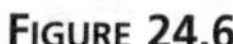

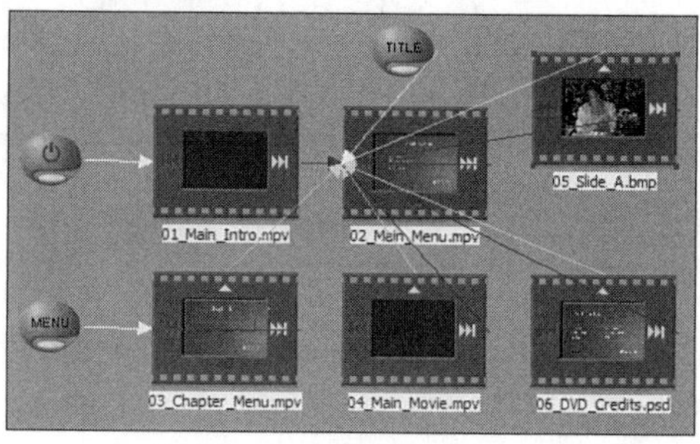

FIGURE 24.6
How your Storyboard window should look after setting all the track links.

Editing Buttons

This is where things get tedious and repetitious—that's the bane of having this much control over your project. In these next few tasks, you will ascribe attributes to the buttons in the two principal menus, plus do some work on the Credits menu.

Task: Prepare for Button Creation and Editing

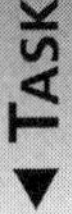

In an effort to simplify this process, in this task you'll open those menus in an expanded window and then switch back and forth between them. This will save several steps later.

Here's how to get that process started:

1. Select the Main Menu icon in the Storyboard window to display that menu in the Preview window.
2. Expand the Preview window by right-clicking its gray border, deselecting Allow Docking, and dragging a corner.
3. As shown in Figure 24.7, move the expanded window and adjust the Storyboard window slider bars so you can see the Main Menu, Chapter Menu, and DVD Credits Menu Storyboard icons.

FIGURE 24.7
Arrange the icons in the expanded Preview window on your desktop so that you can click the Main Menu, Chapter Menu, and DVD Credits Menu.

4. Click the Chapter Menu icon and note how the expanded Preview window display changes to that menu.
5. Switch back to the main menu view by clicking its icon in the Storyboard window.

Task: Create Menu Button Hotspots and Links in Menus

▼ TASK

Now that you've worked out the menu switching routine, it's time to set button hotspot locations, define how users will navigate from one button to the next, set button highlight opacity levels, and link buttons to their associated media. Here's how:

1. Use the Create Button tool, shown in Figure 24.8, to drag button hotspot boundaries around each link. Make sure they don't overlap. Later, you can use the Selection icon (the arrow icon to the left of the Create Button tool) to move the boundary boxes around or to resize them.

> You'll know whether your button hotspots overlap if two or more hotspot boundary boxes suddenly join together into one box. If that happens, use the Selection tool, shown in Figure 24.8, to move or resize the boxes.

> As you create the hotspot boundaries, ReelDVD numbers them and displays the letters `NOP`. *NOP* means *no operation*. The acronym remains onscreen until you link the hotspot to media or a menu.

FIGURE 24.8
Create hotspots using the Create Button tool.

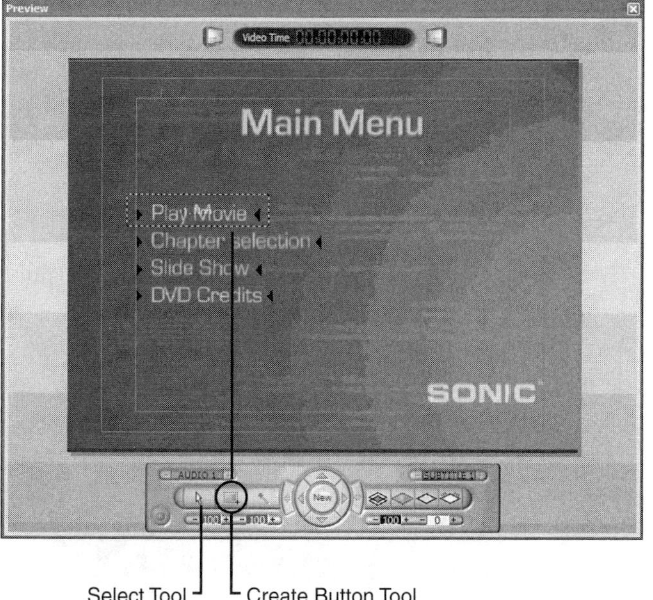

▼

2. Perform the same button hotspot creation process in the Chapter menu. To access it, click its icon in the Storyboard window. Use the Create Button tool to create boxes that match the three squares in the menu. In addition, create a hotspot around the Back button.

You do not need to define a button hotspot for the DVD Credits menu because it is a layered Photoshop file and ReelDVD automatically recognizes any buttons by analyzing transparency layers and associating hotspots with them.

3. Return to the Main Menu Preview window view by clicking the Main Menu icon in the Storyboard window. Now, you'll create links from one main menu button to the next so that, when a viewer presses the Down or Up button on the remote, the correct buttons in the menu become highlighted or active.
4. Select the Link Vertically tool shown in Figure 24.9, click the top button (Play Movie), and then drag down to the next button (Chapter selection). Do the same for the next two buttons, linking each to the next one below it. Then connect the bottom button (DVD Credits) to the top button (Play Movie).

FIGURE 24.9
Use the Link Vertically tool to define the button navigation.

5. Switch to the Chapter menu and follow the same procedure; however, as shown in Figure 24.10, use the Link Horizontally tool to connect the boxes in turn from left to right and link the third box back to the first. Use the Link Vertically tool to connect the left button hotspot to the Back button below it, and vice versa. Return to the Main Menu Preview window.

FIGURE 24.10
Because these buttons run left to right, use the Link Horizontally tool to define this menu's navigation.

Task: Change Button Color and Opacity Settings

TASK

In Hour 23, "Professional DVD Authoring with ReelDVD: Part I," you globally set color and opacity for your entire project's sub-pictures. In this task, you make changes to two individual buttons, specifically reducing their opacity to 0. You won't need to make any color changes, but I'll point out how to do that in case you want to experiment a bit. Follow these steps:

1. As you did earlier in this hour, open the Chapter menu in the Preview window. As shown in Figure 24.11, click the Display Color icon. This shows how the Chapter menu will look if the viewer takes no action—when she does not navigate over any button or press the Enter key on the remote.

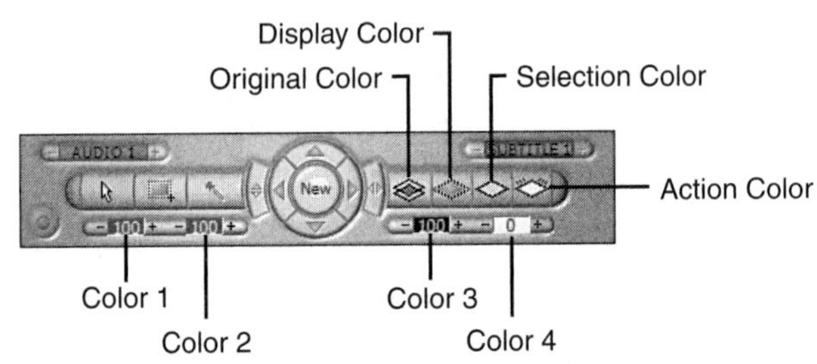

FIGURE 24.11
The four diamond-shaped icons represent the four highlight states: original, display, selection, and action. The numbered, colored boxes represent the four possible colors for your sub-picture button highlight graphics.

> To see how this menu's sub-picture, transparent overlay looks, click the Original Color icon (the leftmost diamond). As shown in Figure 24.12, only the button highlight areas display. Also, they are all solid-color—red or black—shapes or text. As mentioned in the previous hour, these silhouettes (created in a separate graphics program such as Photoshop) merely define the highlight shapes and sizes. ReelDVD digitally replaces them with the highlight color and opacity characteristics you describe. Using two colors in a sub-picture lets you highlight buttons in two different ways in one menu.
>
> Depending on which Vertical or Horizontal Link icon you clicked last, clicking the Original Color icon will also display the respective linking lines.

2. Your task here is simple: Reduce the display color opacity of the boxes to 0 to make them transparent. To do so, click the Display Color icon, which is the second icon from the left in Figure 24.11. This shows how the boxes (and Back button) will look when unselected.

> When you click the Display Color icon, you might see that the Back button becomes black. This further demonstrates how sub-pictures work. The black color is the original color used for the word Back when the graphic artist created this menu. The red color used in the sub-picture (the transparent overlay placed "above" the menu) merely tells ReelDVD that this is a sub-picture highlight element. ReelDVD overlays the red Back on top of the black Back but displays the red Back only as a highlight (depending on the viewer remote control actions) and only using the color and opacity you ascribe to it. If you select 0% opacity (transparent) for a highlight, the original black "back" displays.

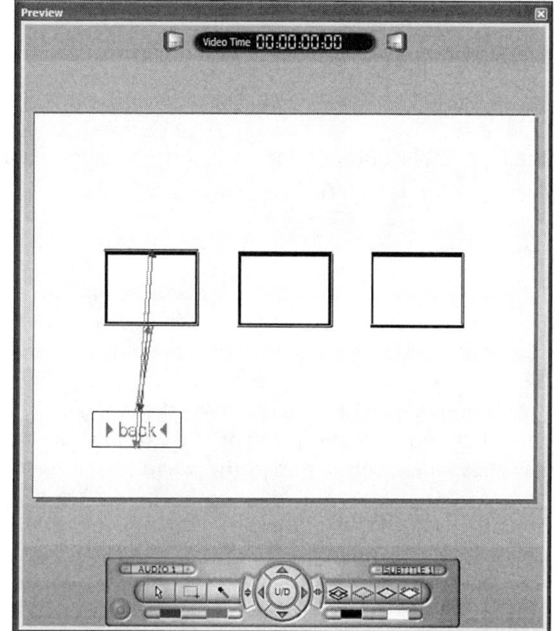

Figure 24.12
If you select the Original Color icon, you'll see the actual sub-picture. Note that this menu uses two colors—red and black—allowing you to create two types of highlights in one menu.

3. The box outlines in the sub-picture are black—sub-picture Color 3. Reduce the Color 3 opacity to 0% by clicking the minus sign (–) 15 times. You'll see the boxes gradually disappear.

If, instead, you want to change the button box outline color without changing the opacity, you could click the Color 3 bar between the minus and plus signs and then select a color from the 16-color pop-up palette. You then could reduce the opacity as well.

If you wanted to actually make the displayed version of the Back button red, you'd need to click the Color 2 icon (because the back sub-picture color is red) and select the color red from the pop-up color palette.

The ReelDVD sub-picture Color Number interface at the bottom of the Preview window can be confusing. Sub-picture colors 1–4 are blue, red, black, and white, respectively. But the interface does not show those four colors; it shows whatever colors you've chosen in the global settings for whatever state icon you've clicked. If you click the other states—selection or action—the colors in the little Color Number boxes change accordingly.

4. Do the same thing to the main menu. Switch to its display in the Preview window by clicking the Main Menu icon in the Storyboard window.
5. Click the Display Color diamond icon (second from the left). Note the small, black triangles next to the four menu options. You want them to disappear when not selected.
6. Reduce the Opacity level for Color 3 (the original sub-picture color for the triangles is black—Color 3) to 0%. The triangles will disappear.

If you click through the other states—Selection and Action—you should see how those triangles will display depending on viewer choices (light gray for Selection and red for Action). But if the button bounding boxes you created are only as wide as the words and do not extend to accommodate the triangle button highlights, you won't see the triangles. If that's the case, use the Selection tool (the arrow icon on the left) to select each box and then drag its sides far enough to display the triangles (see Figure 24.13).

FIGURE 24.13
Make sure your hotspot bounding boxes are wide enough to display the button triangle highlights.

Task: Create Command (or "Button") Links

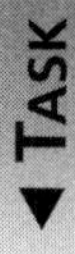

Now you need to connect the button hotspots to their respective media and menus, which is a multistep (but fairly straightforward) process. Essentially, you'll connect a menu to some other object in the Storyboard and then select which item within that destination menu or media you want to link the button. Finally, you'll choose the respective button from the source menu. Here's how to do it:

1. Redock the Preview window by right-clicking somewhere on its gray border, selecting Allow Docking, clicking the window's title border, and dragging it toward the lower-right part of the main ReelDVD interface. A dotted line border will appear indicating that, if you release the mouse button, the Preview window will return to its regular lower-right corner location. Release the mouse button.
2. Click the `03_Chapter_Menu.mpv` Command Link icon (the bent arrow), shown in Figure 24.14, and drag the cursor to the `04_Main_Movie.mpv` icon. The screen shown in Figure 24.15 pops up and displays the three chapters in the main movie.

Sonic Solutions uses the term *command link*, which can be a bit confusing. A more descriptive term would be *button link*.

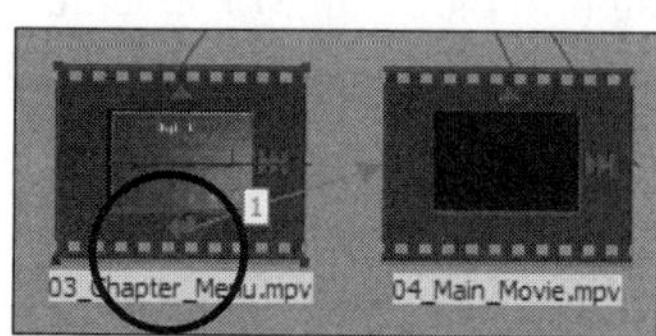

FIGURE 24.14
The command link curved arrow icon lets you connect buttons to media and menus.

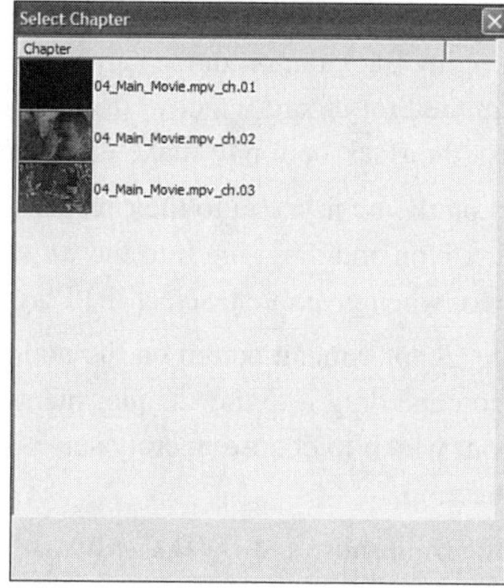

FIGURE 24.15
Dragging the command link icon to an object with video chapters lets you choose to which chapter you want to link.

3. Select the first chapter—the opening title scene—which is a black screen in this window. Next, you will link this chapter scene to the chapter button in the video.

I found this workflow a little confusing the first time I tried it. A more logical approach might have been to first select the *source* menu (the one with the buttons you want to link), use its Command Link to connect to a *destination* menu or media item, select the source button you want to connect, and then select the destination item to complete the link. But, with a little practice you'll quickly get the hang of navigation programming in ReelDVD. This workflow is nonsequential and counterintuitive. It confused me the first time I tried it, and it might confuse you, too. I think the logical sequence should be to select the source menu (the one with buttons you want to link), use its command link icon to connect the menu to a destination menu or media item, select the source button you want to connect, and then select the destination item to which you're connecting it. ReelDVD reverses that a bit, having you select the object in the *destination* that you're linking *to* first and then the button *from* the *source* menu that refers to it.

4. Select the first chapter scene to open the Choose Command Button window, which is a minidisplay of the Chapter menu. Select button one, the one on the left.
5. Click the Command icon, drag it to the Main Movie icon, select a chapter scene, and select its respective button to link the middle button to the second chapter and the right button to the third chapter.
6. Link the Chapter menu's Back button to the main menu. Click the Command icon and drag the cursor to the `02_Main_Menu.mpv` icon. Because no movie chapters exist in the main menu, all you need to do is select the Back button to complete the connection.
7. You can link the DVD Credits Back button in the same way: Click the DVD Credits Command icon in the Storyboard window, drag the command link cursor to the main menu, and select the Back button to make the connection.
8. Connect the four buttons on the main menu to their respective destinations. Start by clicking the command link icon and dragging it to the `04_Main_Movie.mpv` icon. Select the first chapter and, when prompted, select the Play Movie button.
9. Do the same thing for the Chapter menu button on the main menu: Click the Main Menu Command Link icon and drag it to the Chapter menu. Because no video chapters are available from which to choose, merely select the Chapter Menu button to complete the connection.
10. Repeat this process for the Slideshow and DVD Credits. When completed, your Storyboard should look similar to Figure 24.16, which resembles a spider web.

The command (chapter button) link lines have numbers on them that represent the button numbers on the source menu. If you connect several buttons from one menu to another—the three Chapter Menu buttons to the Main Movie, for instance—the highest number button displays. In the case of the link from the Chapter menu to the main movie, that number is 3 (Button 4 is the Back button and has a separate link line running up to the main menu).

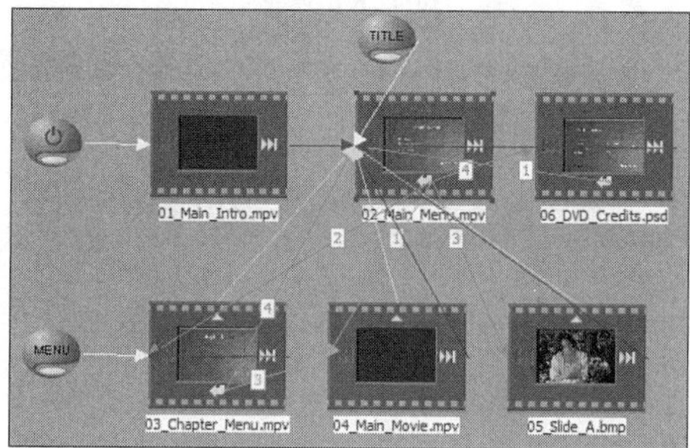

FIGURE 24.16
How your Storyboard window should look after you've created all the media, menu, and button links. The numbers stand for the button numbers.

Proofing Your Project

You have completed this project, so now it's time to test it. ReelDVD has a built-in DVD simulator. It relies heavily on your PC's processor, so it might not play smoothly or with as precise timing as other DVD players. For more precise timing, you can try using CinePlayer or some other DVD player. I'll touch on that option a bit later.

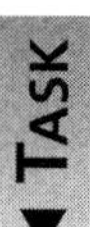

Task: Simulate DVD Playback

Checking your work is a routine process. Here's how to do it:

1. Select the Auto/First Play movie, the `01_Main_Intro.mpv` icon.
2. Undock the Preview window and expand it to get a better view of your project.

Undocking and expanding the window isn't necessary to simulate your project. I simply prefer the larger viewing area.

3. Click the Simulation On/Off button—it's the little red dot in the lower-left corner of the Preview window. It should turn green, indicating it's in Simulation mode (see Figure 24.17).

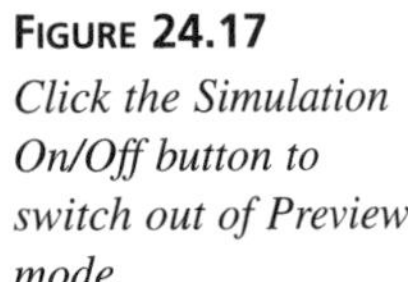

FIGURE 24.17
Click the Simulation On/Off button to switch out of Preview mode.

4. Click the Play button shown in Figure 24.17. The Sonic logo should display, followed by the main menu.

> As you check your project's functionality, ReelDVD displays a text playback log that notes buttons pressed and media viewed. It also likely will state something such as `Menu track '01_Main_Intro.mpv contains fewer subpicture streams than other tracks in this project`. It also might point out that some menus do not have return tracks. Don't worry about that.

5. Use the Up/Down triangles at the center of the control panel to navigate through the four buttons. Click Enter to move to a selection.

When you select the main movie, via either the main or Chapter menu, you'll get the warning box shown in Figure 24.18. This lets you know that ReelDVD will combine—*encode*—the subtitles with the movie. Click Yes.

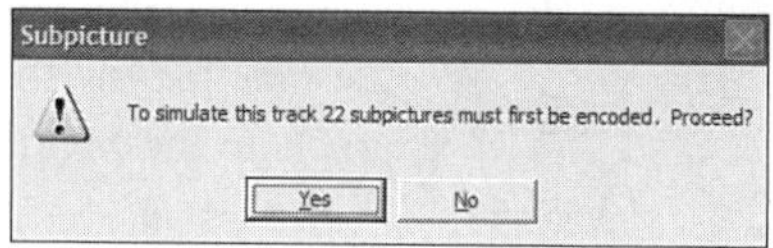

FIGURE 24.18
Viewing the main movie for the first time leads to this prompt to combine the subtitles with the video. Click Yes.

Your goal here is to use the Simulation mode to check for broken links, links to the wrong destinations, incorrect highlight colors or opacities, and to make sure button navigation within each menu is intuitive.

Also, check the Title, Menu, and Return buttons as well as the Skip Forward and Skip Backward keys. See whether what happens make sense. If need be, you can stop the simulation, click a Storyboard window icon, make fixes, and try again.

Emulation Mode

You can test your project fully using more accurate playback software. I won't go into the details here, but it basically involves making a hard disc image of your project via the Project Settings/Layout window. After that's completed, you can test the project using Sonic's CinePlayer or another software DVD player that can play a DVD hard drive image.

Touring DVD Producer

Although I think ReelDVD's flowchart workspace is elegant, it can have one drawback: Projects with many elements may require dozens of linking lines on your flowchart. If you're creating these large projects with ReelDVD and you're not careful about the way you manage the placement of objects on the screen, the multitude of linking lines between them can complicate the flowchart. Fortunately, ReelDVD allows you, with a bit of housekeeping, to arrange these objects in such a way as to keep the flowchart view organized and understandable.

24

DVD Producer minimizes that complexity by using an asset list approach very similar to that used in DVDit! (see Figure 24.19). You simply right-click the Palette window and select the media and menus you want to include in your project. From the Palette, you then can drag and drop assets to the Menu or Movies column.

As you increase your DVD authoring and video production skills, you might want to upgrade to a dual-monitor PC setup. Several PC video and graphics cards from firms such as NVIDIA, ATI, and Matrox offer this feature, which enables you to spread out your workspace over two monitors. In ReelDVD's case, you could use one entire screen solely for the flowchart, thereby expanding your view and reducing the potential for any confusion over the multitude of linked lines.

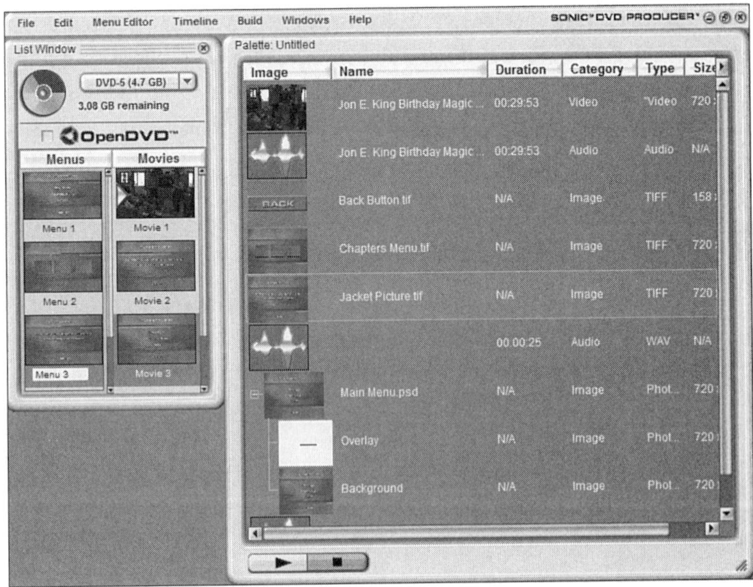

FIGURE 24.19
DVD Producer takes the list approach used in DVDit! and enhances it with some extra ease-of-use features.

Available for $2,500 and up, DVD Producer ships either as standalone authoring software or with a hardware MPEG encoder and an audio input/output box. DVD Producer is a high-end product, and many users choose audio/video hardware to match.

Taking a cue from ReelDVD, DVD Producer includes a timeline, shown in Figure 24.20, as a means to add and edit audio tracks and insert subtitles. You can simply scrub through the timeline to find a new thumbnail to use in a menu button or add a chapter with a single keyboard button stroke.

FIGURE 24.20
The timeline is similar to the one in ReelDVD. It lets you add audio and subtitle tracks plus grab thumbnails for buttons.

As you move up through Sonic Solutions DVD authoring products, one element remains constant: near-universal DVD playback platform compatibility. I can't stress how important this is. The DVD specification is complex, so some authoring software makers use shortcuts to simplify the authoring process. However, those changes frequently lead to burned DVDs that play on only a limited subset of DVD set-top playback machines. On the other hand, if you burn a DVD using a Sonic product, it will play on most DVD players.

Moving several steps up from DVDit!, DVD Producer has a menu editor with numerous helpful features, including these:

- Displays all subpicture overlays as separate windows or composited, displaying all layers at once (see Figure 24.21).
- Lets you define rectangular button hotspots.
- Has a text editor.
- Allows you to use video frames as animated buttons to produce full motion menus.
- Permits you to drag and drop graphics and animations onto the menu to use as buttons. The Back button in Figure 24.21 is an example.

Take a close look at Figure 24.21 and note the numbers in the upper-left corner of each button. Those numerals represent the navigation order as the viewer uses the remote control to maneuver around the menu.

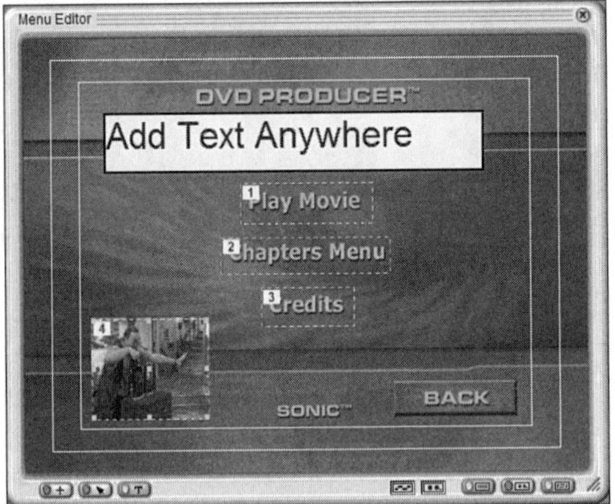

FIGURE 24.21
The Menu editor has everything you need right at your fingertips.

DVD Producer shines in the ease-of-use department. Context-sensitive, right-click menus let you quickly link any menu button to any other element in your project down to individual buttons on different menus.

The Properties menu shown in Figure 24.22 is a powerful tool. It incorporates what Sonic Solutions calls Jump Anywhere technology: You can select any element on any menu, and all the options available for the element display in the Properties window.

You can also set time durations and animation for menus and buttons, easily change subpicture overlay colors and opacities, rework button navigation using drop-down lists, and update links within this all-in-one tool. As you make changes in the sub-picture overlay color scheme, the Menu editor displays them immediately.

FIGURE 24.22
The Properties menu.

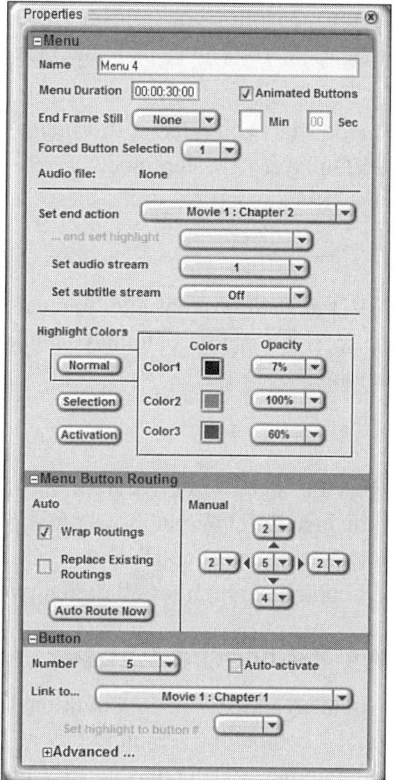

If you keep this window open, you can click any element in the Menu editor window and the Properties window will change its display to that item's feature options. This is an excellent, labor-saving feature.

Here are some other DVD Producer features of note:

- **OpenDVD compliance**—You can take any DVD created in another OpenDVD-compliant product, such as MyDVD, and open it in DVD Producer for further editing. ReelDVD, however, is not yet OpenDVD compliant.
- **Ability to add up to eight audio tracks and 32 subtitle tracks**—This is the maximum permitted by the DVD specification.
- **WriteDirect technology**—Lets you transfer data directly from source files to a wide range of DVD, CD, and digital linear tape formats.
- **Bit Budget Calculator**—Continually tracks the size of your project relative to your selected output media to ensure you don't exceed the desired capacity.
- **Flexible Asset Importing**—DVD Producer accepts all standard graphics file formats as well and AVI and QuickTime video.

- **Built-in Menu Compositor**—enabling you to layer video files, graphics, and text within the authoring application. You can also animate video from your project within the menu.
- **Jacket Picture**—This support lets you select a still image to display whenever the disc is loaded and the DVD player is in stop mode.

DVD Trends

These are exciting times, as DVD technology continues to create opportunities. I don't see this trend waning any time soon. So, before bidding you adieu, here are a few final thoughts on where things are going.

DVD Ubiquity

DVDs will be everywhere. Every PC and TV set will have one, and every new non-linear video editing application will include DVD output as a natural part of the program. Creating slideshows, audio, and home movies on DVD will be as natural as using a VCR. DVDs will become the standard medium for all digital publishing.

Microsoft Windows XP Media Center

Microsoft is taking another stab at creating a standard multimedia PC. Its first foray in the early 1990s failed for lack of a compelling reason to buy it. This new standard, however, integrates the PC with the Internet, a DVD player, a TV tuner, and a remote control. Windows XP Media Center works similarly to a combination TiVo and online Gemstar programming guide. You can watch shows (pausing and rewinding at will), time shift, and operate a CD/DVD player. Media Center shows some real promise, and some major PC manufacturers, including Gateway, HP, and Alienware, have hopped on the bandwagon.

Sonic Solutions, in collaboration with Microsoft, has created a Media Center application called PrimeTime. It has a simple function: You select a TV program, press a button, and burn recorded shows seamlessly to DVD. This is just another example of the coming ubiquity of DVDs.

Integration with the Web

Creating and playing DVDs will take on more immediacy because of Internet connectivity. As you pull in content during the authoring process, you'll be able to access the Web and easily add material and links that will enhance your DVD.

Later, when viewers play your DVD on their Web-connected PCs or set-top appliances, they will be able to access the latest information using those links. Doing a DVD on your

Tahitian vacation? Add some Web links to travel sites. Including material about your favorite musician? Add links to that artist's site. During playback, those links will lead to content that will be more current than what you might have included on the DVD.

This technology is available today as an add-on option to Sonic's top professional authoring app, DVD Creator and DVD Fusion. Called eDVD, it's based on InterActual's web-linking technology and is found on millions of Hollywood titles today. With eDVD, you can add links to your DVD that automatically launch your web browser to provide you with an enhanced viewing experience.

High-Definition TV and DVDs

The history of DVD authoring began with Hollywood studios using high-priced and technical DVD creation workstations to create DVDs.

As PCs have become more powerful and DVD recorders dramatically lower in price, authoring products are now consumer friendly and inexpensive.

That cycle might repeat itself as high-definition TV standards begin to coalesce. Putting HDTV on DVD might be a bit of a challenge at the beginning, but it will quickly take on a consumer-friendly glow as HDTV gains acceptance.

Long Live the DVD Format

DVDs will be around for a long time. What might change is DVD data capacity. Double-sided, double-density discs will become easier to use and author, and blue lasers might finally make an appearance. Because of DVD's near universal acceptance, any new optical format that comes along will need to be backward-compatible with DVD video and CD audio.

The future is less about authoring and more about publishing content on DVD. Any type of software, down to word processors and meeting presentation products, will have DVD output functionality.

Software makers constantly will need to ask, "How do we make it easy for consumers to create DVDs?"

Upcoming products will take whatever digital content you have and offer easy ways to put it on DVDs. They'll do everything for you: make menus, create interactivity, and put up thumbnails.

So, what about high-end products such as ReelDVD and its more expensive siblings? As with nonlinear editors, there will always be prosumer and professional producers looking to differentiate their products with special features. They (and you, I'm guessing) will continue to push the envelope.

Summary

I've wrapped up your DVD authoring instruction by taking you through the simple but elegant drag-and-drop authoring techniques available in ReelDVD. You used them to define button hotspots, set the navigation flow among those buttons, and connect them to their respective media and menus. In addition, you adjusted highlight color and opacity characteristics and finally simulated your project.

Professional-level products give you more control over the DVD authoring process, leading to creative and distinctive DVD titles. In turn it does takes a bit more time and effort to get used to working with professional product versus consumer type DVD authoring applications. With a little practice, though, you'll find that ReelDVD becomes very intuitive and fast to use.

You took a tour of DVD Producer, noting the extra features that go hand-in-hand with its higher price tag. Those features include real-time proofing, design tools for motion and static menus, unlimited undo, text generation and graphic compositing tools, JumpAnywhere navigation technology, and mix and match 4:3 and 16:9 aspect ratio content.

Finally, the next several years look good for DVD authoring. DVDs will become ubiquitous, and DVD authoring tools or modules (some with very narrow functionality) will appear in all sorts of media creation and publishing tools. Even as DVD technology improves with higher-capacity discs, new drives will likely retain backward-compatibility with both DVD and CDs.

You are on the leading edge of this exciting and dynamic technology and are well positioned to use that head start to your advantage.

Workshop

Review the questions and answers in this section to reinforce your ReelDVD authoring techniques. Also, take a few moments to take the short quiz and perform the exercises.

Q&A

Q I try to set a chapter point at a specific location by typing in a time code. But each time I click the Chapter icon, the time displayed shifts by a few frames. What's going on?

A DVDs work with MPEG encoded video. In general, the MPEG compression process compares video frames looking for differences. Instead of storing each frame, it groups similar frames into so-called groups of pictures (GOPs). The DVD spec won't let you put a chapter point in the middle of a GOP, so ReelDVD shifts your selected chapter point to the nearest break between GOPs (and you thought GOP stood for *Grand Old Party*).

Q Sometimes when dragging and dropping media and menu icons in the Storyboard, I inadvertently click a Next/Return/Previous/Command button and create a link between two icons. How do I erase that link?

A To undo an improperly placed link, simple click the source icon's appropriate link button and drag the cursor to an empty space in the Storyboard. The old link will disappear.

Quiz

1. How do you make a video menu play in a continuous loop until the viewer makes a selection on the remote control?
2. You've created a menu using four buttons placed in a square pattern. You want the button navigation to proceed from the top-left, to the top-right, down to the bottom-right, over to the bottom-left, and back up to the top-left. How do you do that?
3. You've set the sub-picture colors and opacities globally but want to make one button's Selection highlight different from the rest in that menu. What do you do?

Quiz Answers

1. Click the Next button on that menu's Storyboard icon and drag the cursor within that icon. Doing so creates an internal loop.
2. After defining the button hotspot boundary boxes, use the Link Vertically and Link Horizontally tools to create the button navigation. Use the Link Horizontally tool and drag from the upper-left to the upper-right; then drag from the lower-right to the lower-left. Use the Vertical Link tool and drag from the upper-right to the lower-right; then drag from the lower-left to the upper-left.
3. This is kind of a trick question. The only way to create a different look for one button's highlight (if there are more than one button on a menu) is in the original graphic art design process. You need to give that one button a different color—from the four available sub-picture colors—than the other buttons in the menu. For instance, this project's Chapter Menu Back button color is red while the other buttons in its sub-picture are black. Therefore, you first must change that button highlight's color in its original graphic file. Then you change that highlight's characteristics. Open that menu in the Preview window, click the Selection diamond icon, click the appropriate Color Number bar, and make the adjustments.

Exercises

1. Create a sub-picture button highlight graphic. To find out how to do that, read the thorough explanation in the ReelDVD Help file under Planning and Preparation, Preparing Assets, Subpicture Assets. It's a fairly simple process. Try adding a graphic file (a photograph will do) to your ReelDVD project, and then drag and drop your sub-picture overlay onto it. Open it in the Preview window and adjust the highlight colors and opacities.
2. Now, go out and make some DVDs! Critique your work. Watch other people navigate through them, and ask those viewers what they think you could do to make your DVDs better. Then make some more DVDs. The more you make, the better your finished products will look and perform. Enjoy.

Index

A

B

E

F

G - H

I - J - K

L

M

N - O

P

Q - R

S

T

U

V

W - X - Y - Z

License Agreement